Examining Energy and the Environment around the World

BRUCE E. JOHANSEN

Global Viewpoints

 ABC-CLIO™

An Imprint of ABC-CLIO, LLC
Santa Barbara, California • Denver, Colorado

Library of Congress Cataloging-in-Publication Data

Names: Johansen, Bruce E. (Bruce Elliott), 1950- author.
Title: Examining energy and the environment around the world / Bruce E. Johansen.
Description: Santa Barbara, California : ABC-CLIO, [2019] | Series: Global viewpoints | Includes bibliographical references and index.
Identifiers: LCCN 2018044168 | ISBN 9781440859298 (print : alk. paper) | ISBN 9781440859304 (ebook)
Subjects: LCSH: Energy industries—Environmental aspects. | Energy consumption—Environmental aspects. | Energy development—Environmental aspects. | Energy policy—Environmental aspects. | Environmental management.
Classification: LCC HD9502.A2 J64 2019 | DDC 333.7—dc23
LC record available at https://lccn.loc.gov/2018044168

ISBN: 978-1-4408-5929-8 (print)
       978-1-4408-5930-4 (ebook)

23 22 21 20 19   1 2 3 4 5

This book is also available as an eBook.

ABC-CLIO
An Imprint of ABC-CLIO, LLC

ABC-CLIO, LLC
130 Cremona Drive, P.O. Box 1911
Santa Barbara, California 93116-1911
www.abc-clio.com

This book is printed on acid-free paper ∞

Manufactured in the United States of America

Contents

Series Foreword

We are living in an ever-evolving world, one that is rapidly changing both in terms of society and in terms of our natural environment. Hot-button topics and concerns emerge daily; the news is constantly flooded with stories of climate change, religious clashes, educational crises, pandemic diseases, data security breaches, and other countless issues. Deep within those stories, though, are stories of resilience, triumph, and success. The Global Viewpoints series seeks to explore some of the world's most important and alarming issues and, in the process, investigate solutions and actionable strategies that countries are taking to better our world.

Volumes in the series examine critical issues, including education, war and conflict, crime and justice, business and economics, environment and energy, gender and sexuality, and Internet and technology, just to name a few. Each volume is divided into 10 chapters that focus on subtopics within the larger issue. Each chapter begins with a background overview, helping readers to better understand why the topic is important to our society and world today. Each overview is followed by eight country profiles that explore the issue at a global level. For instance, the volume on education might have a chapter exploring literacy, honing in on literacy rates and advocacy in eight nations. The volume on war and conflict might dedicate a chapter to women in the military, examining women's military roles in eight countries. The volume on crime and justice might include a chapter on policing, focusing on police infrastructure in eight countries. Readers will have an opportunity to use this organization to draw cross-cultural comparisons; to compare how Brazil is grappling with renewable energy amid a booming economy versus India, for instance, or how Internet access and control differs from Cuba to the United States.

Readers may read through each chapter in the volumes as they would a narrative book or may pick and choose specific entries to review. Each entry concludes with a list of further reading sources to accommodate additional research needs. Entries are written with high school and undergraduate students in mind but are appropriate and accessible to general audiences.

The goal of the series is not to make stark comparisons between nations, but instead to present readers with examples of countries that are afflicted by various issues and to examine how these nations are working to face these challenges.

Preface

This book couples energy and the environment. In the industrial age, first powered by coal, then more often by oil, environmental degradation (aka pollution) nearly inevitably followed resource exploitation and energy production. This holds as well for pollution by carbon dioxide and other heat-trapping gases released into our warming atmosphere. Some carbon dioxide is necessary for the survival of life on Earth (without any at all, we would freeze to death), but the overload created by fossil-fuel combustion is environmentally devastating, and in ways that few people think about, such as acidification of the oceans. Raising the carbon-dioxide level of the air also does the same in the oceans, threatening any living thing housed in a calcium shell. This includes phytoplankton, at the base of the maritime food chain.

The coupling of energy production with environmental damage is not inevitable. Renewable resources (wind and solar, for example) decouple this relationship for generation of energy that is vital to industry and transportation. Even wind power kills some birds, however. It also obstructs sight and makes some noise, although most people would prefer a wind farm to a coal strip mine as a neighbor.

The exploitation of the Earth in pursuit of human comfort, convenience, and profit (with accompanying environmental damage) has long been part of our economies and cultures. In the Bible's book of Genesis, the Christian God sent Adam and Eve out of the Garden of Eden to multiply and subdue the Earth. This commandment circled the Earth more than a millennium later as mercantile capitalism. In a few hundred years after that, human beings have done such an effective job of subduing the Earth that humanity is ruining the atmosphere, soil, and water around the world, meanwhile crowding out many other animal species.

A century ago, environmental preservation received little attention from most people, who thought more in terms of "mother lode" than "Mother Earth." Now that people are witnessing the fact that uncontrolled pollution of the Earth and atmosphere are presenting future generations with a suicide pact, sustainability has come into fashion. The late guerilla ecologist Edward Abbey called our capitalistic economic system "the ideology of the cancer cell," because it is predicated on growth, as it spreads randomly and ultimately may destroy the Earth that sustains it. How can we adjust our motivations, desires, and definitions of what is "good" (out of Eden, read "God") to fit a new world in which *more* is not always *better*?

The keepers of religious doctrine have realized just how effective humankind has been at subduing the Earth and just how damaging that subjugation has become. During the summer of 2015, Pope Francis, who has become well known for directly tackling many controversial issues, made climate change a Vatican priority by issuing an encyclical (essentially a policy statement) detailing how the burdens of global warming worldwide fall disproportionally on the poor. Indigenous peoples around the world bear a disproportionate burden of environmental damage. In the United States today, Native peoples often live on ruined, exhausted land, suffering toxic consequences. Fully one-third of the Superfund sites declared by the U.S. Environmental Protection Agency are on Native American lands.

Pope Francis's encyclical was part of a broader campaign by the pope to advocate protection of the Earth and all of creation. The pope prompted Catholic theologians to reinterpret Genesis to emphasize stewardship over subjugation. "We are the first generation that can end poverty and the last generation that can avoid the worst impacts of climate change," said United Nations secretary general Ban Ki-moon at an international symposium on climate change at the Pontifical Academy of Sciences on April 28, 2015; this was one of the events leading up to the encyclical (Povoledo 2015).

Awareness of economic development's costs animates environmental advocacy. "This, then, is the nemesis that modern Western man, together with his imitators . . . has brought upon himself by following the directive given in the first book of Genesis," wrote the great English historian Arnold Toynbee. "That directive has turned out to be bad advice, and we are beginning, wisely, to recoil from it" (Toynbee 1973).

This is a book about humankind's assault on the environment and what must change in our energy paradigm to sustain human and natural survival. To fit the Global Viewpoints series format, it is presented as a set of national profiles on each of ten subject areas. While national categories are convenient, we must recognize that solving our problems requires international cooperation—and in the case of climate change, quickly. In the longer run, preservation of a habitable Earth by sustainable means is going to require political behavior that advances beyond international competition. This will require humanity to solve conflicts without the wasteful, environmentally ruinous resort to armed conflict. This is not to say that we must abandon diverse languages and cultures, which provide all of us with a rich and beautiful window on our national histories. We can respect all peoples and their cultures without assuming superiority of one over others.

Bruce E. Johansen
Omaha, Nebraska
December 2017

Further Reading

Povoledo, Elisabetta. 2015. "Scientists and Religious Leaders Discuss Climate Change at Vatican." *New York Times*, April 29. Accessed September 24, 2018. http://www.nytimes.com/2015/04/29/world/europe/scientists-and-religious-leaders-discuss-climate-change-at-vatican.html.

Toynbee, Arnold. 1973. "The Genesis of Pollution." *New York Times*, September 16. Reprinted from *Horizon*, n.p.

Introduction

In this book, we present 80 essays on a variety of categories, including climate change, forests, fossil fuels, species extinctions, indigenous peoples, toxic chemicals, other forms of pollution, renewable resources, and water-related issues. All are considered in the context of national profiles (that is, by country). Some examples of country-specific profiles include: global warming's role in Syria's civil war; heat, drought, and floods in India; protests in the United States against the Keystone XL Pipeline; the devastation of tar-sands mining in Canada; declining numbers of penguins in Antarctica; the walrus's loss of ice habitat in the Arctic; toxins in Inuit mothers' milk; the deadly toll of uranium among the Navajo; "Cancer Alley" in Louisiana; the world's dirtiest air in Delhi, India; air pollution in London; Agent Orange's deformations across generations in Vietnam; Chinese resistance to air, soil, and water pollution—and much more.

One section describes issues that affect many countries at once, as well as the two-thirds of the Earth that is covered by oceans, without political boundaries. Some of the worldwide issues include biodiversity's decline, ocean acidity, coral collapse, global warming, thermal inertia's important role in climate change, and the looming extinction of many large primates.

The Worldwide Reach of Toxicity

Some themes lace the entire work. One is the role that a warming atmosphere plays in many environmental maladies worldwide. While that theme is the focus of one section, it plays an important role in many others. Another theme is the extent to which toxicity has pervaded the entire ecosystem from pole to pole, from the highest mountains to the deepest oceanic trenches. Extinction of species also pervades the entire volume, along with its relationship to both human-caused pollution and climate change. Another common thread is the worldwide affliction of indigenous peoples as victims of pollution.

Readers may find the global extent of human development engendered by human industry rather alarming. For example, crustaceans called Hirondellea gigas in the Mariana Trench, six miles deep in the western Pacific Ocean, have higher levels of polychlorinated biphenyls (PCBs) than crabs living in some of China's most polluted rivers. High levels of flame retardants have been detected in crustaceans living in the Kermadec Trench, six miles deep near New Zealand. Animals in these trenches eat any organic material that sifts down from the surface. Living in absolute darkness, they must withstand pressures of seven tons per square inch. They are, quite literally, some of the toughest creatures on Earth.

Alan Jamieson of Newcastle University, United Kingdom, and colleagues encountered high levels of toxicity in these crustaceans despite the fact that production of PCBs and polybrominated diphenyl ethers (PBDEs) has been illegal for about 40 years. These chemicals are not only toxic but also nonbiodegradable, so they continue to recycle through the food chain worldwide. The highest levels of PCBs in the amphipods were 50 times greater than levels found in a survey of crabs in a highly polluted river in China. Furthermore, they wrote, "The pollutants may have reached these remote areas by way of long-range atmospheric and oceanic transport, and through association with particulate matter and sinking carrion that are consumed by animals" (Jamieson et al. 2017).

Extinctions: Past as Prologue

As mass extinctions abetted by human-instigated global warming and toxic pollution have become more common, scientists have sought parallels in the past. According to some scientists, the worst mass extinction in the history of the planet could be replicated in as little as a century if global warming continues at the pace forecast by the Intergovernmental Panel on Climate Change (IPCC). Researchers at England's Bristol University have estimated that a 6°C increase in global temperatures was enough to play a role in the annihilation of up to 95 percent of the species alive on Earth at the end of the Permian period 251 million years ago; this is roughly the same amount of warming expected by the IPCC within a century, if levels of greenhouse gases in the atmosphere continue to rise at present rates (Reynolds 2003, 6). Albert K. Bates wrote, "Sixty-five million years ago, 60 to 80 percent of the world's species disappeared in a cataclysmic mass extinction, possibly caused by an asteroid's impact with Earth. Human population, not an asteroid, will cut the remaining number of species in half again, in just the next few years" (Bates 1990, 137).

The end-Permian mass extinction lasted perhaps 200,000 years. Shu-zhong Shen and colleagues wrote in *Science* (2011, 1367) that "associated charcoal-rich and soot-bearing layers indicate widespread wildfires on land. A massive release of thermogenic carbon dioxide and/or methane may have caused the catastrophic extinction." Drawing an analogue to anticipated conditions in a rapidly warming future world with accelerating ocean acidification and warming, along with declining levels of oxygen, they continued:

> Our studies indicate that both marine and terrestrial ecosystems collapsed very suddenly, and massive release of thermogenic CO_2 as well as methane, is a highly plausible explanation . . . [resulting] in increased continental aridity by rapid global warming, which caused widespread wildfires and accelerated deforestation in the world . . . [which] further enhanced the continental weathering and finally resulted in catastrophic soil erosion. (Shen 2011, 1372)

The wave of mass extinction at the end of the Permian period probably was caused by a series of very large volcanic eruptions that triggered a runaway greenhouse effect that nearly extinguished life on Earth. Conditions in what geologists have termed this "post-apocalyptic greenhouse" were so severe that 100 million years passed before species diversity returned to former levels. Michael Benton, head of Earth sciences at Bristol University, in the United Kingdom, wrote: "The end-Permian crisis nearly marked the end of life. It's estimated that fewer than one in ten species survived. Geologists are only now coming to appreciate the severity of this global catastrophe and to understand how and why so many species died out so quickly" (Reynolds 2003, 6).

The Permian heat wave was felt first and most intensely in the tropical latitudes; loss of species diversity spread from there. Reduction of vegetation, soil erosion, and the effects of increasing rainfall wiped out the lush, diverse habitats of the tropics, which today could lead to the loss of animals such as hippos, elephants, and all of the primates, according to Benton (Reynolds 2003, 6). He added:

> The end-Permian extinction event is a good model for what might happen in the future because it was fairly nonspecific. The sequence of what happened then is different from today because then the carbon dioxide came from massive volcanic eruptions, whereas today it is coming from industrial activity. However, it doesn't matter where this gas comes from; the fact is that if it is pumped into the atmosphere in high volumes, then that gives us the greenhouse effect and leads to the warming with all the other consequences. (Reynolds 2003, 6)

According to a theory first advanced by Anthony Hallam and Paul Wignall (1997), the volcanic eruptions 251 million years ago provoked a number of feedbacks that accelerated global warming of about 6°C. In a chapter of his book *When Life Nearly Died: The Greatest Mass Extinction of All Time,* titled "What Caused the Biggest Catastrophe of All Time?," Benton sketched how the warming (which was accompanied by anoxia, a lack of oxygen) may have fed upon itself:

> The end-Permian runaway greenhouse may have been simple. Release of carbon dioxide from the eruption of the Siberian Traps [volcanoes] led to a rise in global temperatures of 6°C or so. Cool polar regions became warm and frozen tundra became unfrozen. The melting might have penetrated to the frozen gas hydrate reservoirs located around the polar oceans, and massive volumes of methane may have burst to the surface of the oceans in huge bubbles. This further input of carbon into the atmosphere caused more warming, which could have melted further gas hydrate reservoirs. So the process went on, running faster and faster. The natural systems that normally reduce carbon dioxide levels could not operate, and eventually the system spiraled out of control, with the biggest crash in the history of life. (Benton 2003, 276–277)

Greg Retallack, an expert on ancient soils at the University of Oregon, has speculated that the same methane "belch" from the oceans was of such a magnitude that it caused mass extinction via oxygen starvation (anoxia) of land animals. Bob

Berner of Yale University has calculated that a cascade of effects on wetlands and coral reefs may have reduced oxygen levels in the atmosphere from 35 percent to only 12 percent in only 20,000 years. Marine life also may have suffocated in the oxygen-poor water (Hecht 2003). One animal, the meter-long reptile *Lystrosaurus,* survived because it had evolved to live in burrows, where oxygen levels are low and carbon dioxide levels high. According to a report by the *NewScientist,* "It had developed a barrel chest, thick ribs, enlarged lungs, a muscular diaphragm, and short internal nostrils to get the oxygen it needed" (Hecht 2003).

According to Chris Lavers, writing in *Why Elephants Have Big Ears,* a spike of worldwide warming contributed to this mass extinction in part because all of the Earth's continents at the time were combined into one land mass (Lavers 2000, 231). Warming of tropical regions at this time has been estimated at about 11°F, with larger rises near the poles that tended to create a generally warm atmosphere planetwide, "a flattening of the temperature difference between the poles and the equator," a condition that Lavers suspects drastically slowed or shut down ocean mixing, killing many sea creatures (Lavers 2000, 232). "Unstirred," wrote Lavers, "the oceans begin to stagnate. Deep waters gradually lost oxygen, and species began to vanish" (Lavers 2000, 233).

What caused this spike in temperatures? The prime suspect, at least in the beginning, was coal-bearing deposits in the southern reaches of Pangea (the Earth's single land mass), which were oxidized after they were lifted by tectonic activity, releasing large volumes of carbon dioxide when the volcanoes erupted. The level of greenhouse gases in the atmosphere thus increased due to the most concentrated bout of volcanic activity on Earth during the last 600 million years. "This injection of volcanic CO_2," wrote Lavers, "was probably the decisive event that ultimately tipped the biosphere into the new era of the Mesozoic" (Lavers 2000, 235). Today, human beings may be replicating this situation by combusting fossil fields, injecting massive amounts of carbon dioxide and methane into the atmosphere.

Indigenous Environmental Afflictions

In addition to human alteration of the atmosphere, another common thread in this book is the affliction of indigenous peoples with life-threatening pollution. Native peoples share a long history of residing in resource colonies that today host more than a third of the United States' Superfund acute pollution sites. Environmental provocations afflicting Native American peoples in the United States range from uranium to kitty litter—a range that rivals the problems of any Third World nation—from the toll of uranium mining on the Navajos, to the devastation wrought by dioxin, PCBs, and other pollutants on the agricultural economy of the Akwesasne Mohawk reservation in northernmost New York State. As with the Akwesasne Mohawks, some of the most serious problems span international borders. The Yaquis, whose homelands span the U.S.-Mexican border, have been afflicted with some of the same chemicals as the Mohawks on the U.S.-Canadian border.

Many reservation residents suffer from cancers and other illnesses because their lands have been used for several decades as industrial dumps and mine sites. These include acute effects of exposure to dioxins, PCBs, and other persistent organic pollutants, most intensely in the Arctic, where Native consumption of sea life (their traditional diet) has been curtailed and Inuit mothers have been warned not to breastfeed their infants because their milk may be toxic. Toxicity intensifies as it ascends the food chain in an exponential manner in a process known as biomagnification. A breastfed infant is the last stop. The Arctic, which looks so pristine to the untutored eye, is also experiencing the world's most rapid rate of climate change. Ice plays a large role in Inuit culture, but the ice is melting, and many Inuit hunters have been injured or killed by falling through thin ice.

Canada also has become a major source of indigenous environmental contamination and conflict. The Innu of Labrador have been afflicted with sulfide mining, aluminum smelting, and noise pollution from squadrons of military aircraft. Some of the most intense resource exploitation in Canada takes place in remote locations, such as among the Lubicon Cree of northern Alberta, whose lands were so inaccessible in 1900 that treaty makers completely missed them. Today, roads have opened their lands to massive oil drilling and logging.

From the Dine (whom the Spanish called Navajo) to their namesake Dene of Canada's Northwest Territories (and many points between), Native peoples were recruited beginning in the early 1950s to mine "yellow dust"—uranium—and then, over decades, died in large numbers of torturous cancers. Uranium-induced cancers have become the deadliest plague unleashed on Native peoples of North America. By 2014, 350 to 400 former Navajo underground uranium miners had died from maladies caused in large part by exposure to radiation. On Navajo land, its mining and milling are now illegal, as Native people heed their "original instructions," which maintain that uranium is better left in the ground.

Uranium mined from Native American lands supplied a substantial proportion of the fuel for early nuclear power plants as well as the U.S. nuclear arsenal. By the 1970s, many of the early miners were dying of lung cancer. In Washington State, nuclear waste from the Hanford plants afflicted the Yakamas. Uranium mining also has caused a plague of cancer among the Laguna Pueblo. The Navajos succeeded in stopping uranium mining and milling only after several hundred people had died of its effects and many more had suffered the tortures of cancers that once were nearly unknown in their country.

The Role of Thermal Inertia

This book also intends to supply students with readily understandable explanations of scientific concepts that affect our world. Some of the most important scientific themes in ecology are only very rarely discussed in public debate. One of these—perhaps the most vital—is the role of thermal inertia in climate change. Thermal inertia delivers the results of atmospheric change roughly a half-century after our burning of fossil fuels provokes them. The weather today is reacting to greenhouse gas emissions from about 1965. Since then, the world's emissions have risen substantially. With carbon-dioxide levels in the atmosphere already as high as the Pliocene era, 2 to 3 million years ago, scientists have been asking a question that will become more important in coming decades: How long will it be before enough ice melts to raise sea levels to reflect these temperatures?

It's not a matter of "if," but "when," according to a study published in *Science* in July of 2015 (Dutton et al. 2015). This study documented sea-level rises of at least 20 feet (six meters) several times during the last 3 million years and concluded that present-day levels do not reflect carbon-dioxide and other greenhouse-gas levels already surpassed. Generally, because of thermal inertia, temperatures in the air lag any given atmospheric level by 50 years in the air and about 150 to 200 years in the oceans.

To put it briefly, this cake is already being baked. The scientists found that a rise in temperatures of 1°C to 2°C virtually guaranteed a rise of 20 feet. World-wide, several hundred million people around the world live within 20 feet of high tide. Much of the sea-level rise has (and will) come from the melting of ice sheets in Greenland and Antarctica, said lead author Andrea Dutton, a University of Florida geochemist. "This evidence leads us to conclude that the polar ice sheets are out of equilibrium with the present climate," she said ("Global Sea Levels," 2015).

Global warming is a deceptively backhanded crisis in which thermal inertia delivers results a half-century or more after our burning of fossil fuels provokes them. Our political and diplomatic reactions come *after* we see results. Political inertia plus thermal inertia thus presents the human race and the planet we superintend with a challenge to fashion a new energy future *before* raw necessity—the hot wind in our faces—compels action. Global warming is dangerous because it is a sneaky, slow-motion emergency, demanding that we acknowledge a reality centuries in the future with a system of individual, legal, and diplomatic reactions that responds in the past tense.

Dutton and her colleagues compiled their study using computer models and the geologic record to gauge the global ice pack's sensitivity to climate change. They learned that in the past, when average temperatures rose 1°C to 3°C (1.8°F to 5.4°F) warmer than levels prevailing in preindustrial times (before about 1850), sea level peaked at least 20 feet higher than today's levels. "As the planet warms, the poles warm even faster, raising important questions about how ice sheets in Greenland and Antarctica will respond," Dutton said. "While this amount of sea-level rise will not happen overnight, it is sobering to realize how sensitive the polar ice sheets are to temperatures that we are on path to reach within decades" ("Global Sea Levels," 2015).

"It takes time for the warming to whittle down the ice sheets," said Anders Carlson, of Oregon State University's College of Earth, Ocean, and Atmospheric Sciences, a coauthor of the same study. "But it doesn't take forever. There is evidence that we are likely seeing that transformation begin to take place now" ("Global Sea Levels," 2015).

Another paper, by James Hansen et al., also published in 2015, supports the idea that a 2°C rise in temperatures will nearly certainly guarantee a large-scale sea-level rise. In *Atmospheric Chemistry and Physics Discussions*, they wrote:

> There is evidence of ice melt, sea level rise to five to nine meters, and extreme storms in the prior interglacial period that was less than 1°C warmer than today. Human-made climate forcing is stronger and more rapid than paleo forcings. . . . We conclude that 2°C global warming above the preindustrial level, which would spur more ice shelf melt, is highly dangerous. Earth's energy imbalance, which must be eliminated to stabilize climate, provides a crucial metric. (Hansen et al. 2015)

Hansen continued:

> As the evidence accumulates, at some point a scientist must say it is time to stop waffling so much and say that the evidence is pretty strong. In my opinion we have reached that point on the sea-level issue. My conclusion, based on the total information available, is that continued high emissions would result in multi-meter sea level rise this century and lock in continued ice sheet disintegration such that building cities or rebuilding cities on coast lines would become foolish. (Hansen et al. 2015)

The Question of Unburnable Carbon

Another question that pervades this entire volume that has been only rarely discussed outside scientific circles is that of how to handle unburnable carbon. Given the general consensus among world climate scientists and diplomats that global temperatures cannot rise more than 2°C without doing irreparable harm to people and the rest of the planet's flora and fauna, what becomes of the trillions of dollars' worth of recoverable fossil fuels that would have to remain locked away to preserve a habitable Earth? When the threat of a warming climate is truly taken seriously by the fossil-fuel industry, the industry will have to deal with the question of locking away reserves that have been defined as corporate assets that could become worthless.

The value of these "unburnable" assets already has been estimated by scientists, whose calculations appeared in the British journal *Nature* early in 2015. Michael Jakob and Jerome Hilaire, who work with the Potsdam Institute for Climate Impact, wrote, "Cumulative carbon dioxide emissions must be less than 870 to 1,240 gigatons between 2011 and 2050 if we are to have a reasonable chance of limiting global warming to 2°C above average global temperature of pre-industrial times" (Jakob and Hilaire 2015,150). "Reasonable," in this case, is defined as 50 percent, calculated with a model. This target represents one-third to one-fourth of the oil, natural gas, and coal held on company balance sheets as provable reserves. What is provable by companies may be revised in accordance with market prices and development of mining and drilling technology.

A few decades ago, oil from shale and tar sands was known to exist, but exact reserves were largely unknown. By 2010, oil booms had developed in Alberta, North Dakota, and (on a smaller scale) other parts of North America using water sources (hydraulic fracturing, or "fracking") and oil (also called tar) sands mining, as oil prices reached $100 a barrel. As prices fell by more than 50 percent in 2015 and 2016, some of this production shut down because it is unprofitable at that price, only to resume when oil prices increased. The United States and Canada are not the only countries with oil-shale potential but are merely the first to apply the technology to production. "Encouraged by the recent shale-gas production boom in the United States, several world regions, including China, India, Africa, and the Middle East, are seeking to unlock their large endowments or increase existing production," wrote Jakob and Hilaire (2015, 151).

The question of unburnable carbon raises profound social, political, and economic questions in a world in which the fossil-fuel industry is superlatively equipped with payrolls, equipment, and political influence to continue mining and drilling without limit on a 19th-century model, as levels of greenhouse gases and temperatures continue to rise. At what point will climatic changes force a change in how carbon-based fuel reserves are valued? Will it then be too late, given thermal inertia, to forestall planetary disaster?

Christophe McGlade and Paul Ekins, writing in *Nature*, calculated the amount of reserves that cannot be burned to avoid such a disaster: "Our results suggest that, globally, a third of oil reserves, half of gas reserves, and over 80 percent of current coal reserves should remain unused from 2010 to 2050 in order to meet the target of 2°C," they wrote. "We show that development of resources in the Arctic and any increase in unconventional oil production are incommensurate with efforts to limit average global warming to 2°C" (McGlade and Ekins 2015, 187).

In the Middle East, home to much of the Earth's most easily and least expensively recoverable oil, about 40 percent of reserves would be required to remain unexploited to keep temperature rises within the 2°C limit, McGlade and Ekins calculated (2015, 188). Coal, with reserves that are much larger than oil in terms of both energy potential and carbon-dioxide emissions, would have to remain mainly in the ground. China and India would face 66 percent sequester rates, and Africa 85 percent. The United States, Australia, and the territory of the former Soviet Union would be required to leave about 90 percent of coal reserves locked away to provide climate stability. Nearly all (less than 10 percent) oil and gas reserves attributed to unconventional sources (fracking and tar sands) would have to be shut down or remain in the ground to maintain any reasonable chance to stabilize global climate over the long term, according to McGlade and Ekins (2015, 188–189).

Furthermore, McGlade and Ekins wrote:

Our results show that policymakers' instincts to exploit rapidly and completely their territorial fossil fuels are, in aggregate, inconsistent with their commitments to this temperature limit. Implementation of this policy commitment would also render unnecessary continued substantial expenditure on fossil fuel exploration, because any new discoveries could not lead to increased aggregate production. (McGlade and Ekins 2015, 187)

Another calculation of the unburnable carbon question was provided in 2015 by Michael Greenstone, Milton Friedman professor of economics at the University of Chicago, as well as director of the university's Energy Policy Institute (and chief economist of President Barack Obama's Council of Economic Advisers in 2009 and 2010). He wrote in the *New York Times* that recoverable reserves and resources of coal, oil, and natural gas, once combusted, would raise average global temperatures 16.2°F. ("Reserves" are known deposits recoverable at or near market prices; "resources" are known deposits that could be exploited with today's technology above that level.) "If we use all of the fossil fuels in the ground, the planet will warm in a way that is difficult to imagine. Unless the economics of energy markets change, we are poised to use them," Greenstone commented. And there may be more: "Indeed, it is well known that there are ample supplies of coal deeper beneath the Earth's surface that do not yet qualify as resources, and there is increasing evidence that energy from methane hydrates may become relevant commercially" (Greenstone 2015).

Given that fossil-fuel reserves had an estimated value of U.S. $27 trillion at 2014 prices, any effort to sequester substantial amounts of them would produce a financial earthquake in the fossil-fuel industry, according to Jakob and Hilaire, and force the companies to "ask themselves whether they should continue to invest in exploration for, and processing of, oil, gas, and coal, or risk losing billions of dollars of stranded assets" (Jakob and Hilaire 2015, 151).

The Crisis to Come

This work surveys world environmental and energy issues in a context of crises to come unless the ways in which we produce energy are changed. One of the chapters delves into renewable energy, principally wind and solar power. Until recently, wind and solar power had been very minor players in an energy-generation field dominated by coal, oil, and natural gas. Within roughly 15 years, since the turn of the millennium, wind and solar have become much more efficient and cost-competitive. Between 2009 and 2015, the cost of generating electricity with wind declined 61 percent; solar power's cost dropped by 82 percent during the same period (Krugman 2016, A-21).

Solar technology is undergoing a revolution that may eventually allow power to be acquired from nearly any surface on which the sun shines. One such technology is the "artificial leaf." By 2016, scientists were studying a form of artificial photosynthesis that, it was reported in *Science*, "when combined with solar photovoltaic cells, solar-to-chemical conversion rates should become nearly an order of magnitude more efficient than natural photosynthesis" (Liu et al. 2016, 1210). At the point when this technology becomes commercially feasible on a large scale, fossil fuels may be on their way to museum status.

Read, with us, about how the world is changing in humankind's hands and what must be done to reverse what is becoming a disastrous course.

Further Reading

Bates, Albert K., and Project Plenty. 1990. *Climate in Crisis: The Greenhouse Effect and What We Can Do.* Summertown, TN: The Book Publishing Co.

Benton, Michael J. 2003. *When Life Nearly Died: The Greatest Mass Extinction of All Time.* London: Thames and Hudson.

Dutton, A., A. E. Carlson, A. J. Long, G. A. Milne, P. U. Clark, R. DeConto, . . . M. E. Raymo. 2015. "Sea-Level Rise Due to Polar Ice-Sheet Mass Loss during Past Warm Periods." *Science* 349 (July 10). Accessed September 27, 2018. http://doi.org/10.1126/science.aaa4019.

"Global Sea Levels Could Soon Rise 20 Feet as Climate Warms." 2015. *Environment News Service*, July 13. Accessed September 27, 2018. http://ens-newswire.com/2015/07/12/global-sea-levels-could-soon-rise-20-as-climate-warms.

Greenstone, Michael. 2015. "If We Dig Out All Our Fossil Fuels, Here's How Hot We Can Expect It to Get." *New York Times*, April 9. Accessed September 27, 2018. http://www.nytimes.com/2015/04/09/upshot/if-we-dig-out-all-our-fossil-fuels-heres-how-hot-we-can-expect-it-to-get.html.

Hallam, Anthony, and Paul Wignall. 1997. *Mass Extinctions and Their Aftermath.* Oxford: Oxford University Press.

Hansen, J., M. Sato, P. Hearty, R. Ruedy, M. Kelley, V. Masson-Delmotte, . . . K.-W. Lo. 2016. "Ice Melt, Sea Level Rise and Superstorms: Evidence from Paleoclimate Data, Climate Modeling, and Modern Observations That 2°C Global Warming Is Highly Dangerous." *Atmospheric Chemistry and Physics Discussions* 15 (July): 20059–20179. Accessed September 27, 2018. http://doi.org/10.5194/acpd-15-20059-2015.

Hecht, Jeff. 2003. "Suffocation Suspected for Greatest Mass Extinction." *New Scientist*, September 9. Accessed September 27, 2018. https://www.newscientist.com/article/dn4138-suffocation-suspected-for-greatest-mass-extinction.

Jakob, Michael, and Jérôme Hilaire. 2015. "Climate Science: Unburnable Fossil-Fuel Reserves." *Nature* 517 (January 8): 150–152. Accessed September 27, 2018. http://doi.org/10.1038/517150a.

Jamieson, Alan J., Tamas Malkocs, Stuart B. Piertney, Toyonobu Fujii, and Zulin Zhang. 2017. "Bioaccumulation of Persistent Organic Pollutants in the Deepest Ocean Fauna." *Nature Ecology & Evolution* 1 (2017). Accessed September 27, 2018. http://doi.org/10.1038/s41559-016-0051.

Krugman, Paul. 2016. "Wind, Sun, and Fire." *New York Times*, February 1: A-21.

Lavers, Chris. 2000. *Why Elephants Have Big Ears.* New York: St. Martin's Press.

Liu, Chong, Brendan C. Colón, Marika Ziesack, Pamela A. Silver, and Daniel G. Nocera. 2016. "Water Splitting–Biosynthetic System with CO_2 Reduction Efficiencies Exceeding Photosynthesis." *Science* 352 (June 3): 1210–1213.

McGlade, Christophe, and Paul Ekins. 2015. "The Geographical Distribution of Fossil Fuels Unused When Limiting Global Warming to 2°C." *Nature* 517 (January 8): 187–190. Accessed September 27, 2018. http://doi.org/10.1038/nature14016.

Reynolds, James. 2003. "Earth Is Heading for Mass Extinction in Just a Century." *The Scotsman*, June 18: 6.

Shen, Shu-zhong, James L. Crowley, Yue Wang, Samuel A. Bowring, Douglas H. Erwin, Peter M. Sadler, . . . Yu-gan Jin. 2011. "Calibrating the End-Permian Mass Extinction." *Science* 334 (December 9): 1367–1372.

Chapter 1: Climate Change

OVERVIEW

As part of Earth's natural cycle, the greenhouse effect (which scientists call "infrared forcing") is very necessary to life on Earth. Without it, the planet's average temperature would be minus 2°F. It is the added warming provoked by human combustion of fossil fuels that causes a problem. As with chocolate, a little is a good thing; too much is toxic to the system. Fossil fuels provide us comfort and convenience, and altering their use in a fundamental way presents the challenge of the century—and most probably, for several centuries to come.

In 2015, the atmospheric level of carbon dioxide breached 400 parts per million in all areas, at all seasons. Methane and nitrous oxides, the two other principal greenhouse gases, also reached record levels by substantial margins. The same year, world temperatures, stoked by El Niño conditions, surged to a new record as well, above 2014's previous high. Global monthly average temperatures set records for nearly every consecutive month during 2015, 2016, and 2017, even after the El Niño had subsided, often by substantial margins.

"We're moving into uncharted territory at a frightening speed," said World Meteorological Organization secretary general Michel Jarraud (Warrick 2015). "The departures are what we would consider astronomical," said National Oceanic and Atmospheric Administration climate scientist Jessica Blunden. "It's on land. It's in the oceans. It's in the upper atmosphere. It's in the lower atmosphere. The Arctic had record low sea ice. Everything everywhere is a record this month, except Antarctica," Blunden said. "It's insane." Georgia Tech climate scientist Kim Cobb added:

> When I look at the new February 2016 temperatures, I feel like I'm looking at something out of a sci-fi movie. In a way we are: it's like someone plucked a value off a graph from 2030 and stuck it on a graph of present temperatures. It is a portent of things to come, and it is sobering that such temperature extremes are already on our doorstep. (Borenstein 2016)

Accelerating warming of the planet has a way of getting into everything. It is, of course, a worldwide problem that provokes changes in temperature—and more. A casual observer may be surprised at how a change in proportion of a trace gas (carbon dioxide, in this case) can have such wide-ranging influences. When the level of carbon dioxide is raised in oceans, for instance, the acidity level rises, thus endangering creatures that live in calcium shells. The very basis of the oceanic food chain, phytoplankton, is a prime example of a creature that is imperiled by rising levels of carbon dioxide.

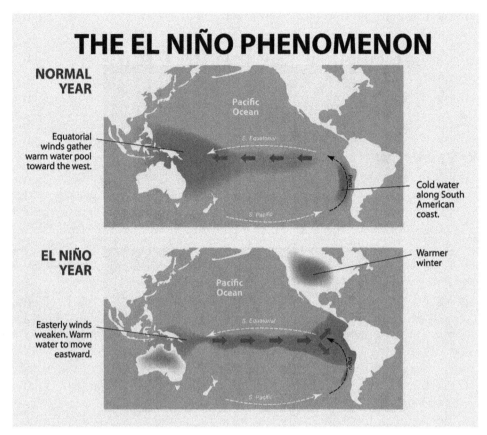

THE EL NIÑO PHENOMENON

NORMAL YEAR

Pacific Ocean

Equatorial winds gather warm water pool toward the west.

S. Equatorial

Peru

S. Pacific

Cold water along South American coast.

EL NIÑO YEAR

Pacific Ocean

Easterly winds weaken. Warm water to move eastward.

S. Equatorial

Peru

S. Pacific

Warmer winter

El Niño warms the ocean and atmosphere in the equatorial Pacific Ocean and shapes weather worldwide. (Designua/Dreamstime.com)

Raise the amount of carbon dioxide, methane, and other heat-retaining gases, and temperatures eventually rise, melting ice and raising sea levels. Human beings have an affinity for the oceans, and a large proportion of us live on or near coastlines. Glance at a map of the world and point out the large cities that will be in peril as sea levels rise a few feet—Shanghai, Kolkata, London, New York City, Miami, and many others. Increase the proportion of greenhouse gases in the atmosphere and change its circulation patterns, expanding convection patterns that meteorologists call "Hadley Cells"; this causes a decline in rainfall over some areas, expanding deserts. Harvests fail, and people go hungry, provoking bitter civil wars. We are adapted to the climate, as are all of Earth's flora and fauna. When the climate changes, everything changes. Ken Caldeira, a researcher at Stanford University's Carnegie Institute of Science, told Chelsea Harvey of the *Washington Post*, "The legacy of what we're doing over the next decades and the next centuries is really going to have a dramatic influence on this planet for many tens of thousands of years" (Harvey 2015).

Rapid Seasonal Change

Climate change is changing the rhythm of the seasons. As David George Haskell, a professor of biology at the University of the South, commented from Sewanee, Tennessee:

> Sexual energies were loosed early this year in Tennessee, then quashed in February. Spring peepers made my ears ring as I walked through wetlands east of Nashville's honky-tonks. These frogs were a month ahead of their normal schedule. But what is normal in a year [when] . . . the season started weeks earlier for plants and animals? Spring has been particularly hasty and irregular this year [2017], but this is no anomaly. In the latter half of the 20th century, the spring emergence of leaves, frogs, birds, and flowers advanced in the Northern Hemisphere by 2.8 days per decade. I'm nearly 50, so springtime has moved, on average, a full two weeks since I was born. And you? We now experience climate change not only through the abstractions of science, but also through lived experience. . . . What we experienced as spring, a predictable appearance of buds and birds, is passing away. Our children will live in uncharted, unnamed seasons. (Haskell 2017)

Further Reading

Haskell, David George. 2017. "The Seasons Aren't What They Used to Be." *New York Times*, March 17. Accessed October 11, 2018. https://www.nytimes.com/2017/03/17/opinion /sunday/the-seasons-arent-what-they-used-to-be.html.

This chapter provides eight country profiles, slices of an enormous subject. In Syria, global warming's role is traced in a civil war that as of 2017 had displaced nearly a fourth of its population. Similarly, the roots of the terror group Boko Haram in Nigeria are traced to failing rains, a spreading Sahara, and agricultural collapse. Australia's increasing heat and aridity threaten crops and water supplies. China struggles with a spread of deserts in its north, even as rising affluence increases fossil-fuel use. At the same time, China has become the world's leader (both as producer and user) of wind and solar power. In India, with a surging population (1.2 billion as of 2017) and rising affluence, demand for all types of fuels has been increasing as temperatures rise and the annual monsoon becomes more inconsistent. Drought intensifies, punctuated by occasional devastating deluges. Greenland adapts to a warming climate as farming spreads, and even Russia has felt the effects of rising temperatures. The United States has the largest armed forces in the world, and we will examine the carbon footprint of war in the conclusion of this chapter.

Further Reading

Borenstein, Seth. 2016. "Beyond Record Hot, February [2016] Was 'Astronomical' and 'Strange.'" Associated Press, March 17. Accessed September 17, 2018. http://phys.org /news/2016-03-hot-february-astronomical-strange.html.

Harvey, Chelsea. 2015. "Scientists Confirm There's Enough Fossil Fuel on Earth to Entirely Melt Antarctica." *Washington Post*, September 11. Accessed September 17, 2018. https://www.washingtonpost.com/news/energy-environment/wp/2015/09/11/scientists-con firm-theres-enough-fossil-fuel-on-earth-to-melt-all-of-antarctica/.

Warrick, Joby. 2015. "Greenhouse Gases Hit New Milestone, Fueling Worries about Climate Change." *Washington Post*, November 9. Accessed September 17, 2018. https://www.washingtonpost.com/national/health-science/greenhouse-gases-hit-new-mile stone-fueling-worries-about-climate-change/2015/11/08/1d7c7ffc-8654-11e5-be39 -0034bb576eee_story.html.

AUSTRALIA

Scorching Heat Waves

Always a desert continent (except in its tropical north), Australia has become even drier and hotter during recent years. Some people there long suspected that global warming is the culprit, and mounting evidence has convinced most of them. For example, a brush fire in Western Australia during January 2017 produced a 3,000-foot-high "fire tornado." Such whirlwinds, which suck up flames, dust, and other debris, have become more common as temperatures rise and drought intensifies, a signature of global warming.

On February 7, 2009, the temperature in Melbourne hit 115.5°F (and as high as 120°F in nearby areas), with relative humidity as low as 6 percent. The next day, more than 600 fires blew up on stiff northwesterly winds, the worst outbreak in the continent's recorded history. Some "fire fronts" reached 300 feet in height, sweeping through and searing small towns, killing 173 people. The fires moved so quickly that they caught some people from behind as they tried to flee in their cars (Kenneally 2009, 46–53). During mid-November 2009, Australia suffered a spring heat wave that set record highs characteristic of mid-summer and pushed fire danger to "catastrophic" in some regions of South Australia and New South Wales. Adelaide had its highest springtime temperatures on record (35°C, 95°F) for eight days in a row. The temperature reached 109°F in Sydney one day.

Heat, Drought, and Wildfires

Reports of heat, drought, and devastating wildfires often dominate the news. In one example of many, during February 2015, two large wildfires, both ignited by lightning, threatened towns in southwestern Australia; one hit just east of Northcliffe, and the other burned more than 60 miles to the north, near Boddington ("Bushfires Menace Towns in Western Australia" 2015). October 2015 (the climatic equivalent of April north of the equator) was Australia's hottest on record. "It is going to be a horror summer," said Trevor Tasker, a firefighter and regional emergency services inspector from Western Australia. "I've never seen conditions like this," Tasker noted. "By the time we knew that fire was alight, it was unstoppable," he said.

"There was nothing anyone could do but get out of the way and let it unleash its fury." One farmer was scorched alive inside his car as he raced to warn others of the wall of flames headed their way (Innis 2015).

The townships of Salmon Gums, Grass Patch, and Norseman, all near Perth, on Australia's southwest coast, recorded temperatures around 107°F on November 17, according to Neil Bennett, a weather forecaster based in Perth at the Bureau of Meteorology. "This is the earliest we have seen the temperatures this high," Bennett said. "But we have had a drying trend going back 30 to 40 years" (Innis 2015).

The year 2013, which started and ended with record heat, was Australia's hottest on record to that time. In addition, one-tenth of Australia's reporting stations experienced record heat between January 1 and 4, 2014. The highest air temperature during that period was in Moomba, Queensland, at 49.3°C (120.7°F) on January 2 ("Heat Wave Stifles Australia" 2012).

During January 2013, Australia was swept by record heat for several weeks, with temperatures reaching 45°C (113°F) in several locations. The Australian Bureau of Meteorology said that the Austral summer set a record for the highest average temperature across the continent, with a high of 40.33°C (104.59°F) on January 7. The next day was almost as hot, with a national average of 40.11°C (104.20°F). The minimum on January 7–8 also set a record, at 32.36°C (90.25°F).

New York Times reporter Matt Siegel wrote from Sydney that "Four months of record-breaking temperatures stretching back to September 2012 have produced what the government says are 'catastrophic' fire conditions along the eastern and southeastern coasts of the country, where the majority of Australians live"

Australia's Sizzling Summers

During the summer of 2016 and 2017, Southeastern Australia sweltered through yet another summer of intense heat, with temperatures as high as 113°F in some parts of the state of New South Wales. "It was nothing short of awful," said Sarah Perkins-Kirkpatrick, of the Climate Change Research Center at the University of New South Wales, in Sydney. "In Australia, we're used to a little bit of heat. But this was at another level" (Fountain 2017).

Summer during the intense El Niño year of 2015 also had arrived in Australia with record late-spring heat (104°F in Sydney). Fire ravaged 580 square miles of wheat fields and killed farmers at work in the south of Western Australia, near Esperance, a town of 14,000 that is 450 miles southeast of Perth. Furious flames were driven by a scorching 50-mile-per-hour wind at temperatures more than 100°F. Farmers were beginning their wheat harvest as exceedingly dry fields, record heat, and strong winds fed the flames.

Further Reading

Fountain, Henry. 2017. "Sydney's Swelter Has a Climate Change Link, Scientists Say." *New York Times*, March 2.

(Siegel 2013a). Temperatures set record highs once again. "If you look at yesterday [January 8, 2013], at Australia as a whole, it was the hottest day in our records going back to 1911," said David Jones, manager of climate monitoring prediction at the Bureau of Meteorology. "From this national perspective, one might say this is the largest heat event in the country's recorded history" (Siegel 2013a).

Southwestern Australia experienced intensifying heat and drought into 2016. On January 6, lightning started a fire in Lane Poole Reserve that enveloped Yarloop, a town about 70 miles south of Perth, destroying 128 homes and 41 other structures, including bridges and community buildings, according to the Department of Fire and Emergency Services. Two Yarloop residents were killed, and four firefighters were injured. Within five days, the fire had burned 276 square miles ("Bushfire Devastates Australian Town" 2016).

Heat made outdoor sports impossible near Melbourne during mid-January 2014, when a hot wind from the interior brought temperatures of 111°F to the Australian Open tennis matches, nearly shutting them down. Press reports described tennis great Maria Sharapova leaving the sizzling court "underneath the shade of an umbrella, an ice vest draped over the back of her neck" (Bishop 2014).

"Climate change is potentially the biggest risk to Australian agriculture," said Ben Fargher, chief executive of the National Farmers' Federation in Australia (Bradsher 2008). Many Australian farmers have switched from rice to wine grapes, which require much less water and produce a pretax profit of about $2,000 an acre, compared to rice at about $240 an acre. In the meantime, strains of rice are being developed that bloom during the cooler early hours of the day. Rice that blooms late in the day is vulnerable to warming.

With irrigation, cultivation of cotton prospered in Australia's Outback. By 2007, however, with Australia suffering record heat and its worst drought on record to that time, production of cotton, with its thirst for water, had fallen sharply. Patrick Barta wrote in the *Wall Street Journal* that in the town of Wee Waa, Australia's self-described "cotton capital," population 2,000, about 250 miles northwest of Sydney, "The Cotton Fields Motel that once was busy with seasonal workers now struggles to fill rooms. Elsewhere in the flat basin [of Australia's largest river system, the Murray-Darling], kangaroos hop along dry levees, and the end of giant water-transport pipes poke out over empty reservoirs" (Barta 2007, A-1).

Australia's cotton production fell by two-thirds between 2001 and 2007 (Barta 2007, A-12). Some fields were down to 2 percent of usual, and reservoirs as large as 240,000 acres had completely dried up. In 2008, the drought in the Murray-Darling river basin continued, following a record-dry June, even after parts of Australia received occasional torrential rains during a strong La Niña that briefly countered some effects of the drought.

Drought in Australia's Agricultural Heartland

By 2009, the Murray-Darling basin (in the continent's southeastern quadrant) had become so dry that hundreds of thousands of river red gum trees, the world's largest such forest, died suddenly. Some former wetlands that had once been swamps became encrusted with cracking dried earth that, absent periodic flushing, reacted with air to form sulfuric acid. These are signs of an epic drought, beyond the cycles that usually affect the area (Draper 2009, 56). On September 23, 2009, Australia's worst dust storm in 70 years produced an orange dawn in Sydney as dust clouds from Australia's interior, suffering its worst drought on record, spread over much of eastern Australia. Visibility was low enough to close Sydney's airport.

The only major industrial country except the United States that long refused to ratify the Kyoto Protocol was Australia, and its conservative government under Prime Minister John Howard was slow to recognize the economic perils of global warming. Largely because 85 percent of its electricity was generated from coal and because it hosts a large number of energy-intensive industries, such as aluminum smelting and steel manufacturing, Australia generated more greenhouse gases per capita in 2004 than any other major country—26.3 tons, trailing only Luxembourg, at 28.1 tons (the United States' comparable figure was 24.1 tons) (Wiseman 2007, 10-A). The Howard government's inability to comprehend the role of climate change in the devastation of Australia's economy probably played a major role in the country's 2007 elections, when he was replaced by Labor Party member Kevin Rudd, who signed the Kyoto Protocol as his first official act after he was sworn into office December 3, 2007. Howard had been quite direct with his disregard of global warming; during September of 2006, Al Gore had come to the country, and the prime minister refused to meet with him.

By the time Howard faced the voters, many Australian urban areas were facing severe water shortages. Inflows behind Sydney's dams from 1991 to 2006 were 71 percent less than their averages from 1948 to 1990 (Pincock 2007, 336). By 2006, polls in Australia indicated that a large proportion of people there ranked a dysfunctional climate as the number-one danger in their lives, ahead of such right-wing mainstays as international terrorism. Howard had been reelected three times as the drought intensified (1998, 2001, and 2004), but by 2007, his string had run out, despite his pledge in June 2007 to implement a cap-and-trade emissions trading scheme.

Australia's Climate Commission, a federal government agency, in 2013 associated extreme weather in that country, especially heat waves and droughts (both of which sparked many wildfires), to global warming worldwide. The fires scorched many heavily populated areas of New South Wales and Queensland, killed six people, and caused $2.43 billion in damage. At times, the heat waves were followed by flooding rains, which often fell with such intensity that water ran off baked, parched earth. While Australian climate scientists had heretofore been reluctant to connect these events with a more general pattern of climate change because Australia is prone to a drought-deluge cycle, the Climate Commission's report, titled *The Angry Summer*, asserted that the "frequency and ferocity of recent extreme weather events indicate an acceleration that is unlikely to abate unless serious steps are taken to prevent further changes to the planet's environment." Tim Flannery, the commission's leader, told the Australian Broadcasting Corporation, "I think one of the best ways of thinking about it is imagining that the baseline has shifted" (Siegel 2013b).

Temperature Records Fall Year by Year

Temperature records have been falling with a regularity heretofore unknown, the report said.

> Included were milestones like the hottest summer on record, the hottest day for Australia as a whole, and the hottest seven consecutive days ever recorded. To put it into perspective, in the 102 years since Australia began gathering national records, there have been 21 days when the country averaged a high of more than 102°F (39°C), and eight of them were in 2013. (Siegel 2013b)

In other words, a preponderance of evidence has established a pattern. The author of the report, Will Steffen, said, "The findings were consistent with an overall global acceleration of weather factors like rising temperatures and heavier rains attributed by scientists to human-caused climate change" (Siegel 2013b).

As thousands of sheep and cattle died in fires during the record heat, climate scientists evoked alarm. "Those of us who spend our days . . . contributing to the scientific literature on climate change are becoming increasingly gloomy about the future of human civilization," Elizabeth Hanna, a researcher at the Australian National University in Canberra, told The Sydney *Morning Herald*. "We are well past the time of niceties, of avoiding the dire nature of what is unfolding, and politely trying not to scare the public" (Siegel, 2013a).

Further Reading

Barta, Patrick. 2007. "Parched Outback: In Australia, a Drought Spurs a Radical Remedy." *Wall Street Journal*, July 11: A-1, A-12.

Bishop, Greg. 2014. "In Wilting Heat, Sharapova Escapes 2nd Round." *New York Times*, January 16. Accessed September 18, 2018. http://www.nytimes.com/2014/01/17/sports/tennis/heat-wilts-players-and-crowd-at-australian-open.html.

Braasch, Gary. 2007. *Earth Under Fire: How Global Warming Is Changing the World*. Berkeley: University of California Press.

Bradsher, Keith. 2008. "A Drought in Australia, a Global Shortage of Rice." *New York Times*, April 17. http://www.nytimes.com/2008/04/17/business/worldbusiness/17warm.html.

"Bushfire Devastates Australian Town." 2016. NASA Earth Observatory, January 12. Accessed September 18, 2018. http://earthobservatory.nasa.gov/IOTD/view.php?id=87302&src=eoa-iotd.

"Bushfires Menace Towns in Western Australia." 2015. NASA Earth Observatory, February 6. Accessed September 18, 2018. http://earthobservatory.nasa.gov/IOTD/view.php?id=85225&src=eoa-iotd.

Draper, Robert. 2009. "Australia's Dry Run." *National Geographic*, April: 34–59. https://www.nationalgeographic.com/magazine/2009/04/murray-darling/. Accessed November 21, 2018.

"Heat Wave Stifles Australia." 2014. NASA Earth Observatory, January 12. Accessed September 18, 2018. http://earthobservatory.nasa.gov/IOTD/view.php?id=82790&src=eoa-iotd.

Innis, Michelle. 2015. "Record Heat Puts Australia at Risk of Intense Fire Season." *New York Times*, November 20. Accessed September 18, 2018. http://www.nytimes.com/2015/11/21/world/australia/australia-fires-record-temperatures.html.

Kenneally, Christine. 2009. "The Inferno." *New Yorker*, October 26: 46–53.

"Land Clearances Turned up the Heat on Australian Climate." 2009. *New Scientist*, May 16. Accessed September 18, 2018. http://www.newscientist.com/article/mg20227084.700-land-clearances-turned-up-the-heat-on-australian-climate.html.

Pincock, Stephen. 2007. "Climate Politics: Showdown in a Sunburnt Country." *Nature* 450 (November 14): 336–338.

Siegel, Matt. 2013a. "Record Heat Fuels Widespread Fires in Australia." *New York Times*, January 9. Accessed September 18, 2018. http://www.nytimes.com/2013/01/10/world/asia/record-heat-fuels-widespread-fires-in-australia.html.

Siegel, Matt. 2013b. "Report Blames Climate Change for Extremes in Australia." *New York Times*, March 4. Accessed September 18, 2018. http://www.nytimes.com/2013/03/05/world/asia/australian-government-blames-climate-change-for-angry-summer.html.

Wiseman, Paul. 2007. "Australia Pushes New Climate Plan." *USA Today*, September 6: 10-A.

CHINA

Global Warming Wild Card

China is the wildest card in the world greenhouse deck. On one hand, the world's most populous country is streamlining energy efficiency and experimenting with new fuel sources. On the other, China is undergoing an industrial revolution with a population of about 1.4 billion, consuming enormous amounts of coal and oil even as its economy becomes more efficient. By 2015, China was consuming half the world's coal production and emitting 50 percent more greenhouse gases than the United States, following a decades-long economic boom.

The scale of industrial development in China in the late 20th and early 21st centuries has no parallel in human history. To gauge the scale of the building boom (and greenhouse-gas generation) in China, consider the amount of cement manufactured and used there between 2010 and 2013: 6.1 gigatons. The United States produced and used 4.4 gigatons of cement during the *entire* twentieth century.

China's Greenhouse-Gas Emissions Surpass the United States'

During China's economic boom, greenhouse-gas emissions grew rapidly. According to the Netherlands Environmental Agency, China's carbon-dioxide emissions increased 8 percent in 2007. This increase represented two-thirds of global growth in greenhouse-gas emissions. By the end of 2007, China's emissions exceeded those of the United States by 14 percent. Per capita, however, United States' emissions were still four times those of China, 19.4 tons to 5.1 tons (Rosenthal 2008).

China's growth has been explosive. In 2014, its economy expanded at 7.3 percent, its *slowest* expansion of gross domestic product (GDP) in 25 years. Coal use—and electricity production from it—grew about 10 percent a year (doubling in about seven years, with compounding), reflecting a similar rise in the country's gross domestic product.

The country's declining growth rate and efforts to install more wind and solar energy by 2015 was causing coal combustion to stabilize, as many coal-fired power plants were operating well below capacity, even as about 150 such plants were still in planning stages or under construction (Wong 2015, A-6). "China already has more coal capacity than it will ever need," said Zhang Boting, vice chairman of the China Society for Hydropower Engineering. "A few years down the road we'll see what a waste these plants are" (Wong, 2015, A-6). At the same time, China has become the world's largest producer and consumer of both solar panels and wind turbines.

Official policies in China now encourage wind and solar power, as government reacts to a warming climate. Eastern China endured a record-breaking heat wave during July and early August 2013. Shanghai set all-time record highs three times in three weeks, peaking at 40.8°C (105.4°F) on August 7, 2013.

China's winter of 2006–2007 was unusually mild, drawing attention to the warming climate. A popular 1,400-year-old ice festival in Harbin, in northeast China, about 400 miles east of the Russian border, literally melted, threatening a tourist attraction that usually draws 5 million people a year. Edward Cody, of the *Washington Post*, wrote:

> The hands had melted off a delicately entwined couple of ballet dancers crafted by an ice-sculpting team from Vladivostok. Eaves fashioned from packed snow drooped into icicles at the Roast Meat Fire House restaurant. Authorities banned people from approaching the ice-cube tower at Ice and Snow World because big chunks kept falling off. . . . Heads are falling from statues and intricately sculpted ice animals are turning into shapeless blobs. (Cody 2007, A-19)

In the midst of the non-winter of 2006–2007, the China Meteorological Administration said that temperatures probably would continue to rise by 7°F to 10.8°F by the year 2100 compared to average temperatures between 1961 and 1990. In Beijing, during the Lunar New Year celebrations, the warmest since authorities began keeping records in 1951, people jogged without jackets in Ritan Park, as boys played basketball in T-shirts. On February 5, the temperature rose to a record of 61°F (Cody 2007, A-19).

Alarms over global warming also have begun to ring in some of China's official agencies. Oceanographers at China's State Oceanic Administration argue that sea-level rise of three feet a century could cause flooding in many of China's coastal areas, home to half of China's large cities and 40 percent of its population.

China's Appetite for Resources

Even as awareness of climate change's perils grows, so does China's industrial base and consumer use of energy. With its expanding industries, more automobiles, and a more affluent population, China scours the world looking for energy in all forms. China's demand for energy and other natural resources can roil world markets. "China's demand has also provided life support to coal producers suffering from declining use in the United States and other industrialized countries," wrote Clifford Krauss and Keith Bradsher in the *New York Times* (2014). By 2013, China was burning 10.1 million barrels of oil a day, one-ninth of the world's supply, but itself produced only 4.2 million. These figures remained relatively stable into 2017.

Before 2015, with Chinese GDP and energy consumption growing at an average of 10 percent a year, global prices for oil and other natural resources soared. When growth slowed to less than 7 percent in 2015, with a lower rate projected in 2016, commodity-futures prices sank worldwide. Chinese oil companies have an eye on U.S. production of oil by hydraulic fracturing ("fracking"). Some in China see shale oil production as a substantial source for natural gas to relieve some of the burden of low-energy "brown" coal that is so abundant in China. However, much of China's shale-gas deposits lie deep underground in complex geological formations, raising safety and environmental issues.

Chinese Car Culture

During the mid-1990s, people in China owned a mere handful of private cars. Private automobile ownership grew by 26 percent between 1996 and 2000 and by 69 percent in 2003 alone. The number of passenger cars on Chinese roads rose from about 6 million in 2000 to about 20 million in 2006. China accounted for 18 percent of global growth in automobile sales between 2002 and 2012. During the 1990s, motor vehicle sales in the Chinese countryside rose from about 40,000 to almost 500,000 per year (Leggett 2001, A-19). Shanghai Automotive Industrial Corporation licensed General Motors technology to build a basic pickup truck for Chinese farmers. The new vehicle, to be called "Combo," is produced in a nonprofit government car factory. This is one of GM's efforts to tap an auto market of a billion consumers and a network of national highways that has been expanding rapidly.

In 2009, for the first time, the Chinese purchased more cars than there were residents of the United States, 12.8 million to 10.3 million. China's car sales increased 42 percent in 2009 compared to 2008, including 72 percent for SUVs, despite the fact that 2009 was a year when most of the world was in recession (Bradsher 2009). Automobile sales in China increased more than 800 percent from 2000 to 2007 (Bradsher 2008). By 2008, more Buicks were being sold in China than in the United States in a market where car size is closely identified with social and economic status. Some wealthy Chinese paid more than $200,000 for a Hummer. By the end of 2016, Chinese owned more than 172 million private cars.

All new cars, minivans, and sport-utility vehicles sold in China starting July 1, 2007 had to meet fuel-economy standards stricter than those in the United States. New construction codes encourage the use of double-glazed windows to reduce air-conditioning and heating costs and high-tech light bulbs that produce more light with fewer watts.

China's highway system may soon surpass that of the United States. The 23,000 miles of highway in 2006 had doubled that of 2001. The Chinese government in 2006 announced plans to build 53,000 freeway miles by 2035. The U.S. Interstate Highway System comprises 46,000 miles. As with the U.S. Interstate System, China's goal is to consolidate the nation and to allow the easy transport of military forces between regions. Policy anticipates that western territories such as Tibet and Xinjiang (meaning "New Frontier") will be fully integrated, ethnically and economically (Conover 2006).

Chinese car culture resembles that of the United States:

> City drivers, stuck in ever-growing jams, listen to traffic radio. They buy auto magazines with titles like *The King of Cars, AutoStyle, China Auto Pictorial, Friends of Cars, Whaam* ("The Car—The Street—The Travel—The Racing"). Two-dozen titles now compete for space in kiosks. The McDonald's Corporation said that it expects half of its new outlets in China to include drive-through lanes. (Conover 2006)

As China's fleet of motor vehicles expanded, its consumption of oil also increased from roughly 2.2 million barrels a day in 1988 to 5.2 million barrels a day in 2003, or roughly 150 percent in 15 years (an average of 10 percent a year). Oil use accelerated after that to almost 7 million barrels a day by 2007 and 11 million a day in 2016. The International Energy Agency issued figures from its office in Paris indicating that increases in Chinese greenhouse-gas emissions between 2000 and 2030 "will nearly equal the increase from the entire industrialized world" (Bradsher 2003, 1).

In addition to its massive industrial expansion, China since the year 2000 has been adding about 7.5 billion square feet of residential and commercial real estate per year, as much as all existing retail shopping centers and strip malls in the United States, according to the United States Energy Information Administration (Kahn and Yardley 2006). An increasing proportion of this space is air-conditioned, increasing energy demand. In addition, most Chinese buildings, even new ones, have little or no thermal insulation and require twice as much energy to heat or cool as the same amount of floor space in similar United States or European climates, according to the World Bank. China has energy efficiency standards, but most new buildings do not meet them (Kahn and Yardley 2006).

Further Reading

Bradsher, Keith. 2003. "China's Boom Adds to Global Warming Problem." *New York Times,* October 22. Accessed on November 21, 2018. https://www.nytimes.com/2003/10/22/world/china-s-boom-adds-to-global-warming-problem.html.

Bradsher, Keith. 2008. "With First Car, a New Life in China." *New York Times*, April 24. Accessed September 27, 2018. http://www.nytimes.com/2008/04/24/business/world business/24firstcar.html.

Bradsher, Keith. 2009. "Recession Elsewhere, but It's Booming in China." *New York Times,* December 10. Accessed September 27, 2018. http://www.nytimes.com/2009/12/10/business/economy/10consume.html.

Cody, Edward. 2007. "Mild Weather Takes Edge Off Chinese Ice Festival; Residents of Tourist City Blame Global Warming." *Washington Post*, February 25: A-19. Accessed September 27, 2018. http://www.washingtonpost.com/wp-dyn/content/article/2007/02/24/AR2007022401421_pf.html.

Conover, Ted. 2006. "Capitalist Roaders." *New York Times Sunday Magazine*, July 2. Accessed September 27, 2018. http://www.nytimes.com/2006/07/02/magazine/02china.html.

Kahn, Joseph, and Jim Yardley. 2006. "As China Roars, Pollution Reaches Deadly Extremes." *New York Times*, August 26. Accessed September 27, 2018. http://www.nytimes.com/2007/08/26/world/asia/26china.html.

Krauss, Clifford, and Keith Bradsher. 2014. "China's Global Search for Energy." *New York Times*, May 22. Accessed September 27, 2018. http://www.nytimes.com/2014/05/22/business/international/chinas-global-search-for-energy.html.

Leggett, Karby. 2001. "In Rural China, General Motors Sees a Frugal but Huge Market: It Bets Tractor Substitute Will Look Pretty Good to Cold, Wet Farmers." *Wall Street Journal*, January 16: A-19.

Rosenthal, Elisabeth. 2008. "China Increases Lead as Biggest Carbon Dioxide Emitter." *New York Times*, June 14. Accessed September 27, 2018. http://www.nytimes.com/2008/06/14/world/asia/14china.html.

Schapiro, Mark. 2014. *Carbon Shock: A Tale of Risk and Calculus on the Front Lines of the Disrupted Global Economy; How Carbon Is Changing the Cost of Everything*. White River Junction, VT: Chelsea Green Publishing.

Wong, Edward. 2015. "Glut of Coal-Fired Plants Casts Doubts on China's Energy Priorities." *New York Times*, November 12: A-6.

GREENLAND

Adaptations to Warming

Melting ice has been changing the economy and culture of Greenland as tourists and energy entrepreneurs arrive from the lower latitudes. For example, a shrimp factory in Narsaq (once the town's largest employer) has closed, as the shrimp departed the nearby warming waters. All but one of the town's eight fishing boats have been retired. The population of Narsaq fell from 3,000 to 1,500 in about ten years, and the suicide rate has risen. "Fishing is the heart of this town," said Hans Kaspersen, 63, a fisherman. "Lots of people have lost their livelihoods" (Rosenthal 2012).

Temperatures in Greenland are rising more quickly than almost any other place on Earth—roughly 5.5°F between 1987 and 2017, five times the world average. In northeastern Greenland, a rise of 14°F to 21°F degrees is anticipated by some climate models by the end of the twenty-first century.

Some of Greenland's 57,000 people are finding that warming temperatures and melting ice offer opportunities. Deposits of gems and minerals have emerged from melting ice, including large deposits of rare earth essential for cell phones, electric cars, and wind turbines. Gold, iron, zinc, and offshore oil also are being sought. More than 160 active licenses had been issued by 2014 from Greenland's Bureau

Residents of Ittoqqortoormiit Village, Greenland, can no longer use traditional dogsleds to hunt, as the sea ice is now dangerously thin or nonexistent. (Adwo/Dreamstime.com)

of Minerals and Petroleum; at the turn of the century, that figure was less than 20. Greenland also is being pressured to relax its "zero tolerance" policy for mining of uranium because melting ice has offered access to deposits that have been known for several decades.

An Influx of Workers

A society comprised mainly of fishermen and hunters may be facing large-scale development, as well as immigration of foreign workers from around the world.

Business entrepreneurs also are knocking on the door in Greenland, looking for more fossil fuels. Four oil companies have applied to explore offshore as mining companies prospect for uranium and gold. Two aluminum companies want to build smelters using glacial meltwater for hydroelectric power. The U.S. Geological Survey estimates that waters off Greenland's northeastern coast may contain as much as 31 billion barrels of oil and gas. More oil may be found on the west coast, enough to tempt Exxon Mobil, Chevron, Canada's Husky Energy and Cairn Energy, and Sweden's PA Resources. In November 2008, Greenlanders approved a self-rule charter that directs mineral royalties to national development. The idea is to leverage the fruits of global warming to wean Greenland off its annual $680 million subsidy from Denmark.

Arctic Ice at Record Low

For a third consecutive year, Arctic ice reached a record low maximum in early March 2017. "This is just another exclamation point on the overall loss of Arctic sea ice coverage that we've been seeing," said Mark Serreze, the director of the National Snow and Ice Data Center. "We're heading for summers with no sea ice coverage at all" (Fountain 2017). Serreze speculated that summer ice could disappear in the Arctic about 2030.

Melting Arctic ice tends to reinforce itself because liquid water is darker than ice, and thus absorbs more heat during months when the sun shines, a feedback loop called the "Arctic amplification." Much of the ice also appears to be thinner than usual, Serreze said, following a season of very unusual warmth. Late in 2016, some areas in the Arctic experienced temperatures more than 35°F above average during short periods. At the North Pole, mean temperatures averaged 23°F above average during November 2016. Similar patterns occurred throughout the winter.

Further Reading

Fountain, Henry. 2017. "Arctic's Winter Sea Ice Drops to Its Lowest Recorded Level." *New York Times*, March 22. Accessed October 11, 2018. https://www.nytimes.com/2017/03/22/climate/arctic-winter-sea-ice-record-low-global-warming.html.

Danger for Hunters

Erosion of Greenland's ice poses practical dangers for indigenous hunters. Winter pack ice is breaking up, and ice fishing, on which many people depend, has become risky and dangerous. DeNeen L. Brown, of *the Washington Post*, described an Inuit hunter's confrontation with glacial ice in Greenland made more dangerous by a warming climate. Aqqaluk Lynge had been chasing a seal. Fear chilled him when the seal dove under the ice and didn't return. Patience is essential when seal hunting, so he waited, but the seal never came back. When an animal begins to act strangely, such as not coming up for breath, something out of order is happening in nature, wrote Brown.

> The iceberg was at his back. Suddenly it began moving like a monster that was waking up. Lynge . . . looked up in alarm, knowing that these floating mountains . . . for all their frozen beauty, are ruthless and deadly. So he decided to get moving . . . but the engine on his motorboat wouldn't start. Just then he noticed the iceberg moving. If the tip is moving, he knew, it could mean that one end is moving up and the other end is moving down. . . . A friend in another boat nearby quickly gave him a tow. Soon they were speeding away from the iceberg, not waiting to look back. Behind them they heard it turning. "We looked back and saw the whole iceberg was collapsing, exploding almost," he said. "We were so afraid." Then it flipped, creating a great tidal wave that crashed hard onto nearby shorelines. By then the two men were out of the wave's path. "When we were finally far away, we could breathe normally again. We were looking back and seeing nothing was left. It exploded underneath the surface of the sea." (Brown 2002, A-30)

New Markets Open

Some of Greenland's people, most of whom live on the coasts near the edge of the ice cap, are reaping benefits from a warming climate in which average winter temperatures rose about 10°F between 1991 and 2007. New pastures are being used to graze sheep and new fields to grow potatoes. The hay-growing season has been lengthening in southern coastal Greenland, where farmers also have planted Chinese cabbage, several types of flowers, and turnips. Greenland farmers raised 22,000 lambs for local consumption during 2006, another market that is thriving on the warming edges of the retreating ice cap. Some of the lamb has been exported to restaurants in Europe.

By 2007, a few food markets in Greenland were selling local cauliflower, broccoli, and cabbage for the first time. Eight farmers grew potatoes commercially. Five grew vegetables, and home gardeners harvested a few strawberries. Greenland Beer's unique selling point is the purity of its glacier-fed water (Native-grown hops may be next). Ewes were having fatter lambs and more of them during growing season, which extended from the middle of May to about September 15, three weeks more than 10 years previously. "Now spring is coming earlier, and you can have earlier lambings and longer grazing periods," said Eenoraq Frederiksen, 68, a sheep farmer whose farm, near Qassiarsuk, is accessible by a harrowing drive across a rudimentary road plowed in the hillside. "Young people now have a lot of possibilities for the future" (Lyall 2007).

Kim Høegh-Dam, who believes that warming coastal waters near Greenland will bring cod that have been abandoning the North Sea and other more southerly waters, has raised more than $1 million to buy cod trawlers and three processing plants. "Global warming will increase the cod tremendously and will bring other species up from the south," he said (Struck 2007, A-1). A government trawler sent to test the cod runs caught 25 tons in one hour, so much that its crew cut its trip short. While some conservations voice concerns about over-fishing (a factor in past declining cod catches), others' mouths water at the prospects not only for cod, but other sea creatures that thrive in relatively cold water, such as shrimp. "The only limiting factor on human endeavor in Greenland is the temperature," Høegh-Dam said, while bouncing on a fast motorboat past icebergs to visit the agriculture station in this country of few roads. Warm the temperature a bit, and new endeavors pop out like lambs from ewes, he believes (Struck 2007, A-1).

Ilulissat, Greenland's third-largest village with 4,500 people, includes one posh hotel, Hotel Arctic, which doubled its size in 2008. From the windows of Hotel Arctic, icebergs moved so quickly that they appear to have formed and re-formed overnight. "Nobody would have predicted 10 to 15 years ago that Greenland would lose ice that fast," glaciologist Konrad Steffen said. "That revises all of the textbooks" (Henry 2008).

Further north along Greenland's coast, fishermen work from boats for longer periods as pack ice forms later and melts earlier. In recent winters, pack ice has failed to form on large areas of the coast, allowing fishing by boat year-round for halibut and other species. Records at an ice patrol station, Daneborg, said that during 1997 the water was open for 80 days. Now it stays ice-free for 140 days.

Further Reading

Brown, DeNeen L. 2002. "Greenland's Glaciers Crumble; Global Warming Melts Polar Ice Cap into Deadly Icebergs." *Washington Post*, October 13: A-30.

Henry, Tom. 2008. "Global Warming Grips Greenland, Leaving Lasting Mark." *Toledo Blade,* October 12. https://www.toledoblade.com/Nation/2008/10/12/Global-warming-grips -Greenland-leaves-lasting-mark.html.

Lyall, Sarah. 2007. "Warming Revives Flora and Fauna in Greenland." *New York Times*, October 28. Accessed September 27, 2018. http://www.nytimes.com/2007/10/28/world /europe/28greenland.html.

Rosenthal, Elisabeth. 2012. "A Melting Greenland: Perils Against Potential." *New York Times*, September 23. Accessed September 27, 2018. http://www.nytimes.com/2012/09/24 /science/earth/melting-greenland-weighs-perils-against-potential.html.

Struck, Doug. 2007. "Icy Island Warms to Climate Change; Greenlanders Exploit 'Gifts from Nature' While Facing New Hardships." *Washington Post*, June 7: A-1. Accessed September 27, 2018. http://www.washingtonpost.com/wp-dyn/content/article/2007/06/06 /AR2007060602783.html.

INDIA

Heat, Drought, Floods, and Electrical Demand

In India, by 2014 the world's third-largest source of greenhouse gases, the government does not expect that emissions will peak until at least 2040. Action to reduce emissions is all "on speculation"—in the future—as the levels of carbon dioxide and methane continue to rise. India's government realizes that as long as these levels rise, temperatures will rise as well in a country that has been suffering record heat. Politically, however, more energy is required to alleviate poverty in a country where 300 million people have no electricity and the primary energy source of many more remains cow dung, which can be converted into biogas.

Glaciers Melt, Drought Expands

The toll of a warming climate is everywhere in India. Even by 2002, the snout of the Himalayan glacier that feeds the mighty Ganga (Ganges River) had developed giant fractures and crevices along a 10-kilometer stretch, indicating massive ice melting. During 15 years of researching such phenomena, Syed Iqbal Hasnain, who heads the Glacier Research Group at Delhi's Jawaharlal Nehru University, had never seen such a rapid deterioration of the frozen massif. He said: "If the rate continues, we could see much of the Gangotri glacier and others in the Himalayas vanish in the next couple of decades" (Chengappa 2002, 40). Hundreds of millions of people who live within the watershed of the Ganges depend on watershed fed by Himalayan glaciers to some degree. Half of India's hydroelectric power is generated from glacial runoff (Lynas 2004, 238). The Indus River supplies 90 percent of the water used in desert areas of Pakistan from glaciers that rapidly lost mass during most of the 20th century and that have continued to do so in the 21st century (Lynas 2004, 238).

Nearly 90 percent of India's water is used in agriculture. Increasing affluence also increases water demand per person as more people buy water-intensive animal protein; increasing manufacturing of consumer goods and generation of electric power also require more water. "The Himalayan glaciers are receding, agricultural yields are stagnating, dry days have increased, patterns of monsoon have become more unpredictable," Jairam Ramesh, the minister of environment and forests at the time, told the *Mint* newspaper in 2009. "So, we are seeing the effects" of climate change (Asokan 2012).

Roughly 15,000 glaciers bordering India, China, and Nepal, many on the Tibetan Plateau (the "Roof of the World"), which supplies the Ganges, Indus, and Brahmaputra rivers, have been melting more quickly than at any time in recorded history, according to research surveying conditions through 2006 published late in November 2008 by Lonnie Thompson and colleagues in *Geophysical Research Letters*. Studying ice melt on the Naimona'nyi glacier, Thompson's team was stunned to learn that all snow built up since 1944 had melted. "We were very surprised not to find the 1962–1963 horizon, and even more surprised not to find the 1951–1952 signal," Thompson said. In more than 20 years of sampling glaciers all over the world, this was the first time both markers were missing. As more heat is trapped in the atmosphere, said Thompson, it holds more water vapor. This humidity condenses as it rises, releasing heat in mountains. "At the highest elevations, we're seeing something like an average of 0.3°C warming per decade," Thompson said (Reilly 2008).

The "Asian Brown Cloud" from low-level soot and other pollution in India and China is changing the color of glaciers in the Hindu Kush, Himalayas, and other mountain ranges, speckling them with soot, changing their albedo, and speeding melting. Jane Qiu, writing in *Nature* (2010), described the work of Xu Baiqing, an environmental scientist who measured black carbon deposited on the glaciers during a half century "and found increased emissions since the 1990s, coinciding with rapid industrial growth in the region." Working with Angela Marinoni and colleagues at the Institute of Atmospheric Sciences (in Bologna, Italy), Baiqing found "high concentrations of aerosols, including black carbon, above 5,000 meters in the Nepalese Himalayas, which caused significant atmospheric warming. They calculate that deposition of black carbon could increase snow and ice melting of a typical Himalayan glacier by 12 to 34 percent by reducing its ability to reflect light" (Qiu 2010, 141).

In India, where glacial water supplies 700 million people, research regarding the melting of 9,000 Himalayan glaciers has lagged due to a lack of funding. Anecdotal evidence, including increasing runoff, indicates accelerated melting. Snow packs have suffered as monsoon rains (which fall as snow at high elevations) have become more sporadic "for reasons that many scientists ascribe to the world's changing climate" (Filkins 2016, 63).

India, Coal Mining, and Pollution

India's city air is sometimes dirtier than even China's from coal effluvia, cooking fires, and vehicle exhaust. New Delhi competes with Beijing and Shanghai for the filthiest urban air in the world. A major cause is coal-fired power, which has been expanding rapidly, as well as cooking fires, mainly in urban areas. The several hundred million people who use cow dung for cooking fuel are major contributors to air pollution.

India's coal is mainly strip-mined, and its high ash content and low heat value mean it delivers twice the pollution per unit of energy as black coal. In addition, "in a country three times more densely populated than China, India's mines and power plants directly affect millions of residents. Mercury poisoning has cursed generations of villagers in places like Bagesati, in Uttar Pradesh, with contorted bodies, decaying teeth, and mental disorders" (Harris 2014).

Gardiner Harris, of the *New York Times,* wrote of Dhanbad, India:

> Decades of strip mining have left this town in the heart of India's coal fields a fiery moonscape, with mountains of black slag, sulfurous air, and sickened residents. The city of Dhanbad resembles a postapocalyptic movie set, with villages surrounded by barren slag heaps half-obscured by acrid smoke spewing from a century-old fire slowly burning through buried coal seams. Mining and fire cause subsidence that swallows homes, with inhabitants' bodies sometimes never found. (Harris 2014)

Many people suffer skin and respiratory problems from continually burning coal fires. The atmosphere is so dirty that smog levels lead to highway shutdowns and airport closures in parts of India. Many people wear masks to screen out the filthy air. Approaching New Delhi by air resembles a descent into a bowl of dun-brown dishwater.

"If India goes deeper and deeper into coal, we're all doomed," said Veerabhadran Ramanathan, director of the Center for Atmospheric Sciences at the Scripps Institution of Oceanography, one of the world's top climate scientists, and former chairman of the United Nations' Intergovernmental Panel on Climate Change (IPCC) (Harris 2014). By 2015, India was the world's third largest emitter of greenhouse gases, and burgeoning consumption of cheap, dirty coal was a major reason.

Coal has a considerable domestic lobby in India. "India's development imperatives cannot be sacrificed at the altar of potential climate changes many years in the future," said India's power minister, Piyush Goyal. "The West will have to recognize we have the needs of the poor" (Harris 2014). Goyal's agency plans to superintend a doubling of India's coal consumption from 565 million tons (in 2013) to more than a billion in 2019, mainly to provide power, regardless of climatic consequences. At the same time, India's federal government has pledged to expand solar power.

Solar Power Grows in India

India is developing solar power even as it expands coal generation. In Madhya Pradesh, India built what was, in 2014, one of the world's largest solar plants on

800 acres of barren soil. However, the air at the solar plant is so dusty that panels must be continually washed. Solar systems financed by small loans may eventually help millions of people in India acquire electric power for the first time and skip the fossil fuel age entirely. By 2015, several companies in India were financing home systems in villages that cost almost $200 (more than 13,000 Indian rupees), but doing so in monthly payments of $3.50 to $5 a month, which equaled what families who support themselves on $2 to $3 a day had been paying for kerosene. One of the companies, Selco (Solar Electric Light Company), has announced that such a business model will provide access to solar-powered electricity to 300 million people in India (and 1.2 billion in the world as a whole) who presently have no power, skipping the fossil-fuel age entirely.

India's Prime Minister Narendra Modi has promised total electrification in India by the end of 2022, mainly by adding coal-fired generation even when smog has reached the worst levels on Earth in the Delhi area. Solar, which provided only 1 percent of India's power in 2015, may reduce reliance on coal but will not replace it.

This kind of financing enables renewable energy to compete with kerosene, which continues to be inexpensive because of government subsidies that cost taxpayers more than $5 billion a year. Kerosene use contributes to carbon emissions as well as skin irritation, respiratory problems, and risk of fire.

Rising Water Levels

Shyamal Mandal lives at the edge of ruin on Ghoramara Island, in the Ganges River delta, which, at two square miles, has shrunk by half in less than 40 years. As described by Somini Sengupta in the *New York Times*:

> In front of his small mud house lies the wreckage of what was once [a] village on this fragile delta island near the Bay of Bengal. Half of it has sunk into the river. Only a handful of families still hang on so close to the water, and those that do are surrounded by reminders of inexorable destruction: an abandoned half-broken canoe, a coconut palm teetering on a cliff, the gouged-out remnants of a family's fish pond. All that stands between [one] home and the water is a rudimentary mud embankment, and there is no telling, he confessed, when it, too, may fall away. "What will happen next, we don't know," he said, summing up his only certainty. (Sengupta 2007)

The rising sea is only one reason why the island is disappearing. Rivers from the Himalayas, emptying into the Bay of Bengal, have swollen with glacial ice melt in recent years, changing the shape and size of islands in the Ganges River delta. In 30 years, about 31 square miles of the islands have gone under water, according to a study by Sugata Hazra, an oceanographer at Jadavpur University in Kolkata, which shares the same delta. Sheikh Suleman, now nearing 60, recalled a time when his harvest of coconuts was so plentiful that his wife would give them freely to their neighbors. Now, he said, she has to beg for a coconut (Sengupta 2007).

More than 600 families were displaced by 2007, as fields were submerged. Two islands vanished entirely. Ironically, rising levels of atmospheric carbon dioxide are playing a major role in the drowning of these islands, though the people who live there use nearly no fossil fuels. According to an account in the *New York Times*, "several hundred families have moved to a displaced people's camp on Sagar, a nearby island" (Sengupta 2007). The islands also are more exposed to intensifying cyclones from the Bay of Bengal in much the same way that the sinking of coastal wetlands of the Mississippi River exposed New Orleans to the storm surge of Hurricane Katrina during 2005.

Further Reading

Asokan, Shyamantha. 2012. "Indian States Fight over River Usage." *Washington Post*, April 1. Accessed September 28, 2018. http://www.washingtonpost.com/world/asia _pacific/indian-states-fight-over-river-usage/2013/04/01/73026ae0-9895-11e2-b68f -dc5c4b47e519_print.html.

Bearak, Max. 2015. "Electrifying India, With the Sun and Small Loans." *New York Times*, January 2. Accessed September 28, 2018. http://www.nytimes.com/2016/01/03/busi ness/energy-environment/electrifying-india-with-the-sun-and-small-loans.html.

Chengappa, Raj. 2002. "The Monsoon: What's Wrong with the Weather?" *India Today*, August 12: 40.

Filkins, Dexter. 2016. "The End of Ice: Exploring a Himalayan Glacier." *New Yorker*, April 4: 59–65.

Harris, Gardiner. 2014. "Coal Rush in India Could Tip Balance on Climate Change." *New York Times*, November 17. Accessed September 28, 2018. https://www.nytimes .com/2014/11/18/world/coal-rush-in-india-could-tip-balance-on-climate-change .html.

Lynas, Mark. 2004. *High Tide: The Truth About Our Climate Crisis*. New York: Picador/St. Martins.

Qiu, Jane. 2010. "Measuring the Meltdown." *Nature* 468 (November 11): 141–142.

Qiu, Jane. 2015. "Himalayan Ice Can Fool Climate Studies." *Science* 347 (March 27): 1404–1405.

Reilly, Michael. 2008. "Tibetan Glaciers Melting at Stunning Rate." Discovery.com. November 24.

Sengupta, Somini. 2007. "Sea's Rise in India Buries Islands and a Way of Life." *New York Times*, April 11. Accessed September 28, 2018. http://www.nytimes.com/2007/04/11 /world/asia/11india.html.

NIGERIA

Heat, Drought, and the Rise of Boko Haram

As with the civil war in Syria (described elsewhere in this section), the social and economic breakdown fostering the rise of the terrorist group Boko Haram has roots in a spreading Sahara Desert, with accompanying heat and drought. While Boko Haram has been depicted in some news media as an isolated group mainly known for kidnapping young girls to keep them from attending schools, its effects on society at large have been much broader.

At the height of the crisis, several million people in northeastern Nigeria had fled the violence sparked by Boko Haram, many of whom became sick and starved. By 2018, more than 8 million people who had fled Boko Haram were in need of humanitarian assistance, including 5.1 million who were severely malnourished, many of them children. The entire agricultural infrastructure of some regions in Northern Nigeria has collapsed.

The uprooting of refugees has been a crisis mainly contained within Nigeria, as the displaced have been crowded into squalid camps or taken up residence in small towns whose economies have collapsed and become too destitute to deal with people in need of food and shelter.

As Leslie Roberts wrote in *Science* (2017, 18), "Food, clean water, and sanitation are scarce or nonexistent, and these conditions create a perfect breeding ground for disease. In a deadly cycle, malnutrition renders children more susceptible to infection and less able to fight it. Epidemics of malaria and measles rage, polio has resurfaced, and child mortality is off the charts."

The unrest in Nigeria is a major contributor to migration across the Mediterranean Sea into Europe, as thousands of desperate people fleeing drought, famine, and jihadi violence cross the Sahara Desert into Libya. This migration has given rise to right-wing political parties in Europe and formed the basis of political conflict and protest movements in Britain, France, Italy, Austria, and Hungary. Protests of immigration from Africa, a large segment of it from Nigeria (people fleeing Boko Haram), played a role in Great Britain's decision to leave the European Union (Brexit) and the revival of neo-Nazi groups in Germany.

In some Northern Nigerian cities, the smuggling of human beings has become the main source of economic activity. Nigeria is Africa's richest country, but distribution of wealth is wildly uneven. Nigeria's economy has grown due to oil extraction and foreign investment, but income inequity also has risen. As Ben Taub wrote in the *New Yorker*, "Some wealthy businessmen travel with paramilitary escorts; police officers demand bribes at gunpoint, and crippled beggars crawl through traffic . . . tapping on car windows and pleading for leftover food" (2017, 38).

Boko Haram (whose name, loosely translated, means "Western education is a sin") exceeds even other terroristic groups in cruelty. It has been notable for an especially brutal kidnapping of more than 200 Nigerian schoolgirls and the massacre of several hundred civilians in the town of Gamboru Ngala. The brutal nature of the group's actions springs at least in part from instability spurred by a warming, drying climate that has deprived many farmers and herders of their homes and livelihoods. In 2009, the United Kingdom Department for International Development issued a warning indicating that climate was contributing to resource shortages in Nigeria because of land scarcity due to desertification, as well as drought-induced water shortages that have provoked crop failures. The situation intensified in the ensuing decade.

According to a study by the U.S. Institute of Peace, a "basic causal mechanism . . . links climate change with violence in Nigeria." The report concluded that climatic shifts, especially drought, play a major role in making land unusable, followed by negative secondary impacts, such as more sickness, hunger, and joblessness. Poor responses to these, in turn, open the door to conflict (Ahmed 2014).

According to Professor Sabo Bako of Ahmadu Bello University, during the 1980s Boko Haram was preceded by the Maitatsine sect of Northern Nigeria, which emerged during the 1980s as victims of several ecological disasters found themselves destitute, leaving them in "a chaotic state of absolute poverty and social dislocation in search of food, water, shelter, jobs, and means of livelihood" (Ahmed 2014).

Many Nigerians joined Boko Haram after having been rendered homeless, having lost all that they owned to severe drought, which has inflicted food shortages on Chad Niger, along Nigeria's northern borders. Boko Haram's recruits are a minority of about 200,000 farmers and herders who have been forced off their farms by intensifying heat and drought. "While a good number of these men were found in major cities like Lagos, pushing water carts and repatriating their earnings to the families they left behind," according to the *Africa Review*, "others were believed to have been lured by the Boko Haram" (Ahmed 2014).

In Northern Nigeria, where activity by Boko Haram has been most intense, 70 percent of the population lived on less than the equivalent of one United States dollar per day per person. David Francis, one of the first reporters to cover Boko Haram from outside of Africa, said that "Most of the foot soldiers of Boko Haram aren't Muslim fanatics; they're poor kids who were turned against their corrupt country by a charismatic leader" (Ahmed 2014).

Drought and heat have reduced arable land in Northern Nigeria and neighboring states and intensified conflict between farmers, who are mostly Christian, and cattle herders, who are overwhelmingly Muslim. This conflict, which also has roots in climate change, has paralleled Boko Haram's terror campaign but appears unrelated. Both are related to ecological breakdown, however. Nigeria also has added 125 million people in 50 years, adding to pressure on the land and demand for food and energy. Lake Chad, on Nigeria's northwestern border with the nation of the same name, has lost 90 percent of its surface area in roughly the same period. Its fishing industry has collapsed, and surrounding farms and fields have been routinely starved for water.

"Lake Chad is dying," President Mahamadou Issoufou of Niger said at the opening of the Paris Climate Conference on November 30, 2015. Once more than 25,000 square kilometers (9,700 square miles), Lake Chad has shrunk to 1,500 square kilometers (580 square miles). Beginning with droughts during the 1970s and 1980s, Lake Chad has dried up almost completely at times, threatening the livelihoods of several million people. "A lack of water will lead to more poverty, hunger, and insecurity," Niger's minister of defense, Karidjo Mahamadou, said (Krinninger 2015).

Erika Eichelberger reported in *Mother Jones* that climate change also has contributed to the devastation by heat and drought. Nigeria as a whole has warmed by about 1.4°F since the mid-20th century, with larger increases in arid northern areas.

Between 2010 and 2013, production of corn in Northern Nigeria fell by 1.6 million metric tons due to unfavorable climactic conditions (Eichelberger 2014). Production of wheat also fell by 15,000 metric tons during the same time period. Many farmers and herders have migrated southward, where rainfall is still adequate, sometimes clashing violently with residents there. "That was when 10 of Yakubu Mama's uncles were slaughtered by farmers from the Eggon tribe," wrote Eichelberger (2014). Mama, a 42-year-old Fulani herdsman, said through an interpreter that the Eggon militia knifed his relatives to death, "one by one." And since the victims' families were too afraid to go and collect the corpses, the bodies were eaten, Mama said, "by pigs and dogs." During this series of incidents during January 2013, about 200 people died. The International Crisis Group estimated that about 8,000 people died in clashes between farmers and herders in Nigeria between 2005 and 2013. In addition, Boko Haram murdered about 5,000 people between 2009 (when it was founded) and 2013 (Eichelberger 2014).

"As is the case with other insurgent movements around the world, economic hardship also helps drive recruitment. Poverty and unemployment in the north have reinforced the Boko Haram narrative that says the government has been corrupted by Western values, and thus cares more about enriching itself than helping Nigerians," according to a report issued in 2011 by the United States Institute of Peace, which is funded by the U.S. Congress (Eichelberger 2014). Poverty intensified by heat, drought, and overpopulation is driving Boko Haram at least as much as its radical Islamic doctrines, according to this report.

According to the Civilian Joint Task Force, the anti–Boko Haram vigilante group made up partly of former Boko Haram members, recruitment "has nothing to do with religion, but a lot to do with economic resources." The terror group "swoop[s] on those who don't have any resources whatsoever," said Mausi Segun, a researcher based in Northern Nigeria (Eichelberger 2014).

Environmental Degradation, Poverty, and Violence

The Nigerian government asserts that quelling violence will require addressing the relationship between environmental degradation, poverty, and violence. Labaran Maku, Nigeria's minister of information, told *Mother Jones* that poverty due partly to crop failures in the north creates "a conducive environment for [terrorism] to prosper" (Eichelberger 2014).

Nigerian government reports assert that spreading deserts are crippling the area around Lake Chad. Eziuche Ubani, who chairs the committee on climate change in Nigeria's House of Representatives, said that it has become "easy" for Boko Haram to recruit Northern Nigerians "because people are already aggrieved and hopeless and depressed because they are losing their source of livelihood" (Eichelberger 2014).

However, Paul Lubeck, the associate director of African studies at the Johns Hopkins School of Advanced International Studies, said that the Nigerian government may be overemphasizing the current effects of climate change on the Boko Haram insurgency to deflect blame for its own inability to stem attacks. "The north is faced with a crisis of unimaginable proportions," he said, "and external solutions are very attractive" (Eichelberger 2014).

Further Reading

Ahmed, Nafeez. 2014. "Behind the Rise of Boko Haram: Ecological Disaster, Oil Crisis, Spy Games; Islamist Militancy in Nigeria Is Being Strengthened by Western and Regional Fossil Fuel Interests." *The Guardian*, May 9. Accessed September 28, 2018. https://www.theguardian.com/environment/earth-insight/2014/may/09/behind-rise -nigeria-boko-haram-climate-disaster-peak-oil-depletion.

Eichelberger, Erika. 2014. "How Environmental Disaster Is Making Boko Haram Violence Worse." *Mother Jones*, June 10. Accessed September 28, 2018. http://www.mother jones.com/environment/2014/06/nigeria-environment-climate-change-boko-haram.

Krinninger, Theresa. 2015. "Lake Chad: Climate Change Fosters Terrorism." DW. Accessed on September 28, 2018. http://www.dw.com/en/lake-chad-climate-change -fosters-terrorism/a-18899499.

Roberts, Leslie. 2017. "Nigeria's Invisible Crisis." *Science* 356 (April 7): 18–23.

Taub, Ben. 2017. "We Have No Choice." *New Yorker*, April 10: 36–49.

RUSSIA

Warming's Chilly Reception

Official reception for global warming has long been chilly in Russia. As a northerly country, some officials often maintain they could use more of it. Melting polar ice aids Russia's access to Arctic resources, including Arctic oil, gas, metals, and materials. Two potential shipping routes, the Northwest Passage off Canada and the Northern Sea Route adjacent to Russia, have been opening as Arctic ice melts. Even given these advantages, however, debits of a warming climate are accumulating in Russia as well as everywhere else. This fact was recognized in 2017 by Russia's president, Vladimir V. Putin, when he came to the defense of the 2015 Paris accord on climate change after U.S. president Donald Trump indicated that he might "cancel" the United States' participation in it.

Heat Waves Scorch Russia

Despite official denial that climate change poses a problem in Russia, evidence has been plentiful. During mid-summer 2010, for example, Russia suffered a withering heat wave that raised temperatures to 100°F in Moscow (where the average mid-summer high is 75°F) and scorched 24 million acres of wheat fields in southern parts of the country, where no rain had fallen since April. The heat kindled peat bog fires around Moscow, "sending coils of smoke into the hazy air. Ignited by a casually flipped cigarette butt or an improperly placed campfire, these fires burrow into the dry peat and then creep along, dozens of feet underground. They pollute the air, ignite above-ground forest fires, and are typically only fully extinguished after a heavy rain in the fall" (Kramer 2010). So much smoke was generated in some areas that airports shut down. On August 4, 2010, the Russian federal government banned wheat exports, playing a major role in a rise of 40 percent in worldwide wheat futures in one day. The fires also spread radioactivity remaining from the 1986 Chernobyl nuclear accident. Large areas that were burned had been coated with fallout following the accident. Moscow's health minister urged residents to stay indoors or leave the city, even if they missed work. By August 9, the damage of the smoky shroud to people's lungs was being measured in packs of cigarettes per day. Estimates ranged from three to eight. The death rate in Moscow rose by 50 percent from early July to early August.

As Moscow endured weeks of 90-plus highs, large numbers of Russians, lacking air-conditioning, drowned in local rivers and lakes. Peat and wildfires around Moscow threw an acrid eye-stinging blanket over the city, adding more carbon dioxide to the atmosphere. The bogs had been drained in Soviet times for harvest of the peat as fuel, leading to the fires. Moscow's 10 million residents were advised to wear face masks. Moscow temperatures hit an all-time record high of 101°F July 29, a reading matched on August 2. St. Petersburg hit 96°F on July 29. Russia also suffered large areas of prolonged drought even as northern Pakistan mourned the deaths of more than 1,000 people in its worst monsoon deluge on record.

As in the western United States and northern Canada, Russia has been ablaze during hotter, drier summers. On one day, July 11, 2012, more than 97 square miles (about 25,000 hectares) of forests were burning in Siberia, according to the Russian Federal Forestry Agency, as uncontrolled wildfires swept through boreal forests in central and eastern Siberia. Smoke from some of the wildfires in Siberia was being pushed across the Pacific Ocean; "on July 8 and 9, 2012," reported NASA's Earth Observatory, "smoke from Siberia arrived in British Columbia, Canada, and caused ground-level ozone to reach record-high levels" ("Wildfires in Siberia" 2012).

The summers of 2012 and 2013 brought Russia two of its most severe wildfire seasons, following intense heat waves across parts of Siberia. A persistent high-pressure ridge (called a "blocking high" by meteorologists) pushed temperatures to 90°F as far north as Norilsk, near the Arctic Circle. The average daily high in Norilsk in July is 61°F. While most of Siberia's wildfires had occurred south of latitude 57° north, along the southern edge of the taiga (a broad swath of continental forest), in 2012 and 2013, the heat waves allowed ignition and continuous burning to latitude 65° north, in areas where such intense fires had heretofore been unknown.

The higher the temperature, the easier fires can burn, because less energy is necessary to raise temperature to the point of ignition. With temperatures soaring in northern Russia, it was easier for previously active fires to continue burning and for lightning to spark new ones. Russia's north latitudes have warmed much more rapidly since the 1970s than the world average, at an average of about .51°C per decade, compared to about .17°C globally—three times as fast. A study by Anatoly Shvidenko, of the International Institute for Applied Systems Analysis, "expect[s] a doubling in the number of forest fires in Russia's taiga forests by the end of the century, as well as increases in the intensity of those fires" ("Heat Intensifies" 2013).

Jon Ranson, a NASA scientist on a field expedition in Siberia, witnessed the fires on a flight between St. Petersburg and Krasnoyarsk:

> We left St. Petersburg near sunset, with the sun low on the horizon. Smoky sunsets create very red skies, and the colors were pretty spectacular. Dozens, or maybe hundreds [of fires], colored the sky on July 6. . . . It is sobering to realize that in two years so close together that the taiga has suffered such extreme fires. Is this the result of climate change? Or a freak occurrence? What I know, for sure, is that these fires appear to be consuming a lot of forest. They must be releasing a whole lot of carbon into the atmosphere, and what happens here does affect the rest of our world. ("Wildfires in Siberia" 2012)

Shipping Routes Open in the Arctic

By 2011, shipping routes along northern Canada and Russia opened in the summer, as the Arctic ice cap steadily melted. Russian tugboats were traversing the country's northern coast rather routinely, as described by Andrew E. Kramer of the *New York Times*:

> Rounding the northernmost tip of Russia in his oceangoing tugboat this summer, Capt. Vladimir V. Bozanov saw plenty of walruses, some pods of beluga whales, and in the distance a few icebergs. One thing Captain Bozanov did not encounter while towing an industrial barge 2,300 miles across the Arctic Ocean was solid ice blocking his path anywhere along the route. Ten years ago, he said, an ice-free passage, even at the peak of summer, was exceptionally rare. (Kramer, 2011)

As ice melted, these shipping channels widened. With the tugboats came explorers for oil and other resources that long had been locked under the ice at all seasons. ExxonMobil signed a contract to explore the Russian sector of the Arctic Ocean. Rosneft, the Russian state oil company, signed a partnership with ExxonMobil. The irony here is that melting ice caused by combustion of fossil fuels creates an opportunity to find, and burn, more of them.

At the same time, Russian president Vladimir V. Putin frequently endorsed melting of Arctic ice as an opportunity, regardless of the climatic consequences, as he observed that the Arctic is the shortcut between the largest markets of Europe and the Asia-Pacific region, and an "excellent opportunity to optimize costs" (Kramer 2011). The Northeast Passage cuts 4,000 to 5,000 miles off a more southerly route between Western Europe and Japanese or Chinese ports, with attendant savings in time, payroll, and fuel, saving shippers hundreds of thousands of dollars per trip. Ships also use the route without contracting expensive services from icebreakers.

The Russians occasionally have cut a path along this route for a century by searching for favorable summer conditions and remaining as close to land as possible because it is the only route across trackless northern Siberia. Ships have competed with each other to break the speed record for traversing the Northeast Passage. During the summer of 2011, a tanker carrying natural gas condensate made the passage in six and a half days, faster than the previous record of eight. After the Soviet Union collapsed in 1991, the route opened to freight haulers and prospectors from several nations. Larger ships with deeper drafts began to ply the route as more ice receded year by year. In 2009, two commercial cargo ships made the trip; the next year, 18 did so. After that, the route became rather routine. Cruise ships also have been offering Arctic views on routes from Murmansk to Anadyr, another Russian port not far from Alaska on the Bering Sea.

Russian Permafrost Melts

Some of the permafrost that covers as much as 65 percent of Russia has been melting; scientists there expect the permafrost boundary to recede 150 miles northward during the next quarter-century. Already, the diamond-producing town of Mirny, in Yakutia, has evacuated a quarter of its population because their houses melted into the previously frozen soil. Parts of the Trans-Siberian Railway's track have twisted and sunk due to melting of permafrost, causing delays of service of several days at a time.

By 2002, melting permafrost had damaged 300 apartment buildings in the Siberian cities of Norilsk and Yakutsk. "Assuming that the region [Siberia] continues to warm at the modest rate of 0.075°C per year, Lev Khrustalev, a geocryologist at Moscow State University, estimated that by 2030, all five-story structures built between 1950 and 1990 in Yakutsk, a city of 193,000 people, could come crashing down unless steps are taken to strengthen them and preserve the permafrost" (Goldman 2002, 1494).

Russian climate experts described Siberia's problems with melting permafrost during an international conference on climate change in Moscow held in early fall 2003. Georgy Golitsyn, director of Moscow's Institute of Atmospheric Physics, said that by the end of the 21st century, temperature increases in Siberia could be twice the worldwide rise of 1.4°C to 5.8°C anticipated by the IPCC (Meuvret 2003). "Extreme weather events might happen more frequently, [with] the melting of permafrost, which is already noticeable, and damage risks to buildings, roads, and pipelines. Pipelines are always having some trouble," he said (Meuvret 2003).

"Here in Moscow and in European Russia, really cold episodes are becoming quite rare. And in Siberia, very heavy frosts [snows] have almost disappeared. Instead of minus 40°C or minus 50°[C], which were quite frequent, now these occur just occasionally, and they usually experience minus 30°[C]," Golitsyn said (Meuvret 2003). The disastrous flooding of the Lena River basin in 2001 gave a foretaste of the problems to come, he noted. "The winter had been normal, but . . .

the soil was frozen, and then in May a heat wave came with temperatures up to 30°[C] when the snow had not melted yet. . . . This is the type of catastrophe we might have more frequently" (Meuvret 2003).

Further Reading

Goldman, Erica. 2002. "Even in the High Arctic, Nothing Is Permanent." *Science* 297 (August 30): 1493–1494.

"Heat Intensifies Siberian Wildfires." 2013. NASA Earth Observatory, August 2. Accessed September 28, 2018. http://earthobservatory.nasa.gov/IOTD/view .php?id=81736&src=eoa-iotd.

Kramer, Andrew E. 2010. "Russians and Their Crops Wilt Under Heat Wave." *New York Times*, July 19. Accessed September 28, 2018. http://www.nytimes.com/2010/07/20 /world/europe/20russia.html.

Kramer, Andrew E. 2011. "Warming Revives Dream of Sea Route in Russian Arctic." *New York Times*, October 18. Accessed September 28, 2018. http://www.nytimes .com/2011/10/18/business/global/warming-revives-old-dream-of-sea-route-in-rus sian-arctic.html.

Meuvret, Odile. 2003. "Global Warming Could Turn Siberia into Disaster Zone: Expert." Agence France Presse, October 2.

"Northwest Passage Nearly Open." 2007. NASA Earth Observatory, August 27. Accessed November 5, 2018. https://earthobservatory.nasa.gov/images/7993/north west-passage-nearly-open.

"Wildfires in Siberia." 2012. NASA Earth Observatory, July 12. Accessed September 28, 2018. http://earthobservatory.nasa.gov/IOTD/view.php?id=78515&src=eoa-iotd.

SYRIA

Drought and Civil War

Syria's civil war and the rise of the Islamic State in Iran and Syria (ISIS) has been associated by scientists with prolonged drought in the region (beginning in 2007), which, in turn, has roots in global warming. The drought intensified after that year. According to Elizabeth Kolbert, writing in the *New Yorker* (2015, 23), "The country [in 2008] experienced its driest winter on record. Wheat production failed, many small farmers lost their herds, and prices of basic commodities more than doubled." Within months, as the drought intensified, hundreds of thousands of people abandoned their homes and farms in the countryside and moved to Damascus, Aleppo, Homs, and other cities, crowding them into urban areas already strained by the arrival of more than a million refugees from ongoing war in Iraq. By 2017, more than 4.5 million Syrian refugees had moved to Turkey, Lebanon, and Jordan, as well as several European countries. By 2016, NASA had issued a report using tree rings to determine that the drought in this region from 1998 to 2012 was the worst in at least 900 years ("NASA" 2016).

Researchers from the University of California Santa Barbara and Columbia University described civil unrest linked to the collapse of farming in Syria and the migration of 1.5 million farmers to cities, with related poverty that provoked civil unrest beginning in 2007. Because of the civil war, weather records after that time are scarce. Droughts have become more frequent and intense in Syria; three of Syria's longest droughts have occurred during the last 30 years, as temperatures have risen and winter precipitation has declined.

"There are various things going on, but you're talking about 1.5 million people migrating from the rural north to the cities," said climate scientist Richard Seager, a coauthor of the study in the *Proceedings of the National Academy of Sciences*. "It was a contributing factor to the social unraveling that occurred that eventually led to the civil war" (Borenstein 2015). The study's lead author, Colin Kelley, said that climatic change combined with oppression by the Assad regime, immigration of at least 1 million refugees from Iraq, and political instability across the region to cause the civil war. However, said Seager, this is the "single clearest case" ever presented by scientists of climate change playing a part in conflict because "you can really draw a blow-by-blow account with the numbers" (Borenstein 2015).

The Most Severe Drought on Record

Kelley and colleagues summarized the situation in the *Proceedings of the National Academy of Sciences* (2015). They wrote that prior to the beginning of the conflict in 2011, agricultural infrastructure in the area known as the greater Fertile Crescent was damaged by its most severe drought on the instrumental record. The drought had a "catalytic effect," they wrote, "contributing to political unrest," combining natural variability and a long-term drying trend, with unusual severity provoked by rising temperatures (Kelley et al. 2015).

Similar to the drought in Nigeria (discussed earlier in this chapter) that played a role in the development of Boko Haram, conditions in Syria led to the collapse of agriculture infrastructure and mass migration of refugees inside the country, as well as to Europe, where resistance provoked right-wing political movements. Kelley and colleagues wrote:

> Precipitation changes in Syria are linked to rising mean sea-level pressure in the Eastern Mediterranean, which also shows a long-term trend. There has been also a long-term warming trend in the Eastern Mediterranean, adding to the drawdown of soil moisture. No natural cause is apparent for these trends, whereas the observed drying and warming are consistent with model studies of the response to increases in greenhouse gases. Furthermore, model studies show an increasingly drier and hotter future mean climate for the Eastern Mediterranean. Analyses of observations and model simulations indicate that a drought of the severity and duration of the recent Syrian drought, which is implicated in the current conflict, has become more than twice as likely as a consequence of human interference in the climate system.

The Syrian civil war has become a vivid illustration of a country in which climate-related crises have been fundamental to violent conflict. Another paper in the *Proceedings of the National Association of Sciences* traced conflict and climate generally to societies with existing social and political conflicts that fracture along ethnic lines. The authors of this paper assigned climate change a role as a triggering mechanism:

> This overall state of affairs is likely to be exacerbated by anthropogenic climate change and in particular climate-related natural disasters. Ethnic divides might serve as predetermined conflict lines in case of rapidly emerging societal tensions arising from disruptive events like natural disasters. Here, we hypothesize that climate-related disaster occurrence enhances armed-conflict outbreak risk in ethnically fractionalized countries. (Schleussner et al. 2016)

Climate Change a Trigger for Conflict

"This debate comes up time and again—is climate change really something like a trigger for violent conflict?" said Hans Joachim Schellnhuber, director of the Potsdam Institute for Climate Impact Research (Harvey 2016). He applied climatic criteria to a list of armed conflicts between 1980 and 2010, analyzing each disaster, and found a significant link between climate disasters and the outbreak of violent conflict specifically in countries with high degrees of ethnic fractionalization. Schellnhuber's work found that about 23 percent of armed conflicts in highly ethnically divided nations coincided with climate-related disasters. "We cannot explain the full complexity of the emergence of violent conflict, but here we have found something really robust, a factor that really matters," Schellnhuber said (Harvey 2016).

Syria is one dramatic example where millions of people in a single nation have become refugees because of war with origins in environmental crisis. During the 21st century, the phrase "environmental refugee" has become more familiar around the world. Lowland residents who could be forced out of their homes by a three-foot rise in sea levels during the present century include 26 million people in Bangladesh, 70 to 100 million in China, 20 million in India, and 12 million in the Nile Delta of Egypt (Gelbspan 1997, 162). In Egypt, a one-meter sea-level rise could cost 15 percent of the country's gross national product, including much of its agricultural base (Edgerton 1991, 72–73). A 14-inch rise in sea levels could flood 40 percent of the mudflats that ring Puget Sound, obliterating a significant habitat for shellfish and waterfowl (Gough 1999, 48). These locations are only a few of many examples, because the Earth's junctures of seas and rivers have been important crossroads for human trade (as well as fertile farming areas) throughout human history. Many river deltas are densely populated and very vulnerable to even a small amount of sea-level rise.

The World Bank released a report in 2013 listing the ten cities in the world that may incur the highest bills for damage from sea-level rise. They are: Miami, New York City, New Orleans, Tampa, and Boston, all in the United States, as well as Guangzhou, China; Mumbai, India; Nagoya, Japan; Shenzhen, China; and Osaka, Japan. The Bank surveyed prospective damage in 136 large coastal cities and concluded that damage could rise to $1 trillion a year. Directed by World Bank economist

Stephane Hallegatte, with the Organisation for Economic Co-operation and Development (OECD), the study warns that "coastal cities face a high risk from increasingly costly flooding as sea levels rise amid climate change. Their current defenses will not be enough as the water level rises" ("10 Coastal Cities" 2013). Many of these cities are at risk not only because of rising seas, but also subsiding land.

Further Reading

Borenstein, Seth. 2015. "Syria's Civil War Linked Partly to Drought, Global Warming." Associated Press, March 3.

Edgerton, Lynne T., and Natural Resources Defense Council. 1991. *The Rising Tide: Global Warming and World Sea Levels.* Washington, DC: Island Press.

Gelbspan, Ross. 1997. *The Heat Is On: The High Stakes Battle Over Earth's Threatened Climate.* Reading, MA: Addison-Wesley Publishing Co.

Gough, Robert. 1999. "Stress on Stress: Global Warming and Aquatic Resource Depletion." *Native Americas* 16, no. 3/4 (Fall/Winter): 46–48.

Harvey, Chelsea. 2016. "How Climate Disasters Can Drive Violent Conflict Around the World." *Washington Post*, July 25. Accessed September 28, 2018. https://www.washingtonpost.com/news/energy-environment/wp/2016/07/25/how-climate-disasters -can-drive-violent-conflict-around-the-world/.

Kelley, Colin, Shahrzad Mohtadi, Mark A. Cane, Richard Seager, and Yochanan Kushnir. 2015. "Climate Change in the Fertile Crescent and Implications of the Recent Syrian Drought." *Proceedings of the National Academy of Sciences* 112, no. 11 (March 2): 3241–3246. Accessed September 28, 2018. http://doi.org/10.1073/pnas.1421533112.

Kolbert, Elizabeth. 2015. "Unsafe Climates." *New Yorker,* December 7: 23–24.

"NASA: Recent Mideast Drought Worst in 900 Years." 2016. *Omaha World-Herald*, March 4: 7-A.

Schleussner, Carl-Friedrich, Jonathan F. Donges, Reik V. Donner, and Hans Joachim Schellnhuber. 2016. "Armed-Conflict Risks Enhanced by Climate-Related Disasters in Ethnically Fractionalized Countries." *Proceedings of the National Association of Sciences* 113, no. 33 (July 25): 9216–9221. Accessed September 28, 2018. http://doi.org/10.1073 /pnas.1601611113.

"10 Coastal Cities at Greatest Flood Risk as Sea Levels Rise." 2013. Environment News Service, September 3. Accessed September 28, 2018. http://ens-newswire .com/2013/09/03/10-coastal-cities-at-greatest-flood-risk-as-sea-levels-rise/.

UNITED STATES

The Carbon Footprint of War

Given the lethal toll of war, even the military opposes its use in international diplomacy until all other avenues have been exhausted. Most war is waged in a self-justifying gale of hatred-fueled nationalistic rage that disregards all "collateral damage" as nasty but necessary. The environmental degradation of war is no state secret. Vietnam was being sprayed with Agent Orange (with consequent birth defects that have now spread through several generations) even as citizens of the United States celebrated their first Earth Day.

War has become progressively more mechanized—and carbon-intensive—with the passage of time. As the Pentagon's own reports now tell us, we have no national security without climate security. War has become the ultimate environmental oxymoron. The Pentagon has recognized climate change as a national-security threat and found it useful for pitching an increase in defense spending. Every four years, the U.S. Defense Department reviews its mission as rationale for its budget. In 2010, climate change found its way into this document as a "preparedness mission . . . occurring in multiple and unpredictable combinations." The report was cited by Defense Secretary Robert Gates to help justify the department's $553 billion budget (as of 2011). It called climate change "an accelerant of instability" ("Pentagon Budget" 2011: 2-A).

Even as the Pentagon's own reports acknowledge that global warming poses a national-security risk, the carbon footprint of war is discussed only rarely. War has been sewn so tightly into our national psyche that even the most ardent environmentalists rarely question what war costs the atmosphere. Carbon dioxide and other greenhouse gases have no opinion on the morality of war. They merely retain heat, whatever its source. Thermal inertia delivers the results of atmospheric change roughly a half-century after our burning of fossil fuels provokes them. The weather today is reacting to greenhouse gas emissions from about 1965. Since then, the world's emissions have risen substantially. The Cold War, the Vietnam War, the Gulf War, the Iraq War, and the Afghanistan War have all played their part.

War's Carbon Intensity

Modern machine-enabled war waged at long distances is very carbon-dioxide intensive. Compared to the visible and all-too-obvious death and environmental mayhem caused by warfare, the often-silent, long-range toll of its carbon footprint may not be easily visible but is hardly harmless. U.S. armed forces, which maintain as many as 1,000 bases in other countries, consume about 2 million gallons of oil per day, half of it in jet fuel. Fuel economy has not been a priority in modern fossil-fueled warfare. Humvees average four miles per gallon, while an Apache helicopter gets half a mile per gallon.

Consumption of fossil fuels has increased over time, with waste apace. The U.S. Air Force alone uses half the oil consumed by the Department of Defense, burning through 2.6 billion gallons of fuel in six months during 2006 while prosecuting wars in Iraq and Afghanistan. The armed forces at that time were using as much fuel per month in limited wars as they consumed through similar periods during World War II, a conflict of truly global scope, between 1941 and 1945.

Horses to Jets

A little more than a hundred years ago, at the beginning of World War I in Europe, the main motive force in battle was the horse and shoe leather, as troops in Europe marched off to battle on foot or horseback. World War I quickly witnessed a

A U.S. Air Force Boeing B-52G Stratofortress bomber of the 1708th Bomb Wing takes off on a mission during Operation DESERT STORM, adding large amounts of carbon dioxide to the atmosphere. (U.S. Department of Defense)

dramatic escalation in war's carbon-dioxide production with the advent of aerial bombardment, as well as increasing use of tanks. War is often a powerful technological motor and carbon-consumption innovator. World War II began with quarter-century-old biplanes and ended with jet-propelled fighters, a massive increase in fuel consumption.

The mechanization of the military provided many more opportunities to ramp up carbon-dioxide production during the world wars of the early 20th century. World War II's Sherman tank, for example, got 0.8 miles per gallon. Seventy-five years later, tank mileage had not improved: the 68-ton Abrams Tank got 0.5 miles per gallon. Fighter jets' typical fuel consumption was on the order of 300 to 400 gallons per hour at full thrust or 100 gallons per hour at cruising speed during hundreds of hours of training or combat missions. Blasting to supersonic speed on its afterburners, an F-15 fighter can burn as much as four gallons of fuel per second. According to Gar Smith, writing in *Earth Island Journal*, the B-52 Stratocruiser, with eight jet engines, consumes 86 barrels of fuel per hour—that's 3,784 gallons, or 63 gallons per minute (Smith 1990–1991).

How many decades of riding a bike to work would be required to offset one F-15 flying for an hour? Assuming that our hypothetical bike replaces a car that gets 25 miles per gallon, a daily commute of five miles would use a gallon a week. That's 350 weeks, roughly seven years, to fuel a fighter jet at full thrust for one hour.

During the 1950s and 1960s, U.S. B-52s were in the air at all times on the theory that an airborne fleet would prevent the Soviet Union from obliterating the entire U.S. nuclear-armed armada on the ground. Each of these B-52s burned thousands of gallons of fossil fuel per hour while aloft. That's 73 bike commuters' annual fuel savings per hour a B-52 is aloft.

Calculating War's Carbon Footprint

The carbon footprint of war is important and intriguing, but impossible to exactly calculate because of its size, scope, and complexity. The carbon footprint for a bag of potato chips, a carton of milk, or a pair of athletic shoes may be calculated by measuring the energy used at each step of manufacturing and transportation during the entire creation and sale of a given product. Calculating the carbon footprint of a single consumer product is a complex chore, but it's attainable. When we are really serious about carbon footprints, we will know the amount of greenhouse gases generated by each platoon sent to war, each bomb dropped, each tank deployed. However, today we know the carbon footprint of a bag of British potato chips from a Tesco grocery store in England, but war—that elephant in the greenhouse—remains unmeasured. The United States launched the Iraq War on the pretext of protecting vital oil supplies, even as it consumed oil at a phenomenal rate. At the start of the Iraq war in 2003, the United Kingdom Green Party estimated that the United States, Britain, and the minor parties of the "coalition of the willing" were burning the same amount of fuel in the Iraq war (40,000 barrels a day) as the 1.1 billion people of India at that time. The U.S. Air Force used 2.6 billion gallons of jet fuel every six months, 10 percent of the U.S. domestic market. By the end of 2007, according to a report from Oil Change International by Nikki Reisch and Steve Kretzmann, the Iraq War had put at least 141 million metric tons of carbon dioxide equivalent into the air, as much as adding 25 million cars to the roads. The Iraq War by itself added more greenhouse gases to the atmosphere than 60 percent of the world's nations. Greenhouse-gas emissions from the war in Iraq (or elsewhere abroad) are not reported in the national inventories compiled by the United Nations Framework Convention on Climate Change, a gigantic loophole.

So how would one begin to sketch the carbon footprint of a war? Here is a preliminary sketch:

- First, add in all the energy used to produce the weapons, transport, and other provisions that are consumed in the war.
- Add the emissions produced getting soldiers, supplies, and civilian contractors to the theater of war and home again—in the case of a war pursued thousands of miles from home, often by air transport, quite a bit. Add the cost of running their armed personnel carriers, heating and cooling their lodgings, and so forth, as well as the greenhouse gases added by the conduct of combat itself.
- Add the carbon and other greenhouse gases added to the atmosphere by fires initiated by bombings and other explosions. In Iraq, pay special attention to

intentional sabotage of oil pipelines and suicide bombings, as well as improvised explosive devices.

• Add the carbon cost of tending the wounded. In this war, Iraq's emergency room spans nearly half the world, from airborne surgery to the Landstuhl Regional Medical Center in Germany and hospitals in the United States.

• Add anything that this list omits for any special circumstances of any given war.

A "Green" Military?

A "green" military isn't one with high-mileage tanks or bombers flying on biofuel, but a worldwide refit of the military's mission that requires more than a change of technology. It requires a redefinition of nationalism to conform to the needs of the Earth system in our time in which the military becomes a service organization that reacts to environmental threats in a future where war as we know it today is illegal on environmental grounds.

Peacemakers in our time are often assumed to be naïve dreamers. Given the environmental circumstances, however, a timely end to war is not naïve, but necessary. Armies of the future will study the best ways to solve international conflicts without armed conflict and the monumental pollution that accompanies the death and destruction.

The greenhouse-gas emissions of war should be regulated on a worldwide basis, and the United States, the world's premier military power, should take the lead in decarbonizing international relations. With the carbon footprint of war adding to its cost in blood and treasure, this tally of greenhouse-gas emissions should convince us that no war is worth what we pay for it. A plausible argument may be made that the Earth can no longer afford fossil-fueled war.

Further Reading

Bryce, Robert. 2005. "Gas Pains." *Atlantic Monthly*, May: 34–35.
Johansen, Bruce E. 2009. "The Carbon Footprint of War." *The Progressive*, October: 27–28.
Johnson, Chalmers. 2004. "America's Empire of Bases." *TomDispatch*, January 15.
Johnson, Chalmers. 2007. "737 U.S. Military Bases = Global Empire." Information Clearing House, February 19.
Karbuz, Sohbet. 2007. "U.S. Military Oil Consumption—Facts and Figures." *Energy Bulletin*, May 20: 2. Accessed September 28, 2018. http://karbuz.blogspot.com/2007/05/us-military-energy-consumption-facts.html.
"Pentagon Budget Reflects Growth of Security Needs." 2011. McClatchy Newspapers in *Omaha World-Herald*, February 13: 2-A.
Sanders, Barry. 2009. *The Green Zone: How a Greening Culture Cannot Ignore the Military*. Oakland, CA: AK Press.
Smith, Gar. 1990–1991. "How Fuel-Efficient Is the Pentagon? Military's Oil Addiction." *Earth Island Journal* (Winter). Environmentalists Against War. Accessed September 28, 2018. http://www.envirosagainstwar.org/know/read.php?itemid=593.

Chapter 2: Deforestation, Reforestation, and Desertification

OVERVIEW

The world's tropical forests continue to lose coverage at an accelerating rate, with record losses of about 40 million acres in each of 2016 and 2017, according to satellite data compiled by Global Forest Watch. Causes of deforestation compound one another, from clearance for agriculture to expansion of urban areas to intensifying drought related to changes in atmospheric circulation as global temperatures rise. "These new numbers show an alarming situation for the world's rain forests," said Andreas Dahl-Jorgensen, deputy director of the Norwegian Government's International Climate and Forest Initiative (Plumer 2018).

Climate change is playing a role in turning large amounts of fertile land to desert—enough within the next generation to create an "environmental crisis of global proportions," as well as large-scale migrations, especially in parts of Africa and Central Asia, according to the United Nations. "The costs of desertification are large," said Zafar Adeel of the United Nations University (Rosenthal 2007). "Already at the moment there are tens of millions of people on the move," Adeel said. "There's internal displacement. There's international migration. There are a number of causes. But by and large, in sub-Saharan Africa and Central Asia this movement is triggered by degradation of land," said Adeel (Rosenthal 2007). Overuse of limited water resources has been made worse by climate-change driven drought. In addition, populations are rising, and many rivers are being diverted for irrigation and short-term gain.

"Today, those migrants who are escaping dry lands are mostly moving around far from the developed world," said Janos Bogardi of the United Nations University in Bonn, Germany, a technical adviser on the report. "Those who end up on boats to Europe are the tip of an iceberg," he noted (Rosenthal 2007). "The numbers we now find alarming may explode in an uncontrollable way," Bogardi said, eight years before Europe was inundated by a flood of refugees fleeing from both war and drought in Syria, Iraq, Afghanistan, and parts of Africa. "Because if you look at land use now and dry land, there is the potential that we are nearing a tipping point" (Rosenthal 2007).

About a third of the land on Earth is now desert—almost 20 million square miles—and the percentage rises every year. The Intergovernmental Panel on Climate Change projects that rainfall could decrease by 15 to 20 percent in the

The illegal harvesting of trees is a problem worldwide, negatively influencing the climate. (Apisit Wilaijit/Dreamstime.com)

Middle East and 25 percent in North Africa as weather becomes more extreme (Bilger 2011, 112).

Forests worldwide absorb more than 25 percent of the carbon dioxide that human beings put into the atmosphere—about the amount emitted by all of humanity's cars and trucks—equal approximately to the volume absorbed by the oceans, which are growing acidic enough to imperil sea life with calcium-based shells. Humankind has been reducing forest cover for centuries, usually to make way for agriculture and cities. In our time, deforestation has been most intense in Indonesia and the Amazon Valley of Brazil but extends to many other areas with expanding populations as well, such as China, India, and parts of sub-Saharan Africa.

Trees are roughly 50 percent carbon. Alive, they remove carbon dioxide from the atmosphere and replace it with oxygen. Trees that are dying or dead (or being burned as fuel) become sources, not absorbers ("sinks") of carbon dioxide and other greenhouse gases. The destruction of forests around the world is no trivial matter for the atmosphere's carbon budget. While the atmosphere contains about 750 billion tons of carbon dioxide, forests contain about 2,000 billion tons. Roughly 500 billion tons is stored in trees and shrubs and 1,500 billion tons in peat bogs, soil, and forest litter (Jardine 1994).

By 2007, about one-fifth of the human-caused greenhouse gases being released into the atmosphere came from deforestation, during which carbon stored in trees enters the air. Indonesia has been clearing more forests than any other country. In some areas, such as the province of Riau (on the island of Sumatra), more than half the forests have been felled in a decade, many for palm oil plantations. The burning and drying of Riau's peatlands also release about 1.8 billion tons of greenhouse gases a year (Gelling 2007).

The Earth had less than 5 million square miles of old-growth forest remaining by about 2013. An estimated deforestation rate of 62,000 square miles a year, the amount being logged in recent years, could reduce that figure to zero within two human lifetimes. More than 50 percent of the world's forests have been destroyed within the past 100 years. An area equivalent to 57 soccer fields falls each minute (Webb 1998).

Africa has been plagued with expanding deserts as the Sahara pushes the populations of Morocco, Tunisia, and Algeria northward toward the Mediterranean. At the southern edge of the Sahara, in countries from Senegal and Mauritania in the west to Sudan, Ethiopia, and Somalia in the east, the demands of growing human populations and livestock numbers are converting land into desert. Iran also is doing battle with desertification. Mohammad Jarian, who directs Iran's Anti-Desertification Organization, reported in 2002 that sandstorms had buried 124 villages in the southeastern province of Sistan-Baluchistan (Brown 2006).

During the first 20 years of the 21st century, about 60 million people are expected to leave the Sahelian region of Africa, a region of northern Africa that borders the fringe of the Sahara Desert. United Nations secretary-general Kofi Annan said in 2002 that in northeast Asia, "dust and sandstorms have buried human settlements and forced schools and airports to shut down . . . while in the Americas, dry spells and sandstorms have alarmed farmers and raised the specter of another Dust Bowl, reminiscent of the 1930s." In southern Europe, "lands once green and rich in vegetation are barren and brown," he said ("Global Climate Shift" 2002).

Australian government researcher Leon Rotstayn has compiled evidence indicating that air pollution probably has contributed to catastrophic drought in the Sahel. Sulfate aerosols, tiny atmospheric particles, have contributed to a global climate shift, he said. "The Sahelian drought may be due to a combination of natural variability and atmospheric aerosol," said Rotstayn. "Cleaner air in [the] future will mean greater rainfall in this region," he continued ("Global Climate Shift" 2002).

"Global climate change is not solely being caused by rising levels of greenhouse gases. Atmospheric pollution is also having an effect," said Rotstayn, who is affiliated with the Commonwealth Scientific and Industrial Research Organisation (CSIRO), the Australian government's climate-change research agency. Using global climate simulations, Rotstayn found that sulfate aerosols, which are concentrated mainly in the Northern Hemisphere, make cloud droplets smaller. This makes clouds brighter and longer lasting, so they reflect more sunlight into space, cooling the Earth's surface below ("Global Climate Shift" 2002). As a result, the tropical rain belt, which migrates northward and southward with the seasonal movement of the sun, is weakened in the Northern Hemisphere and does not move as far north. This change has had a major impact on the Sahel, which frequently has experienced devastating drought since the 1960s. "Rainfall was 20 to 49 percent lower than in the first half of the 20th century, causing widespread famine and death" ("Global Climate Shift" 2002).

This chapter will examine topics such as wildfires in Australia, Portugal, and Russia, as well as deforestation from Brazil and Honduras to Indonesia and Myanmar. Finally, the chapter will take a look at an effort to combat desertification in China, with its Green Great Wall.

Further Reading

Bilger, Burkhard. 2011. "The Great Oasis." *New Yorker*, December 10 and 26: 110–121.

Brown, Lester R. 2006. "The Earth Is Shrinking: Advancing Deserts and Rising Seas Squeezing Civilization." Earth Policy Institute, November 15.

Gelling, Peter. 2007. "Forest Loss in Sumatra Becomes a Global Issue." *New York Times*, December 6. Accessed September 28, 2018. http://www.nytimes.com/2007/12/06/world/asia/06indo.html.

"Global Climate Shift Feeds Spreading Deserts." 2002. Environment News Service, June 17. Accessed September 28, 2018. http://ens-news.com/ens/jun2002/2002-06-17-03.asp.

Plumer, Brad. 2018. "Tropical Forests Suffered Near-Record Tree Losses in 2017." *New York Times*, June 27. Accessed September 28, 2018. https://www.nytimes.com/2018/06/27/climate/tropical-trees-deforestation.html.

Rosenthal, Elisabeth. 2007. "Likely Spread of Deserts to Fertile Land Requires Quick Response, U.N. Report Says." *New York Times*, June 28. Accessed September 28, 2018. http://www.nytimes.com/2007/06/28/world/28deserts.html.

Webb, Jason. 1998. "World Forests Said Vulnerable to Global Warming." Reuters. November 4.

AUSTRALIA

Wildfires Run Rampant

Australia, always a desert continent, is getting even drier as temperatures warm globally, aggravated by El Niño ocean circulation patterns. The continent's forests (mainly in the north and southeast) have been scorched by wildfires driven by hot winds and record temperatures.

The scorching summers of 2015 and 2016 had Australians recalling another El Niño summer: searing drought and heat waves during the summer of 2008–2009. According to an Associated Press account published in the *New York Times*, "Towering flames razed entire towns in southeastern Australia and burned fleeing residents in their cars as the death toll rose [to more than 180], making it the country's deadliest disaster" ("Death Toll" 2009). Several thousand homes were destroyed in early February as temperatures rose above 100°F several days in Melbourne and Sydney; temperature reports were as high as 117°F in the scorched interior of Victoria State, the highest on record.

Thousands of volunteer firefighters and Australian Army troops battled the wind-whipped flames. Some entire small villages (including Marysville and several hamlets in the Kinglake district, 50 miles north of Melbourne) were wiped out by racing flames. In Kinglake, according to the same Associated Press report:

> Just five houses out of about 40 remained standing. . . . Street after street was lined by smoldering wrecks of homes; roofs collapsed inward, iron roof sheets twisted from the heat . . . burned-out hulks of cars dotted roads. . . . The landscape was blackened as far as the eye could see. Entire forests were reduced to leafless, charred trunks, farmland to ashes.

During the fires, around 3:00 a.m. on the morning of January 29, 2009, in the northern suburbs of Adelaide, strong northwesterly winds mixed hot air aloft to the surface. At RAAF Edinburgh, a regional airport, the temperature rose to 41.7°C (106°F) at 3:04 a.m., an event without known precedent in southern Australia.

Some brushfires in Australia start with lightning, but others have been human caused, such as a fire that burned more than 47,000 hectares in New South Wales. The State Mine fire started on Marrangaroo Army Range in the Blue Mountains of New South Wales near the city of Lithgow on October 16, according to the Environment News Service ("Australian Army" 2013). A spokesman for the South

---

Australia: A Record Hot Summer

While heat waves over Australia's inland desert are a regular aspect of climate there, records set in February 2017 were remarkable and an indication of an upward trend over time.

"With overheated bats dropping from trees and bushfires burning out of control, temperatures smashed records in many areas," reported the NASA Earth Observatory. "On February 12, 2017, air temperatures rose to 46.6°C (115.9°F) in the coastal city of Port Macquarie, New South Wales, breaking the city's all-time record by 3.3°C (5.9°F). Two days earlier, the average maximum temperature across all of New South Wales hit a record-setting 42.4°C (108.3°F)—a record that was broken the next day when it rose to 44.0°C (111.2°F)" ("Australia" 2017). One town in New South Wales, Mungindi, near the border with Queensland, reported 52 consecutive days with highs more than 35°C (95°F), a record for the province.

A study by the Climate Council of Australia (Steffen et al. 2014) has asserted that heat waves (three or more consecutive days of unusually high temperatures) have grown significantly longer, more intense, and frequent between 1971 and 2008.

Further Reading

"Australia: A Really Unusual Heat Wave." 2017. NASA Earth Observatory, February 21. Accessed October 11, 2018. http://earthobservatory.nasa.gov/IOTD/view.php?id=89683&src=eoa-iotd.

Steffen, Will, Lesley Hughes, and Sarah Perkins. 2014. "Heatwaves: Hotter, Longer, More Often." Climate Council of Australia Ltd. http://www.climatecouncil.org.au/uploads/9901f6614a2cac7b2b888f55b4dff9cc.pdf.

Australian Country Fire (ACF) Service confirmed that "Defence had been responsible for starting a fire at the Cultana training area, south of Port Augusta, on Saturday and that it was 'part of an exercise.'" Australia's largest environmental group said the fires are in part the result of the "new climate reality" and that Australians must start cutting carbon pollution now. ACF's Paul Sinclair added, "Sydney is, quite literally, on fire right now. Homes and cars are being lost as bushfires rage, fueled by a record-setting heat wave which NSW Fire Commissioner Greg Mullins says he's never seen the likes of so early in the year" ("Australian Army" 2013).

"Coincidentally, last week also saw the release of the Intergovernmental Panel on Climate Change's report unequivocally telling us if we don't start acting to rein in pollution, we can expect more of it," said Sinclair ("Australian Army" 2013). "In Australia, the problem we face is that we have been polluting unheeded for decades, and now we are beginning to pay the price," he said. "That price is a risk to our health and our well-being—access to clean drinking water, access to affordable food. The price is more frequent, more devastating bushfires, droughts, and floods" ("Australian Army" 2013).

The Pattern Extends to Chile

Wildfires also have been worsening in other parts of the Southern Hemisphere, due to atmospheric wind patterns called "Hadley Cells" that are expanding due to rising temperatures. A similar pattern is expanding the Kalahari Desert in southern Africa, along with Cape Town's water supply. During January 2017, Chile experienced its worst fire disaster in 50 years, losing more than 500,000 hectares to more than 100 wildfires. Several firefighters died in the stubborn fires that roared out of densely wooded steep terrain. The fires were stoked by windy, hot, and dry summer weather. Chilean President Michelle Bachelet declared a national state of emergency. One fire consumed more than 10,000 hectares—26 square kilometers—in one night.

Cristian Orellana, a forestry manager in central Chile, said, "The heat was such that even planes couldn't fly over to drop water" ("'Extreme' Chilean Wildfires" 2017). The fire razed the town of Santa Olga, turning its 4,500 inhabitants into refugees who huddled in makeshift army tent camps. The Environment News Service reported, "The fires can reach extreme temperatures, sparking devastating firestorms spread by strong winds. . . . Trees are blown off their roots and the fires carbonize everything in their path[s], leaving only white ash scattered across the hills" ("'Extreme' Chilean Wildfires" 2017). Entire vineyards in Chile's famous wine region were reduced to ash.

Fires Worsen in Australia

Fires continued to scorch large parts of Australia through the summer of 2016–2017 in what was becoming nearly an annual pattern. The British Broadcasting Corporation reported on February 12, 2017 that almost 100 fires were burning across New South Wales alone, a third of them uncontained, as temperatures hit records as hot as 47°C (117°F) and a dry wind desiccated the earth. More than 2,500 firefighters were trying to contain the flames. Officials said the conditions were worse than during the 2009 "Black Saturday" fires in the state of Victoria, which killed 173 people. "This is the worst day we have seen in the history of New South Wales when it comes to fire danger ratings and fire conditions," Shane Fitz-simmons, the state's rural fire chief, told reporters ("Nearly 100 Bushfires" 2017). Emergency warnings were issued for several areas, with residents told to evacuate if they could or seek shelter and avoid bush or grassland where it was too late to leave.

One blaze, the Sir Ivan fire east of Dunedoo (northwest of Sydney), was spreading quickly amid difficult and dangerous conditions, the rural fire service said. The Associated Press reported that "one fire alone burned through 50,000 hectares [124,000 acres]. New South Wales Rural Fire Services Commissioner Shane Fitz-simmons said . . . that two firefighters were hospitalized. One suffered burns to the hands and face, while another had a laceration to the hand" ("Homes Destroyed" 2017). Homes, machinery, and farms were swallowed as fast-moving flames jumped containment lines. Dry, hot northwesterly winds from Australia's interior, some gusting to 50 miles an hour, stoked the flames.

Further Reading

"Australian Army Started Bushfire Blazes, Climate Also Blamed." 2013. Environment News Service, October 23. Accessed September 28, 2018. http://ens-newswire .com/2013/10/23/australian-army-started-bushfire-blazes-climate-also-blamed/.

"Death Toll in Australian Bushfires Climbs to 84." 2009. Associated Press in the *New York Times*, February 9. Accessed September 28, 2018. http://www.nytimes.com /2009/02/09/world/asia/09australia.html.

"'Extreme' Chilean Wildfires Worst in Decades." 2017. Environment News Service, February 6. Accessed September 28, 2018. http://ens-newswire.com/2017/02/06 /extreme-chilean-wildfires-worst-in-decades/.

"Homes Destroyed in Fierce Australian Wildfires." 2017. Associated Press in Fox News World, February 12. Accessed September 28, 2018. http://www.foxnews.com /world/2017/02/12/homes-destroyed-in-fierce-australian-wildfires.html.

Kenneally, Christine. 2009. "The Inferno." *New Yorker*, October 26: 46–53.

"Nearly 100 Bushfires Raging in Australia's New South Wales State." 2017. British Broadcasting Corporation, February 12. Accessed September 28, 2018. http://www.bbc .com/news/world-australia-38948669.

"Scientist: Warming Threatens Koalas." 2008. Associated Press in *Omaha World-Herald*, May 8: 7-A.

BRAZIL

The Amazon Is No Longer the World's Lungs

Urbanization and industrialization continue in the Amazon Valley, large parts of which are no longer rain forest. It has become Brazil's wild northwest, South

A Brazilian shanty town on the Amazon River reflects the increasing population in the area, a major cause of deforestation. (Paura/Dreamstime.com)

America's fastest-growing region in terms of human presence, the "world's last great settlement frontier" in the words of Brian J. Godfrey, a geography professor at Vassar College and coauthor of *Rainforest Cities* (Romero 2012). All of this has implications for the world's carbon cycle, as both climate change and human agency convert a long-time carbon-dioxide "sink" into a net source of greenhouse gases. Deforestation in the Amazon Valley already, as of 2015, is one of Earth's largest sources of greenhouse-gas emissions.

Forest cover in Brazil continued to decline at record rates in 2016 and 2017, according to satellite data compiled by Global Forest Watch, despite efforts by government to curtail illegal logging. Corporations with large operations there, such as Cargill, have pledged to farm in a more sustainable manner, but despite all of these efforts, smaller farmers continue to clear land and large tracts have been burned because of recurrent drought. "The big concern is that we are starting to see a new normal, where fires, deforestation, and climate change are all interacting to make the Amazon more flammable," said Mikaela Weisse, a research analyst with Global Forest Watch (Plumer 2018).

On December 1, 2015, Survival International reported:

Wildfires are raging through the Brazilian Amazon, destroying vast areas of forest on the eastern fringes of the "earth's lungs" [which traditionally absorb carbon dioxide and produce oxygen]. The fires are reportedly being started by illegal loggers, in retaliation for tribal peoples' [the Awá tribe's] efforts to defend their territories and keep the invaders out. They threaten one of the few remaining areas of pre-Amazon forest in Brazil, the last environment of its type in the world. ("COP21" 2015)

More than half of the Awá's land has been destroyed by fire.

Deforestation Slows, Then Accelerates Again

In recent years, Brazil has been reducing forest loss but in an uneven fashion. Between 1995 and 2005, forest loss in the Brazilian Amazon averaged 19,500 square kilometers per year—roughly the area of Israel. By 2013, that rate had been cut by 70 percent, even as beef and soya production continued to grow. A combination of measures was applied: corporate commitments coupled with strong laws, satellite surveillance and robust enforcement, restrictions on access to credit for farms and ranches in counties with high deforestation, the creation of protected areas and indigenous reserves, and improvements in land tenure and governance. Brazil's federal government worked closely with the beef and soya industries, NGOs, and international partners. In 2008, for example, Norway committed US\$1 billion to Brazil because it wanted to demonstrate practical new ways to protect forests globally. "Even so," according to one report, "Brazil's progress is fragile—deforestation in the Amazon has increased over the past 18 months [2014–2015]" (Victor and Leape 2015, 440).

After a decade during which Amazon deforestation stalled, by 2016 it was accelerating again. Growing worldwide demand for several crops (most notably soybeans) was increasing the size of farms and the burning of jungle habitats. According to the *New York Times*:

> In the Brazilian Amazon, the world's largest rain forest, deforestation rose in 2015 for the first time in nearly a decade, to nearly 2 million acres from August 2015 to July 2016. That is a jump from about 1.5 million acres a year earlier and just over 1.2 million acres the year before that, according to estimates by Brazil's National Institute for Space Research. . . . In Bolivia . . . deforestation appears to be accelerating as well. (Tabuchi et al. 2017)

About 865,000 acres of land—an area nearly the size of Rhode Island—have been deforested, on average, annually for agriculture since 2011, according to estimates from the nongovernmental Bolivia Documentation and Information Center.

Cities Spread as the Amazon Valley Dries

The Parauapebas area, for example, has changed in a generation from an obscure frontier settlement to an urban area of 2 million people "with an air-conditioned shopping mall, gated communities, and a dealership selling Chevy pickup trucks" (Romero 2012). Manaus's population has grown to more than 1.7 million. By 2015, the population of the area as a whole had surpassed 25 million. Of nineteen Brazilian cities that have doubled in population in a decade, 10 are in the Amazon. "More population leads to more deforestation," said Philip M. Fearnside, a researcher at the National Institute for Amazon Research (Romero, 2012). Construction of hydroelectric dams is drawing workers to the Amazon Valley as well, as are open-pit iron-ore mines near Pará and Parauapebas, as well as other locations, meeting heavy demand in several other nations, most notably China.

According to one observer, "Already the Amazon desert is emerging in Rondônia. As far as the eye can see, there is only red, cracked ground. The red dust fills the air and sky itself turns red. Ten years ago this area was jungle, then it was slashed and burned to form farmland. But the jungle soil is notoriously poor, so it was used for pasture" (Brandenburg and Paxson 1999, 225).

The Mato Grosso region of the Amazon, formerly a rain forest, is becoming a dryland savanna. The destruction of the forest reinforces itself because rain clouds form more easily above moist forests. Deforestation also degrades soil quality because most of the Amazon's nutrients come from decaying vegetation. "By removing the forest you remove the nutrients," said Yadvinder Malhi of the University of Oxford. The deforestation of the Mato Grasso ("Great Forest" in Portuguese) is being aggravated by logging; about 17 percent of this region's forest already has been cleared (Brahic 2009).

The Amazon is very sensitive to drought, according to a 30-year study published March 6, 2009, in *Science*; the study provides the first scientific evidence that a drying environment causes increasing carbon loss in tropical forests as it kills trees ("Amazon Carbon" 2009). "For years the Amazon forest has been helping to slow down climate change. But relying on this subsidy from nature is extremely dangerous," said Professor Oliver Phillips, from the University of Leeds, who was the lead author of the research. "If the Earth's carbon sinks slow or go into reverse, as our results show is possible, carbon dioxide levels will rise even faster" ("Amazon Carbon" 2009).

Deforestation Enhances Drought

Just as a lush tropical rain forest augments precipitation, so too can deforestation enhance drought.

> Vegetation affects precipitation patterns by mediating moisture, energy, and trace-gas fluxes between the surface and atmosphere. When forests are replaced by pasture or crops, evapotranspiration of moisture from soil and vegetation is often diminished, leading to reduced atmospheric humidity and potentially suppressing precipitation. Climate models predict that large-scale tropical deforestation causes reduced regional precipitation. (Spracklen et al. 2012)

What is more, drought in the Amazon and other rain forests combines with rapid climate change, habitat loss, and attacks by invasive species to reduce biodiversity in some of what used to be the richest habitats on land anywhere on Earth. Thiago Rangel wrote:

> These human-induced processes may have boosted the background rate of species extinction by 100 to 1,000 times. However, species do not go extinct immediately

when their habitat shrinks, climate changes beyond their tolerance limit, or an invasive species spreads. It may take several generations after an initial impact before the last individual of a species is gone. (Rangel 2012, 162)

The Amazon's capacity as a carbon sink has declined. R. J. W. Brienen and colleagues wrote in *Nature* that rates of net increase in biomass declined by one-third since the 1990s (Brienen et al. 2015). Until recently, forests, most notably those of the Amazon Valley, have been absorbing a large amount of the carbon dioxide produced by human activities. Signs are that this important carbon sink may have reached its maximum capacity (Hedin 2015, 295).

> While this analysis confirms that Amazon forests have acted as a long-term net biomass sink, we find a long-term decreasing trend of carbon accumulation. This is a consequence of growth rate increases leveling off recently, while biomass mortality persistently increased throughout, leading to a shortening of carbon residence times. Potential drivers for the mortality increase include greater climate variability, and feedbacks of faster growth on mortality, resulting in shortened tree longevity. (Brienen et al. 2015, 344)

In 2010, the Amazon Valley experienced its second "100-year drought" in five years. "This is what's quite alarming—that we've seen these two very unusual events," said Simon Lewis, a University of Leeds (Great Britain) forest ecologist. "And those two unusual events are consistent with those predictions that suggest that the Amazon may be severely impacted over the next few decades by these droughts" (Joyce 2011).

Human activity is pushing the Amazon Valley toward drought in many ways, of which fossil-fuel effluvia is only one. Rising sea-surface temperatures in the Atlantic Ocean change circulation patterns in the atmosphere in ways that can induce drier weather in parts of the valley, as changes in climate patterns from more frequent El Niño episodes (which also may be provoked by rising temperatures) do the same. Land-use changes associated with the spread of logging, ranching, and farming also cause parts of the shrinking rain forest to dry out, as world demand surges for tropical timber, soybeans, and free-range beef, among other products. Roads expand to link production centers in the Amazon Valley to ports on South America's Pacific coasts, allowing exports to the rapidly developing economies of China, India, and other parts of Asia. Demand for "green" ethanol meanwhile causes more Amazon forest to be cleared for sugarcane fields. In an area where 25 to 59 percent of rainfall is "recycled" by the forest, once roughly 30 to 40 percent of the rain forest goes, the area as a whole could pass a threshold, a "tipping point" into a drier climate on a long-range basis (Malhi et al. 2008, 169).

Per Cox of the Center for Ecology and Hydrology in Winfrith, United Kingdom, anticipates that increased drought frequency could devastate about 65 percent of the Amazon's forest cover by the end of the 21st century (Cox et al. 2004, 137).

Further Reading

"Amazon Carbon Sink Threatened by Drought." 2009. NASA Earth Observatory, March 5.

Brahic, Catherine. 2009. "Parts of Amazon Close to Tipping Point." *New Scientist*, March 5. Accessed September 28, 2018. http://www.newscientist.com/article/dn16708-parts -of-amazon-close-to-tipping-point.html.

Brandenburg, John E., and Monica Rix Paxson. 1999. *Dead Mars, Dying Earth*. Freedom, CA: Crossing Press.

Brienen, R. J. W., O. L. Phillips, T. R. Feldpausch, E. Gloor, T. R. Baker, J. Lloyd, . . . R. J. Zagt. 2015. "Long-Term Decline of the Amazon Carbon Sink." *Nature* 519 (March 19): 344–348.

"COP21: Amazon Fire Destroying Rare Forest Home of Uncontacted Tribe." 2016. Survival International, December 1. Accessed November 26, 2018. https://www.survivalinter national.org/news/11033.

Cox, P. M., R. A. Betts, M. Collins, P. P. Harris, C. Huntingford, and C. D. Jones. 2004. "Amazonian Forest Dieback Under Climate-Carbon Cycle Projections for the 21st Century." *Theoretical and Applied Climatology* 78 (June): 137–156.

Hedin, Lars O. 2015. "Biogeochemistry: Signs of Saturation in the Tropical Carbon Sink." *Nature* 519 (March 19): 295–296.

Joyce, Christopher. 2011. "'Alarming' Amazon Droughts May Have Global Fallout." National Public Radio in NASA Earth Observatory, February 7. Accessed September 28, 2018. http://www.npr.org/2011/02/07/133462608/alarming-amazon-droughts-may -have-global-fallout.

Malhi, Yadvinder, J. Timmons Roberts, Richard A. Betts, Timothy J. Killeen, Wenhong Li, and Carlos A. Nobre. 2008. "Climate Change, Deforestation, and the Fate of the Amazon." *Science* 319 (January 11): 169–172.

Plumer, Brad. 2018. "Tropical Forests Suffered Near-Record Tree Losses in 2017." *New York Times*, June 27. Accessed September 28, 2018. https://www.nytimes.com/2018/06/27 /climate/tropical-trees-deforestation.html.

Rangel, Thiago F. 2012. "Amazonian Extinction Debts." *Science* 337 (July 13): 162–163.

Romero, Simon. 2012. "Swallowing Rain Forest, Cities Surge in Amazon." *New York Times*, November 24. Accessed September 28, 2018. http://www.nytimes.com/2012/11/25 /world/americas/swallowing-rain-forest-brazilian-cities-surge-in-amazon.html.

Spracklen, D. V., S. R. Arnold, and C. M. Taylor. 2012. "Observations of Increased Tropical Rainfall Preceded by Air Passage Over Forests." *Nature* 489 (September 13): 282–285.

Tabuchi, Hiroko, Claire Rigby, and Jeremy White. 2017. "Amazon Deforestation, Once Tamed, Comes Roaring Back." *New York Times*, February 24. Accessed September 28, 2018. https://www.nytimes.com/2017/02/24/business/energy-environment/deforesta tion-brazil-bolivia-south-america.html.

Victor, David G., and James P. Leape. 2015. "Global Climate Agreement: After the Talks." *Nature* 527 (November 26): 439–441.

CHINA

The "Green Great Wall"

China has been cutting its forests for at least 2,000 years. The Yellow River received its color (and its name) from eroded soil laid bare by over-logging. After World

War II, much of China's old growth was stripped to grow food, supply fuel for steel mills, and other reasons. Until the 1970s, the countryside was being stripped as sandstorms, called "yellow dragons," raked Beijing. In 1978, however, China decided to replant a "great wall" of greenery. During 1981, the National People's Congress by law required each citizen more than 11 years of age to plant at least three poplar, larch, or eucalyptus saplings each year. The numbers have been astounding; roughly 56 billion trees were planted between 1999 and 2009. Today, however, this grand reforestation project is threatened by climate change that has provoked drought, heating, and spreading deserts.

The Green Great Wall is a popular name for the Three-North Shelter Forest Program, also called the Three-North Shelterbelt Program. Three-North is a reference to the three regions of China involved in the program: Northeast, North, and Northwest. By 2050, this artificial forest is planned to stretch over 400 million hectares. China's Great Green Wall has been designed to install almost 90 million acres of new forest, the largest ecological restoration project in human history. By 2010, China's Communist Party announced with considerable fanfare that the country already had the largest human-made forest in the world, more than 500,000 square kilometers, providing China with roughly 20 percent forest cover. By 2050, government planners foresee an arc of greenery 4,480 kilometers (or about 2,800 miles) long, from Xinjiang province in China's far west to Heilongjiang in the east.

Planting Trees as Deserts Expand

China faces a tall order as deserts expand over many areas in its north and west. In Inner Mongolia's Kubuqi Desert, which adjoins the Gobi Desert, reaching close to Beijing, "We are on the front line of a huge Chinese dust bowl advancing east," said Byong Hyon Kwon, once a South Korean ambassador to China who has become an advocate of the Green Great Wall (Trafford 2014). China has been losing roughly 3,600 square kilometers (1,400 square miles) a year to the Gobi Desert. Dust storms, which have been increasing in severity, have been blowing away topsoil across the northern tier of provinces. The storms also have affected agriculture in North and South Korea as well as Japan, and traces of them have reached the West Coast of the United States.

Aerial seeding has been used in areas with enough precipitation to (at least in theory) allow the trees to take root on their own. In drier areas, farmers are given cash incentives to plant trees and shrubs. Sand-tolerant vegetation is also used to stabilize advancing sand dunes, with varying rates of success. Some of the forests are planted quite carefully:

> 1.5-meter-square frames are set in the sand and wired together to make a grid that is heavy enough not to blow away. Trees are then planted inside the squares. Poplars are chosen in part because their roots, which grow like spider webs, can sprout more baby trees. If the tree survives, it should reach a man's height in about four years. The volunteers also plant Salix, a shrub that grows in sandy soil. (Trafford 2014)

Tree planting has become a mass spectacle. Each spring, as many as 3 million Communist Party members, civil servants, and others plant trees as part of a nationwide campaign to showcase China's environmental consciousness. Former U.S. vice president, Nobel Prize winner, and best-selling author Al Gore was impressed, declaring that China had planted two and a half times as many trees per year as the rest of the world combined, "the largest tree-planting program the world has ever seen" ("China's Great" 2010). China's official goal is to have 42 percent of its land covered with forests by 2050 (Land 2014).

While Gore was impressed, some scientists in China expressed concerns that the massive project was not all it was cracked up to be. According to an account in the London *Guardian*, "Jiang Gaoming, professor at the Chinese Academy of Sciences' Institute of Botany and vice secretary-general of the China Society of Biological Conservation, said the Green Great Wall has, in some places, accelerated ecological degeneration by putting pressure on precious water resources in arid and semi-arid regions." Jiang also said that trees planted during the Great Green Wall project are non-native. "Native trees actually play a much bigger role in preventing desertification," he said, because their root systems are better established ("China's Great" 2010). In addition, recently planted forests are less efficient than

---

Deserts Spread in North China

Researchers based at Peking University and the Chinese Academy of Sciences disclosed in 2015 that lakes on the Mongolian Plateau, northwest of Beijing, have been shrinking rapidly. Using several decades of satellite images, the researchers said that lake surface area had decreased 30 percent in a quarter century because of warming temperatures and declining precipitation, as well as increasing mining and agricultural activities. Several lakes have completely dried up. Results were published in *Proceedings of the National Academy of Sciences* (Tao et al. 2015, 2281–2286).

Xinkai Lake, near the Russian border in northern Inner Mongolia, was sizable in 2001 but dry by 2006. Coal mining has been a major drain of water; the number of mines increased from 156 in 2000 to 865 in 2010. Inner Mongolia by 2014 was China's second-largest coal-producing area, as well as a major supplier of rare-earth minerals. Farmers also have been irrigating to an extent that reduces water levels. According to NASA's Earth Observatory (quoting Tao et al.), "The amount of irrigated cropland in Inner Mongolia has increased from 6,600 square kilometers in the late 1970s to 30,003 square kilometers in 2010" ("Shrinking Lakes" 2015).

Further Reading

"Shrinking Lakes on the Mongolian Plateau." 2015. NASA Earth Observatory, April 8. Accessed October 11, 2018. http://earthobservatory.nasa.gov/IOTD/view.php?id=85665 &src=eoa-iotd.

Tao, Shengli, Jingyun Fang, Xia Zhao, Shuqing Zhao, Haihua Shen, Huifeng Hu, . . . Qinghua Guo. 2015. "Rapid Loss of Lakes on the Mongolian Plateau." *Proceedings of the National Academy of Sciences* 112, no. 7 (February): 2281–2286.

older, more complex growth at removing greenhouse gases from the atmosphere. Grasslands also often can sequester more carbon dioxide and methane than young forests.

An article in *Yale Environment 360* said that after more than 30 years of mixed results, some scientists have questioned "the long-term viability of significant aspects of China's reforestation push," most notably its use of non-native species with heavy water demands in a climate that has always been dry and is becoming more so (Luoma 2013).

China's Green Great Wall will not adequately address desertification because it deals only with its surface manifestations. The roots of spreading deserts lie in global warming, including worldwide changes in atmospheric circulation patterns, which require long-term global-scale solutions. The Green Great Wall may stall the advance of deserts in China for a time, but it is not a cure. Jiang Fengguo, director at the Soil and Water Conservation Supervision Station in Hexigten Banner, Inner Mongolia Autonomous Region said that the Green Great Wall may stall desertification for a time but probably will not reverse it. "There will still be problems. Desertification still exists, and the continuing deterioration of the . . . environment has not been reversed," he said ("China's Great" 2010).

This "solution" has several problems, despite its overblown claims. The numbers are indeed impressive (roughly 56 billion trees planted between 1999 and 2009, comprising 5.88 million hectares of forest) until one delves under the public-relations gloss. A substantial number of these trees were "planted" by airplanes and subsequently died. Others were planted by hand with a great deal of care in planned shelterbelts but were unsuited to their surroundings and sucked up groundwater at rates that were not sustainable in a dry climate that shows every sign of further drought. Some of the forests that did take root are biologically relatively sterile monocultures.

The "yellow dragons," Beijing's dust storms, have persisted even after more than 30 years of tree planting, as they "thicken the skies over Beijing with dust and send people with asthmatic lungs and weak hearts to the hospital," aggravating air pollution from coal-fired power plants that gives that city and others in China the dirtiest air on Earth (Land 2014).

Drought and Atmospheric Circulation

Some of China's enduring drought stems from changes in worldwide atmospheric circulation compelled by worldwide climate change. Even though warmer air generally holds more moisture, not everyone will see more precipitation in a globally warmed world. Many deserts already are expanding in a worldwide pattern influenced by atmospheric circulation patterns that meteorologists call "Hadley Cells."

Near the equator, warm, moist air rises, cools, and unleashes downpours. In the upper troposphere, the air spreads north and southward toward both poles, descending at about 30° north and south latitude, creating deserts. As temperatures rise, the Hadley Cells reach further north and south of the equator. While precipitation patterns are also influenced by other factors (such as ready access, or lack thereof, to ocean-borne moisture), rainfall is strongly influenced by Hadley Cells. Rising air portends instability, low pressure, and storminess; descending air generally produces high pressure and clear skies. In a warmer world, Hadley Cells expand, which causes deserts to expand, a process that is already evident from news reports around the world.

Droughts in regions where Hadley Cells favor descending air now span the globe, from parts of China to Spain, Iraq, Afghanistan, the Murray-Darling Basin of Australia, and the United States' Southwest, including California, Nevada, New Mexico, Arizona, and Texas. In China, the Gobi Desert, also within the northern reaches of Hadley Cell range, has been expanding. In Iran, Lake Urmia, once plied by cruise ships, has lost nearly all of its water, and water rationing has been proposed for Tehran.

Even as it seeded billions of trees, in a 2006 report China told the United Nations Convention to Combat Desertification that 2.63 million square kilometers (a quarter of its land mass) was covered with desert, compared with 18 percent in 1994. Between the early 1980s, when mass tree planting began, and 1994, even the Chinese government admitted that its grasslands had been shrinking by about 15,000 square kilometers annually. Every year, 30 percent of the newly planted trees die and have to be replaced. In 2009, one-quarter of 53,000 hectares planted died, some due to drought, others because of severe winter storms.

Thirsty Trees and Groundwater Levels

Major stands of thirsty trees can be counterproductive in the long run. For example, in Mingin, a relatively arid area in Northwestern China, planting large numbers of non-native trees with high water demands played a role in a groundwater decline of 12 to 19 meters. Land erosion, over-farming, and pollution also have played a role in making soils in some areas unsuitable for intense planting of trees. Lack of diversity also makes the new forests more prone to diseases.

Jiang Gaoming, an ecologist at the Chinese Academy of Sciences, has characterized this grand design as a "fairy tale" (Luoma 2013). According to *Earth Science Reviews*, Beijing University scientist Shixiong Cao and five coauthors said that on-the-ground surveys have shown that over time, as many as 85 percent of the plantings will fail (Luoma 2013). David Shankman, a geographer at the University of Alabama, said that "over years or decades the plantings have tended to eventually deplete local soil moisture and die *en masse* simply because the planted species are not native to the region and don't tolerate local conditions" (Luoma 2013).

Jianchu Xu, senior scientist at the World Agroforestry Centre and a professor at the Kunming Institute of Botany, Chinese Academy of Science, said that native perennial grasses, "with their extensive root systems would be better protectors of topsoil" (Luoma 2013).

> In what could be a hopeful turn, China's State Forestry Administration has indicated that it has gotten the message. The nation's lead forestry agency has begun

collaborating on projects aimed specifically at restoring native species. The agency is working with the Climate Community and Biodiversity Alliance (CCBA), whose members include Conservation International, the Nature Conservancy, and the Rainforest Alliance. (Luoma 2013)

Further Reading

"China's Great Green Wall Grows in Climate Fight." 2010. *The Guardian*, September 23. Accessed September 29, 2018. http://www.theguardian.com/environment/2010/sep/23/china-great-green-wall-climate.

Land, Graham. 2014. "The Great Green Wall: Reforesting China." Asian Correspondent.com, February 3. Accessed September 29, 2018. http://asiancorrespondent.com/119175/reforesting-china/.

Luoma, Jon R. 2017. "China's Reforestation Programs: Big Success or Just an Illusion?" *Yale Environment 360,* January 17. Accessed September 29, 2018. http://e360.yale.edu/feature/chinas_reforestation_programs_big_success_or_just_an_illusion/2484/.

Trafford, Abigail. 2014. "Can China's Great, Green Wall Stop Its Creeping Deserts?" *Washington Post* in *The Star*, February 1. Accessed September 29, 2018. http://www.thestar.com/news/world/2014/02/01/can_chinas_great_green_wall_stop_its_creeping_deserts.html.

HONDURAS

Tropical Mountain Forests Devastated

Tropical mountain forests, such as those in Honduras, depend on predictable, frequent, and prolonged immersion in moisture-bearing clouds. Deforestation upwind of lowland forests alters surface energy budgets in ways that influence dry-season cloud fields (Lawton et al. 2001, 584). Cloud forests form where mountains force trade winds to rise above the condensation level, the point of orographic cloud formation. "We all thought we were doing a great job of protecting mountain forests," said Robert O. Lawton, a tropical forest ecologist at the University of Alabama in Huntsville, Alabama. "Now we're seeing that deforestation outside our mountain range, out of our control, can have a big impact" (Yoon 2001, F-5).

Deforestation Shapes Climate

In Honduras, the clearing of forests, sometimes many miles from the mountains, alters this pattern, often raising the elevation at which clouds form. Lawton and colleagues used Landsat and Geostationary Operational satellite images to measure such changes. They found that their "simulations suggest that conversion of forest to pasture has a significant impact on cloud formation" (Lawton et al. 2001, 586). Patterns found in Honduras resemble those in other tropical areas, including parts of the Amazon Valley. "These results suggest that current trends in tropical land use will force cloud forests upward, and they will thus decrease in area and become increasingly fragmented—and in many low mountains may disappear altogether" (Lawton et al. 2001, 587).

Deforestation's effects probably extend further than most observers have here-tofore believed. "Mountain forests . . . may be affected by what's happening some distance away," said Lawton. Each year about 81,000 square miles of tropical for-ests are cleared, said Gary S. Hartshorn, president of the Organization for Tropical Studies, a consortium of rain forest researchers at Duke University (Polakovic 2001, A-1). Thus, the weather in the lush cloud forests of Honduras and surrounding countries is changing because of land-use changes, including deforestation, many miles away. As trees on coastal plains are removed and replaced by farms, roads, and settlements, less moisture evaporates from soil and plants, in turn reducing clouds around forested peaks 65 miles away.

Climate Changes Worldwide

According to a study on Honduras published in *Science,* "these results suggest that current trends in tropical land use will force cloud forests upward and they will thus decrease in area and become increasingly fragmented and in many low mountains may disappear altogether" (Polakovic 2001, A-1). "It's incredibly ominous that over such a distance deforestation can alter clouds in mountains. This is a very serious concern," said Hartshorn. "This is confirmation of what we have predicted for a long time," said Stanford University ecologist Gretchen Daily. "The implications are very serious for the tropics and other parts of the world" (Polakovic 2001, A-1).

Using data collected from satellites and computer models, scientists examined how forest clearing along the Caribbean coastline, where more than 80 percent of lowland forests have been cleared for farms and towns, influences weather down-wind in the Cordillera de Tilarán mountain range. Evaporation from lowland veg-etation is a principal source of moisture for the 4,000-to-5,000-foot mountains during the dry season of January to mid-May.

As Gary Polakovic explained in the *Los Angeles Times*:

> The researchers found that the moisture content of the clouds over the mountains has declined by about half since intensive land clearing began in the 1950s. Also, the cleared land is warmer, pushing the base of clouds nearly a quarter of a mile higher on some days, meaning they pass over the mountain range dropping little moisture. In contrast, clouds were more abundant over forested lowlands just across the bor-der in Nicaragua, where forest still blankets much of the coastal plain. (Polakovic 2001, A-1)

Tropical rain forests are typically the last areas colonized by people, a final refuge of biodiversity in places where lowlands have been cleared and developed. "Many cloud forest organisms have literally nowhere to go," said Dr. Nalini M. Nadkarni, an ecologist at Evergreen State College in Olympia, Washington. "They're stuck on an island of cloud forest. If you remove the cloud, it's curtains for them" (Yoon 2001, F-5).

Many watersheds fed by wet highland forests also are threatened. "We always knew that if you have a town and a mountain behind it, you protect the mountain forests to protect the water," said James O. Juvik, a tropical ecologist at the University of Hawaii at Hilo. "This says, even if you leave the cloud forest intact like good conservationists, if you clear the lowland forests, you can diminish the cloud forest and affect your water flow" (Yoon 2001, F-5).

Five years before flooding from Hurricane Mitch devastated Honduras in 1998, J. Almendares, a Honduran medical doctor, warned readers of the British medical journal *Lancet* that deforestation was making the country more vulnerable than ever to deadly flooding. Almendares presented evidence that "desiccation and soil erosion caused by cattle grazing and sugarcane and cotton cultivation have altered the regional hydrological cycle" (Almendares and Sierra 1993, 1401). These changes have led to fewer rainy days but more intense downpours.

Deforestation, Flooding, and Hurricane Mitch

Almendares presented temperature statistics from one deforested area of Honduras indicating that the average ambient air temperature had risen 7.5°C between 1972 and 1990 (Almendares and Sierra 1993, 1403). In neighboring Nicaragua, during the final months of the Sandinista revolution (1979), similar temperature rises were reported in Managua after dictator Anasazio Somoza ordered the wholesale destruction of trees in the city to deny Sandinistas places to hide during gun battles.

Central America had 200,000 square miles of forest in 1900. By the 1980s, only 36,000 square miles survived, and the rate of deforestation was increasing, especially in the Miskito region of Nicaragua and Honduras. A 1998 report by United Nations agencies and nongovernmental organizations documented a regional deforestation rate of 958,360 acres (1,500 square miles) a year (Weinberg 1999, 51).

Honduras lost a third of its forests between 1964 and 1990. Honduran forests continued to be felled at an annual rate of 80,000 hectares a year during the 1990s, a rate that, if sustained, would strip a quarter of remaining forested land per decade. In the meantime, people who can no longer wrest a living from denuded (or corporate-controlled) land have been moving to Honduran cities, where malaria became endemic.

During October 1998, Hurricane Mitch made landfall on Central America's Miskito Coast, with winds as strong as 178 miles an hour, dropping as much as three feet of rain. Crossing the mountains, Mitch turned northwest, moving slowly through El Salvador, Guatemala, and southern Mexico. By the time the storm reached the sea again, it had killed 10,000 people and left nearly 3 million others homeless. In Honduras, thousands of people who survived the storm lost their jobs in devastated banana plantations.

The devastation wrought by Mitch raised questions in Central America not only about potential increases in hurricane severity due to global warming, but also about changes in land use across the region which make many areas more prone to severe flooding during heavy rains. The same questions were raised in Caracas, Venezuela, after devastating floods there late in 1999.

Bill Weinberg described the conditions that have made hurricane flooding so devastating in Central America:

> Fire and flood fuel each other in a vicious cycle. Landless peasants colonize the agricultural frontier or clear forested slopes for their *milpas* (fields). The more forest is destroyed, the more the hydrologic cycle is disrupted; with no canopy for transpiration, local rainfall and cloud cover decline; aridity makes the surviving forest vulnerable to wildfires. Then, when the rains do come, sweeping in from the Caribbean on the trade winds, there are no roots to hold the soil and absorb the water. Millennia [of] accumulated wealth of organic matter is swept from the mountainsides in deluges of mud. Tlaloc, the revered Nahua rain god who the Maya called Chac Mool, brings destruction instead of abundance. (Weinberg 1999, 51–52)

In Nicaragua, close to the Honduras border, 2,000 people died in the municipality of Posoltega as 10 communities were buried in mudslides when the Casita volcano crater collapsed. Three-quarters of a million people were left homeless in the area. Posoltega lies in an area that has been almost completely deforested. Hurricane Mitch's trail of death and damage in eastern Nicaragua also was intensified by deforestation of the upper Rio Coco watershed. More than 20 inches of rain caused the river to rise more than 60 feet within a few days.

Preservation of forests became an environmental issue in Honduras as well as surrounding nations following Hurricane Mitch's devastation. Local native peoples and ecologists united the same year to evict Korean-owned timber giant Solcarsa, which had won a government contract in a large area of the Miskito rain forest. Honduras and Nicaragua have lost 60 percent of their forest cover within two generations.

Hurricane Mitch left Honduras with an enduring legacy—a plague of rats. Almost a year after the disaster, their numbers exploded from an overabundance of dead meat—both human and animal—following the hurricane. The rats overran fields and homes, decimating crops.

Former Honduran president Rafael Callejas attributed the extreme death and damage wrought by Hurricane Mitch to "mudslides that were the result of uncontrolled deforestation and therefore could have been prevented" (Weinberg 1999, 54). Ecological activism can be as dangerous in Honduras as the winds, rains, and mudslides of a major hurricane. A few weeks before Mitch made landfall, Carlos Luna, a local opponent of logging in the central mountains near Tegucigalpa, the country's capital and largest city, was gunned down by unknown assailants.

The main cause of tropical deforestation—perhaps two-thirds of it—is slash-and-burn agriculture. Much of the deforestation is caused, in turn, by small farmers and ranchers who are threatened by the spread of commercial farming and ranching enterprises. Driven out of more populous areas, subsistence farmers are forced to destroy large amounts of tropical forest to create new farmland. According to a study by the *Global Futures Bulletin*, the pattern is the same in much of Africa, Asia, and Latin America. Increasing human population and pressure on the land is the ultimate cause of the deforestation of slash-and-burn agriculture, according to the report.

Further Reading

Almendares, J., and M. Sierra. 1993. "Critical Conditions: A Profile of Honduras." *Lancet* 342 (December 4): 1400–1403.

Lawton, R. O., U. S. Nair, R. A. Pielke Sr., and R. M. Welch. 2001. "Climatic Impact of Tropical Lowland Deforestation on Nearby Montane Cloud Forests." *Science* 294 (October 19): 584–587.

Polakovic, Gary. 2001. "Deforestation Far Away Hurts Rain Forests, Study Says; Downing Trees on Costa Rica's Coastal Plains Inhibits Cloud Formation in Distant Peaks. 'It's Incredibly Ominous,' a Scientist Says." *Los Angeles Times*, October 19: A-1.

Preston, Douglas. 2018. "Deep in the Honduran Rain Forest, an Ecological SWAT Team Explores a Lost World." *New Yorker* (May 21). https://www.newyorker.com/science /elements/deep-in-the-honduran-rain-forest-an-ecological-swat-team-explores-a-lost -world.

Weinberg, Bill. 1999. "Hurricane Mitch, Indigenous Peoples, and Mesoamerica's Climate Disaster." *Native Americas* 16, no. 3/4 (Fall/Winter): 50–59.

Yoon, Carol Kaesuk. 2001. "Something Missing in Fragile Cloud Forest: The Clouds." *New York Times*, November 20: F-5.

INDONESIA

Farms Rise, Forests Fall

Decades of uncontrolled exploitation have resulted in massive deforestation across Indonesia. The deforestation rate between 1984 and 1998 was around 1.6 million hectares each year, and recent deforestation has been closer to 2.0 to 2.4 million hectares each year, one of the highest deforestation rates in the world. By 1997 Indonesia had lost 72 percent of its original forest cover, and more than half of Indonesia's remaining forests were threatened (Bryant et al. 1997). Twenty years later, more than 85 percent of original forest cover was gone, a trend that has been accelerating due not only to burning (for open land for agriculture) but also to several droughts.

In 2015, Indonesia experienced one of its worst fire seasons on record during El Niño conditions that aggravated drought. In 2016, Indonesia's government imposed a moratorium on conversion of peatland that releases enormous amounts of carbon dioxide when burned. By 2017, fire loss on protected peatlands had declined by 88 percent, part of which was provoked by enforcement, as well as wetter weather. "Still," wrote Brad Plumer in the *New York Times*, "the real test of success may come when the next El Niño hits" (Plumer 2018).

During 2015, at least 2.6 million hectares of Indonesian forest was burned, a swath of land as large as Sicily. During the autumn, "The fires blanketed much of South-East Asia in a noxious haze and released a vast plume of greenhouse gases" ("Despite Tough Talk" 2016). The government issues edicts that go unenforced as the conversion of forests to farmland continues.

Forests Become Furniture

Indonesia's remaining tropical forests are quickly being drawn into the global economy as merchandise. The last of central Java's once-great teak forests has ended up as lawn chairs in the United States, Europe, Japan, Australia, China, and other countries. Dozens of workshops around the forest are turning the teak logs into chairs, mainly for export. Those chairs, and other manufactured products, spelled the end of the majestic forests that once blanketed large parts of Indonesia. Their disappearance also has hastened the extinction of innumerable animal and plant species indigenous to this country. "We are facing a cataclysm," said Togu Manurung, the director of Forest Watch Indonesia, an environmental organization that documents the destruction of the country's forests (Gargan 2001, A-7). With a boom in palm-oil production by 2015, forests have been felled and replaced by palm plantations.

Citing rapid deforestation, the Indonesian Forum on the Environment in 2001 called for a moratorium on industrial logging for two to three years, or until the forest industry and forest management could be reformed. The group proposed that "the logging moratorium would be phased in . . . with the objective of stopping the illegal and destructive logging and promoting sustainable forest management. Illegal logging can most easily be stopped during a period when no industrial logging is allowed" ("Portrait" 2001). In the short term, according to this proposal, forest-based industries would be forced to rely on plantation-produced or imported wood "or else close their operations" ("Portrait" 2001). The proposed moratorium was never enforced, and deforestation continued at accelerating rates. During the 1990s, the rate of Indonesia's deforestation accelerated from 2.47 million acres to 4.2 million acres annually (Gargan 2001, A-7). Free-boot capitalism was even more rapacious in the forests than the former "New Order" of the Suharto dictatorship.

Borneo's Tanjung Puting National Park, designated by the United Nations as a biosphere reserve, is being systematically and illegally logged, according to Forest Watch, Telepak Indonesia, and Indonesia's Ministry of Forestry and Estate Crops. Suripto, the secretary general of the forestry ministry (like many Indonesians, he goes by one name), charged during the year 2000 that lumber companies and sawmills owned by Abdul Raysid, a member of Indonesia's Parliament, were illegally processing ramin logs. The ramin is the most valuable tree in the national park, whose blond, straight-grained wood is used extensively in furniture, wood moldings, window blinds, and pool cues. Suripto's findings were ignored by authorities, and the logging continued (Gargan 2001, A-7).

Moratorium Announced, but Not Enforced

Indonesian forestry minister M. Prakosa and trade and industry minister Rini M. S. Suwandi announced a moratorium on the export of logs and wood chips during October 2001 to "safeguard the conservation of Indonesian forests" ("Indonesia: Low Expectations" 2001). The main result of the ban may not be the conservation of forests, however, but a supply of raw material for domestic wood-processing industries. According to the World Rainforest Movement *Bulletin*, "The new ban will be a boost to the country's highly indebted wood industry and the timber tycoons whose businesses have suffered from a lack of raw materials. A total of 128 companies are under the control of the Indonesian Bank Restructuring Agency (IBRA). A previous commitment to close half of these companies has not been fulfilled" ("Indonesia: Low Expectations" 2001). Additionally, according to the WRM *Bulletin*, the export ban is not likely to significantly reduce timber smuggling "while Indonesia's notoriously corrupt police force, government apparatus, and courts continue as before" ("Indonesia: Low Expectations" 2001).

Now the question for Indonesian forests becomes one of enforcement, since much present logging already is illegal and the power of the central government over the country's many islands has been waning. Illegal logging activity has been increasing in many Indonesian forests despite governmental edicts. According to the Environment News Service:

> The Environmental Investigation Agency reported in 1999 that Tanjung Puting National Park was full of logging camps and an extensive network of wooden rails had been built to drag out the timber. In the east of the park, a logging road was built to truck out the illegal timber. Steel barges were observed loaded with illegal wood, and investigators tracked the timber to local sawmills and factories. ("Indonesian President" 2002)

In Gunung Leuser National Park, according to the same report, investigators witnessed loggers with chainsaws operating in the Suaq Balimbing research area, which provides prime orangutan habitat and is the only place where these apes have been observed using tools ("Indonesian President" 2002).

The Penan are one of the few surviving nomadic peoples of the rain forest in Sarawak (the northern side portion of Borneo, controlled by Malaysia). Of roughly 9,000 Penan alive today, only about 300 continue to live as nomads in the traditional way. They live in a forest laced by rivers, as well as one of the world's most extensive network of caves and underground passages. The Penan are experiencing one of the highest rates of logging on earth. Some Penan also are threatened by a massive dam project. The proposed Bakun dam will flood 70,000 hectares of land, displacing indigenous peoples, wildlife, and large areas of rain forest.

The Penan Homeland Diminished

The Penan are one of several indigenous tribes known collectively as the Dayak. For centuries before the arrival of European colonizers, the Dayak lived in communal long houses, surviving by hunting and slash-and-burn agriculture in the lush rain forest. Aside from occasional raids by notorious head-hunters, life was generally peaceful.

The Penan live amid awe-inspiring scenery. The peaks of Batu Lawi, a Penan sacred site, are said to be places of spirit and mystery.

Rapid deforestation began in the Penans' homeland during the 1960s. An Internet web page of the Sarawak Peoples Campaign described a "frenzy of logging that has gripped Malaysia . . . a rate of forest destruction twice that of the Amazon and by far the highest in the world" ("The Penans," n.d.). In the Baram River drainage alone, more than 30 logging companies, equipped with as many as 1,200 bulldozers, worked 1 million acres of forest on lands traditionally belonging to the Kayan, Kenyah, and Penan. More than 98 percent of Sarawak's timber is exported in the form of raw logs, virtually all of which is destined for Asian markets ("The Penans," n.d.).

Japanese companies supply the bulldozers and heavy equipment that is utilized to extract the logs. Japanese interests provide the insurance and financing for the ships that carry the raw logs to be processed in Japanese mills and sold as lumber to construction firms often owned by subsidiaries of the same companies that first secured the wood in Sarawak. Once milled in Japan, the wood produced by the oldest and perhaps richest tropical rain forest on Earth is used principally for packaging material, storage crates, and furniture. Roughly half of it is used in construction, mostly as plywood cement forms that are used once or twice and then discarded ("The Penans," n.d.).

During 1987, having appealed in vain to the government for more than seven years to end the destruction of their traditional homelands, the Penan declared:

> We, the Penan people of the Tutoh, Limbang, and Patah Rivers regions, declare: stop destroying the forest or we will be forced to protect it. The forest is our livelihood. We have lived here before any of you outsiders came. We fished in clean rivers and hunted in the jungle. We made our sago meat and ate the fruit of the trees. Our life was not easy but we lived it contentedly. Now the logging companies turn rivers to muddy streams and the jungle into devastation. Fish cannot survive in dirty rivers and wild animals will not live in devastated forest. You took advantage of our trusting nature and cheated us into unfair deals. By your doings you take away our livelihood and threaten our very lives. You make our people discontent[ed]. We want our ancestral land, the land we live off, back. We can use it in a wiser way. ("The Penans," n.d.)

The statement concluded, "We are a peace-loving people, but when our very lives are in danger, we will fight back. This is our message" ("The Penans," n.d.).

On March 31, 1987, armed with blowpipes, a group of Penan erected a blockade across a logging road in the Tutoh River basin. In April, about 100 people blockaded a road though their territory at Uma Bawang. By October, Penan from 26 settlements had joined the protest. Blockades have continued since that time, as whole villages moved onto logging roads, building makeshift shelters directly on the right-of-way. Often the protests lasted for months, and when they finally were suppressed by government forces, new ones sprang up in other areas. At their peak, the blockades halted logging in half of Sarawak.

The government and the logging companies have reacted to these peaceful protests by changing the law to make blockading of roads a criminal offense. Hundreds of indigenous people were harassed, arrested, and imprisoned as a result.

Environmental activists have been accused of undermining the nation's economic security and have had their passports confiscated. In the meantime, loggers continued to strip the forests.

Further Reading

Bryant, D., D. Nielsen, and L. Tangley. 1997. *The Last Frontier Forests: Ecosystem on the Edge.* Washington, DC: World Resources Institute.

"Despite Tough Talk, Indonesia's Government Is Struggling to Stem Deforestation." 2016. The Economist, November 26. Accessed September 29, 2018. http://www .economist.com/news/asia/21710844-weather-helping-little-despite-tough-talk -indonesias-government-struggling-stem.

Gargan, Edward A. 2001. "Lust for Teak Takes Grim Toll; Illegal Logging Decimating Indonesia's Majestic Forests." *Newsday*, June 25: A-7.

"Indonesia: Low Expectations on Log Export Ban." 2001. World Rainforest Movement *Bulletin* 53 (December). Accessed September 29, 2018. https://wrm.org.uy/articles-from -the-wrm-bulletin/section2/indonesia-low-expectations-on-log-export-ban/.

"Indonesian President Calls for Logging Halt." 2002. Environment News Service, May 27. Accessed September 29, 2018. http://ens-news.com/ens/may2002/2002-05-27-19. asp#anchor2.

Parry, Richard Lloyd. 2001. "The Hunt for Bruno Manser." *The Independent*, September 23: 18–22.

"The Penans of Sarawak." n.d. Sarawak Peoples Campaign. Accessed September 29, 2018. http://www.rimba.com/spc/spcpenanmain1.html.

Plumer, Brad. 2018. "Tropical Forests Suffered Near-Record Tree Losses in 2017." *New York Times,* June 27. Accessed September 29, 2018. https://www.nytimes.com/2018/06/27 /climate/tropical-trees-deforestation.html.

"Portrait of Indonesian Forestry: Supply, Demand, and Debt: A Call for Moratorium on Industrial Logging." 2001. The Indonesian Forum on the Environment, April 24.

MYANMAR

Forced Labor in the World's Last Teak Forest

Forest-dwelling peoples in Myanmar (Burma), notably the Karen, have been impressed into slave labor to harvest the world's last sizable forests of teak by the country's military rulers. The Burmese military government also has been using roads originally built for gas exploration to allow logging access in indigenous homelands comprising the "panhandle" in the country's far south. Boycotts of Burmese teak have been organized by environmental activists in the United States and Europe, where the valuable wood is most often used in Scandinavian-style furniture. Large amounts of teak also are sold (much of it harvested in Burma but processed in Thailand and China) for luxury yachts and other pleasure boats. In some cases, Burma teak also is being used as flooring and outdoor furniture.

In addition to the world's last teak forests, Burma territory once included some of the largest virgin rain forests that remained in mainland Asia. Many of these

Workers haul teak out of a forest in Myanmar. (AndiGrieger/iStockphoto.com)

forests were home to rare species, such as the Asian rhino and the Asian elephant, among others, as well as the aforementioned forest-dwelling peoples.

History of Burmese Teak Logging

Burmese teak logging began in earnest during the British colonial period. British demand for ships made of the durable wood consumed most of the commercially viable teak in India and eventually in Thailand. Destruction of teak forests also has provoked flooding and drought in parts of these countries. By the late 1990s, commercial-scale teak harvest was restricted almost entirely to Burma. In 1994, it was estimated that Burma held 80 percent of the world's remaining natural teak (Stevens 1994).

At the beginning of the 20th century, forests covered 80 percent of Burma. In 1948, 72 percent of the country was forested. By 1988, Burma's forest cover had decreased to about 47 percent of its land area. By the early 1990s, a decade later, Burma's forest cover was estimated to be roughly 36 percent. By the turn of the millennium, Burma was experiencing the third-highest rate of deforestation in the world, after Brazil and Indonesia, at roughly 8,000 square kilometers a year (Rainforest Relief 1997).

A military coup in Burma during 1962 initiated a reign of terror and oppression. In 1988, after millions of Burmese rallied for democracy, the military junta formed the State Law and Order Restoration Council (SLORC) to strengthen its domination of Burmese government. The SLORC, composed of several high-level generals, ordered the killing of at least 3,000 (other estimates are much higher) dissident Burmese demonstrators in 1988 during widespread unrest.

The junta later called general elections, during which its opponents won more than 82 percent of the seats in Parliament. The military then ignored these results and refused to yield power. The SLORC generals consolidated their rule with forced labor, rape, torture, forced relocation, and intimidation (Rainforest Relief 1997). The same methods were used against the Karen, Shan, Karenni, Mon, and Chin indigenous peoples as the junta sought to raise foreign exchange by harvesting the world's last sizable stands of teak, as well as other valuable hardwoods across Burma.

Teak Logging and Deforestation

According to an analysis by Rainforest Relief, teak logging, like most tropical deforestation, causes extreme degradation of tropical forests. Teak trees grow in dense stands throughout the forest. Logging roads plowed to gain access to these stands play a fundamental role in allowing further deforestation (as well as lower water tables) for primary forests in Burma, Laos, Cambodia, and Thailand.

Roads also serve as conduits for other invaders of indigenous lands, including prospectors for various metals and minerals, squatters looking for land, and (especially in Burma's case) troops charged with forcing the indigenous population into a slave-labor force for the harvesting of salable teak.

Intense logging in indigenous areas of Burma has aggravated flooding during seasonal monsoons. According to a report by Tim Keating of Rainforest Relief:

> As forests are cleared, rain runs off instead of being absorbed by the forest and recirculated into the environment and atmosphere slowly. This also leads to periods of drought, both of which adversely affect local peoples' ability to grow food. Erosion is exacerbating flooding and [has] caused silting of rivers, affecting fish populations . . . having a negative impact on local people. (Keating 1997)

Similar effects are evident among other hardwoods, as well as evergreen trees in Burma's far north.

By the early 1990s, the sale of teak and other hardwoods had become the second-largest legal source of income for the Burmese military (not counting its illegal trade in heroin). In 1992 and 1993, Burma extracted nearly 1 million cubic tons of teak logs with state-owned or contracted operations, up from 700,000 in 1983 (Keating 1997). China is the largest importer of teak logs from Burma, followed by Thailand. A large proportion of this teak is processed for export as lumber, furniture, and other consumer items (including luxury yachts and other pleasure boats) for sale in the United States and Europe.

Eyewitnesses in Burma have reported, according to *Drillbits and Tailings* (from reports by the international human-rights group EarthRights and the U.S. Embassy in Rangoon), that coerced labor has been used to build a Burmese natural-gas pipeline. The consortium building the Yadana pipeline includes the French oil company Total, the U.S. firm Unocal, the Petroleum Authority of Thailand (which is run by the Thai government), and MOGE, Burma's state-owned oil-and-gas company.

The Yadana and Yetagun pipelines cross the southeastern Tenasserim region of Burma, pumping gas from offshore wells in the Andaman Sea to power plants in neighboring Thailand ("New Evidence" 2000). The pipelines have been built from the Yadana gas field off the coast of southern Burma, through the Tenasserim rain forest, then into Thailand.

Teak Logging and Oil Extraction

Deserters from the Burmese Army have reported that they were ordered on a primary mission to secure the area on behalf of the oil companies. "Securing" the area, according to the eyewitnesses, "included the construction of a string of military bases, intimidation and terror against villagers, and forcible relocation of entire villages" ("New Evidence" 2000). According to the same reports, military housing was built with forced, unpaid labor, including young children. These reports were denied by the government.

Unocal employee Joel Robinson told the U.S. Embassy in Rangoon that Unocal had hired members of the Burmese military as security guards for the Yadana pipeline. Soldiers also were said by Robinson to have been hired as supervisors for construction of helicopter landing sites. The military, in turn, forced unpaid laborers to build the helipads. The Total company later provided some of the laborers with small amounts of money, much of which later was taken from them by soldiers ("New Evidence" 2000).

On May 22, 2000, the same day that the U.S. Embassy's report was released, more than 100 demonstrators rallied at Unocal's annual meeting in Brea, California, to protest the treatment of the Burmese villagers. According to an account in *Drillbits and Tailings*, as drummers in skull masks performed with mock bones on upturned Unocal oil barrels outside, a shareholder resolution sought to scale Chief Executive Officer Roger Beach's salary according to the company's ethical performance, including its behavior on its Burma project. The proposal won approval of 16.4 percent of the company's shareholders, which is high for a human-rights-based measure on a corporate ballot ("New Evidence" 2000). The demonstration was organized by the Burma Forum of Los Angeles and endorsed by the Los Angeles County Federation of Labor.

Activism Against Coerced Labor

Activism in the United States in opposition to coerced labor in Burma has been spearheaded by Ka Hsaw Wa, 29, director of EarthRights International. The name means "White Elephant," a *nom de plume* that he uses to avoid reprisals from Burma's military regime. Ka Hsaw Wa was born in Burma and was tortured by the junta for his environmental and human rights convictions before his subsequent escape to the United States. Ka Hsaw Wa, who is Karen, joined a massive demonstration at the age of 18 in Burma in 1988, demanding human rights, democracy, and an end to military rule. He was arrested and tortured for three days ("Earth Day" 2000). He was then forced to flee his home and go into hiding in the forests near the Thai border. Slipping anonymously back and forth across the border between Burma and Thailand, Ka Hsaw Wa documented environmental and human-rights abuses associated with construction of the Yadana Natural Gas Pipeline. He has called the pipeline "the most notorious project in Burma" because it is damaging the Tenasserim forest, which is home to many ethnic groups as well as many rare plants and animals ("Earth Day" 2000). Ka Hsaw Wa is a winner of the $125,000 1999 Goldman Prize for Asia. He also has been honored with the $20,000 tenth annual Conde Nast Traveler Environmental Award. In 1999, he received the Reebok Human Rights Award.

"The Tenasserim rain forest is one of the largest intact rain forests in mainland Southeast Asia," said Ka Hsaw Wa. "The people of Burma do not want the pipeline, but Unocal wants it and builds it anyway. They have sent the brutal Burmese military dictatorship to protect their investment" ("Earth Day" 2000). The Tenasserim rain forest is on the Isthmus of Kra, between the Andaman Sea and the Gulf of Thailand—land that has long been occupied and used by the Karen and other indigenous peoples.

Ka Hsaw Wa described how the local people are forced to work for the pipeline, to carry ammunition and grow food for the soldiers. "Many villagers were forced to move from their land. Women have been raped and many have been killed," he said ("Earth Day" 2000). "The same soldiers hired by the Unocal company who are oppressing people in the pipeline regions are also destroying the environment. They are catching wild animals, they are cutting down trees to sell, to grow food and other projects," Ka Hsaw Wa continued ("Earth Day" 2000).

Corporate representatives of Unocal maintain that the company is improving the lives of the local people. According to a statement on the company's website:

Unocal does not defend the actions and policies of the government of Myanmar [Burma]. We do defend our reputation and the integrity of the Yadana project. Our hope is that Myanmar will develop a vital, democratic society built on a strong economy. The Yadana project, which has brought significant benefits in health care, education, and economic opportunity to more than 40,000 people living in the pipeline area, is a step in the right direction. ("Earth Day" 2000)

Further Reading

"Earth Day with Ka Hsaw Wa: Everybody Belongs." 2000. Environment News Service, April 7. Accessed November 6, 2018. http://www.ens-newswire.com/ens/apr2000/2000-04-07-01.html.

Keating, Tim. 1997. "Forced-Labor Logging in Burma." Draft. 2nd in the *Rainforest Relief Reports* series, in cooperation with the Burma UN Service Office of the National Coalition Government of the Union of Burma (June). Accessed September 29, 2018. http://www.rainforestrelief.org/documents/Teak_Is_Torture.pdf.

McGeehan, Adam. 2017. "It's Official, No More Teak out of Myanmar." *Escape Artist*, April 18. Accessed September 29, 2018. http://www.escapeartist.com/blog/its-official-no-more-teak-out-of-myanmar/

"New Evidence Reveals Unocal's Complicity in Abuses in Burma." 2000. *Drillbits and Tailings* 5, no. 8 (May 31). Accessed September 29, 2018. http://groups.yahoo.com/group/graffis-l/message/11105.

Rainforest Relief. 1997. "Teak Is Torture and Burma's Reign of Terror; Mon, Karen, and Karenni Indigenous Peoples Threatened in Burma." Teak Week of Action Press Release, March 29. http://wgbis.ces.iisc.ernet.in/envis/doc97html/ecoteak330.html.

Stevens, Jane. 1994. "Teak Forests of Burma Fall Victim to Warfare." *The Oregonian* [Portland, OR]. March 16.

PORTUGAL

Wildfires and Renewables

Portugal is being scorched by wildfires nearly every summer, a problem that it shares with many other countries, from the United States' Western states and Alaska to Canada, Russia, Australia, and elsewhere. In Portugal, the prevalence of fires—and their damage—has helped spark a countrywide conversation and action plan for dealing with the changing climate that includes a drive to substitute wind and solar power for fossil fuels.

The expansion of wildfires may be the most pervasive evidence of a warming climate. For example, by 2014, scientists were reporting that "the effects of human-induced climate change are being felt in every corner of the United States, with water growing scarcer in dry regions, torrential rains increasing in wet regions, heat waves becoming more common and more severe, wildfires growing worse, and forests dying under assault from heat-loving insects" (Gillis 2014). Half of the United States Forest Service's budget by 2015 was going into fighting fire, which caused some who work there to complain that they were running a fire department, not a forest service.

Worst Fires on Record

Fires in Portugal during the summer of 2017 were the worst on record. On June 17, 2017, lightning ignited a deadly wildfire that spread across the mountainous areas of Pedrógão Grande, a city in central Portugal northeast of Lisbon. The raging fires were clearly visible to NASA satellites in space for several nights.

"Fires across Portugal's forested landscape during the warm, dry summer months are not uncommon," reported NASA's Earth Observatory. "In 2016, hundreds of fires raged on the mainland and also on the Portuguese island of Madeira" ("Wildfires Light" 2017). The fires of 2017 were notable for a high death toll of 62 people because drought and heat were exceptionally intense, causing fires to spread so quickly that half of the people who died were caught from behind by the inferno and roasted alive in their cars as they tried to escape. The worst previous fire season was in 1966, when 25 people died, all of them Portuguese who were fighting the fires near Sintra.

Patrícia De Melo Moreira, a photographer for Agence France-Presse, was on the scene with firefighters when several blazes enveloped people fleeing in their cars. The flames were so hot that the asphalt highway under them "was completely destroyed—melted," she said (Minder 2017).

The fire brought "a dimension of human tragedy that we cannot remember," Prime Minister António Costa said during a visit to the scorched area. The flames spread along four fronts at the same time with "great violence," said Jorge Gomes, the secretary of state for internal administration (Minder 2017). In addition to burning people in their cars, the fire engulfed several houses as about 1,600 firefighters fought the flames, assisted by airplanes and helicopters. Low humidity, high temperatures, and a parched landscape combined to make the fire unusually violent and unpredictable. "We know fire behavior has changed and continues to change, yet we continue to be surprised every time, when we shouldn't be," said Don Whittemore, a former assistant fire chief in Colorado who has studied wildfire behavior. "The notion that firefighters will be able to put out, suppress, or make safe a wildfire is becoming less and less of a reliable notion" (Minder 2017).

From 1993 to 2013, Portugal, a relatively small nation, experienced more forest fires than France, Spain, Italy, or Greece, according to a 2016 report from the European Environment Agency. Portugal sometimes reaches temperatures of more than 100°F during the summer as hot, dry adiabatic (downslope) winds from the inland Iberian Peninsula dry forests and make them prone to fire. These winds are similar to those in the Santa Anas that create hazardous fire conditions in Southern California and the offshore winds that sometimes blow from inland Australia to the coastal urban areas of Sydney and Melbourne. In Portugal, dense woodland and hilly terrain are covered with pine and eucalyptus trees that contain oil, which burns easily. During recent years, the forests have transitioned to pine and eucalyptus (which are useful commercially) from oak, which burns less easily. Portugal also has an unusually large number of fires that start due to arson or accidents.

The Pedrógão Grande fire of 2017 was not sparked by arson or accident, however, but by a "dry thunderstorm" that contained lightning but no rain that reached the ground. Police were able to find the tree that was struck (in the town of Escalos Fundeiros), starting the fire. Dry thunderstorms, which also may produce strong winds that spread the flames, also play a role in igniting fires in many other parts of the world. Fires in Portugal increased not only because of hot, dry weather, but because recession had caused cuts in forestry management. The World Wildlife Fund (WWF) in Portugal said that the government should improve forestry management, which "is more effective and financially more efficient than the huge mechanisms used every year to combat forest fires" (Minder 2017).

Fires, Heat, and Global Climate Change

An Associated Press analysis said that global climate change has become a factor in these fires: "Ecologists say southern Europe, like many other parts of the world, is experiencing longer and hotter summers and shorter and drier winters." Greenpeace said that 2017 was the hottest start to summer in 40 years (Hatton and Giles 2017). The European Environment Agency report issued in December 2015 said that climate change is "expected to have a strong impact on forest fire regimes in Europe." According to the report, "Although most of the wildfires in Europe are ignited by humans (either accidentally or intentionally), it is widely recognized that weather conditions and the accumulation of fuel play dominant roles in affecting the changes in fire risk over time" (Hatton and Giles, 2017).

Portugal's epidemic of wildfires has helped focus public attention on the need to reduce carbon-dioxide emissions. By 2010, nearly 45 percent of Portugal's electricity was coming from renewable sources, up from 17 percent in 2005. The transition also has sparked some gripes about rising power rates, up 15 percent during the same period (Rosenthal 2010, A-1). Roughly half of Portugal's renewable energy comes from wind, and the other half from hydropower; sometimes they operate together, as wind turbines pump water uphill at night, which flows downhill by day, creating electricity to meet customer demand. Portugal's two-way grid also uses rooftop solar panels in a minor role. The grid uses fossil fuels for backup (efficient natural-gas-fueled plants are the biggest players), and different components of load are constantly balanced to match weather conditions with demand. The power grid has been overhauled to transfer power from remote but wind-rich areas to cities.

The energy transition has placed little stress on Portugal's national budget because construction costs have been largely absorbed by declining costs of imported fossil fuels, especially oil, which doubled in cost while the grid was changing. Imported fuel had accounted for half of Portugal's trade deficit. Portugal plans to use renewables for 60 percent of its electrical power and 31 percent of its total energy by the year 2020 (Rosenthal 2010, A-8).

By 2007 in Portugal, a new solar array spread across 150 hilly acres near Serpa, about 120 miles southeast of Lisbon. This project was old-fashioned photovoltaic, constructed by General Electric Energy Financial Services and PowerLight Corporation of the United States in partnership with the Portuguese company Catavento ("Portugal Celebrates" 2007). Solar power continued to spread across Portugal after that. In 2016, for the first time, Portugal kept its lights on for four consecutive days (May 7–11) on renewable power alone.

Unlike frequently misty Germany, southern Portugal is among the sunniest places in Europe, with as many as 3,300 hours of sunlight annually, nine hours during an average day. This solar array, when finished, produced enough power to supply 8,000 homes. It replaced 30,000 tons' worth of annual greenhouse-gas emissions ("Portugal Celebrates" 2007). This plant's photovoltaic system uses silicon solar cell technology to convert sunlight directly into electricity, producing 20 gigawatts per year. "This project is successful because Portugal's sunshine is plentiful, the solar power technology is proven, government policies are support-ive, and we are investing . . . to help our customers meet their environmental challenges," said Kevin Walsh, managing director and leader of renewable energy at GE Energy Financial Services ("Portugal Celebrates" 2007).

Further Reading

Gillis, Justin. 2014. "Climate Change Study Finds U.S. Is Already Widely Affected." *New York Times*, May 6. Accessed September 29, 2018. http://www.nytimes.com/2014/05/07/science/earth/climate-change-report.html.

Hatton, Barry, and Ciaran Giles. 2017. "Portugal, a Country Helplessly Prone to Forest Fires." Associated Press in ABC News, June 19. https://phys.org/news/2017-06-portugal-country-helplessly-prone-forest.html.

Minder, Raphael. 2017. "Portugal Fires Kill More Than 60, Including Drivers Trapped in Cars." *New York Times*, June 18. Accessed September 28, 2018. https://www.nytimes.com/2017/06/18/world/europe/portugal-pedrogao-grande-forest-fires.html.

"Portugal Celebrates Massive Solar Plant." 2007. Associated Press in the *Sydney Morning News*, November 1.https://www.smh.com.au/technology/portugal-celebrates-massive-solar-plant-20070329-4tk.html.

Rosenthal, Elisabeth. 2010. "Portugal Gives Itself a Clean-Energy Makeover." *New York Times*, August 10: A-1, A-8.

"Wildfires Light Up Portugal." 2017. NASA Earth Observatory, June 20. Accessed September 29, 2018. https://earthobservatory.nasa.gov/IOTD/view.php?id=90427&src=eoa-iotd.

RUSSIA

Fires in the Taiga (Boreal Forests)

As in the western United States and northern Canada, Russia's vast forests have been ablaze in recent years. A report by Greenpeace International suggested that between 50 and 90 percent of the Earth's existing boreal forests are likely to disap-pear if atmospheric levels of carbon dioxide and other greenhouse gases double. These forests comprise a third of the Earth's remaining tree cover, about 15 mil-lion square kilometers, across Russia (where they are called "taiga"), Canada, the United States, and Scandinavia, as well as parts of the Korean Peninsula, China, Mongolia, and Japan. Large forests also cover many mountain ranges outside of these zones. In total, boreal forests cover about 10 percent of the world's land area. As world temperatures have risen, fires in boreal forests have emitted increasing amounts of mainly carbon dioxide and water vapor, as well as several other gases and particles. Incomplete combustion also generates about half of the atmosphere's carbon monoxide, an odorless, poisonous gas.

Fires in Siberia

During 2010, heat and drought provoked severe fires in forests and peat bogs across western Russia, fouling the air in Moscow and other cities. Two years later, another wave of fires struck the taiga further east, in Siberia. According to a report from NASA's Earth Observatory, on June 18, 2012, a total of 198 wildfires were burning across Russia (Ranson 2012). They had charred an area of 32 square miles, or 8,330 hectares, mainly in central Russia, where firefighters battled out-of-control fires for several months. The fires caused local authorities to declare a state of emergency in several regions: the Khanty-Mansiisk autonomous area, the Tyva Republic, the Sakha Republic, Krasnoyarsk, Amur, Zabaikalsky, and Sakhalin. The Russian government said, "Many of the fires started when people lost control of agricultural fires and campfires. However, lightning sparked some of the blazes as well" ("State of Emergency" 2012)

On one day, July 11, 2012, more than 97 square miles (about 25,000 hectares) of forests were burning in central and eastern Siberia, according to the Russian Federal Forestry Agency, as several uncontrolled wildfires spread through boreal forests. Smoke from some of the wildfires in Siberia was being pushed across the Pacific Ocean. "On July 8 and 9, 2012," reported NASA's Earth Observatory, "smoke from Siberia arrived in British Columbia, Canada, and caused ground-level ozone to reach record-high levels" ("Wildfires in Siberia" 2012a, 2012b).

On September 11, 2012, fires burning in and near Tomsk in south central Siberia burned throughout the summer. The NASA Earth Observatory reported, "Thick smoke billowed from numerous wildfires near the Ob River and mixed with haze and clouds that arrived from the southwest." More than 17,000 wildfires had burned more than 30 million hectares (74 million acres) through August 2012, according to researchers at the Sukachev Institute of Forest in the Russian Academy of Sciences. By comparison, 20 million hectares had burned in 2011, roughly the average between 2000 and 2008 ("Wildfires in Siberia" 2012a, 2012b). A Greenpeace report indicated that global warming's toll on the boreal forests had begun by the early 1990s. The report warned that decaying forests may provide an extra

---

Cattle-Killing Wildfires in March

Jack Healy of the *New York Times* reported from Ashland, Kansas, in March 2017 that "ranching families across this countryside are now facing an existential threat to a way of life that has sustained them since homesteading days: years of cleanup and crippling losses after wind-driven wildfires across Kansas, Oklahoma, and the Texas panhandle killed seven people and devoured homes, miles of fences, and as much as 80 percent of some families' cattle herds." And this was mid-March, not mid-summer, hardly what used to be wildfire season (Healy 2017). By April 1, 2017, wildfires had scorched 2.1 million acres in the United States, six times the average.

Further Reading

Healy, Jack. 2017. "Burying Their Cattle, Ranchers Call Wildfires 'Our Hurricane Katrina.'" *New York Times*, March 20. Accessed October 11, 2018. https://www.nytimes.com/2017/03/20 /us/burying-their-cattle-ranchers-call-wildfires-our-hurricane-katrina.html.

boost to rising carbon-dioxide levels, provoking global warming to feed upon itself and become more severe.

Satellites Survey Wildfires

Satellites provide the best way to monitor wildfire emissions over large areas, such as the Russian taiga, that are sparsely populated and contain few ground-based instruments. Christine Wiedinmyer, a scientist at the National Center for Atmospheric Research, has developed a model that analyzes information about vegetation (such as the percentage of tree cover and the type of forest) to describe emissions from forest fires. "In September 2012, according to NASA, Wiedinmyer used her model to calculate Russian fire emissions for every year dating back to 2002. She found that the amount of carbon monoxide produced in 2012 was significantly more than what was produced in 2010 and the second most in a decade. Through August 31, the model showed that Russian wildfires had released an estimated 48 teragrams of carbon monoxide since the beginning of 2012. By comparison, the model estimated fires yielded just 22 teragrams of carbon monoxide in all of 2010" (Wiedinmyer 2011, 625).

According to the NASA Earth Observatory:

> High temperatures play an important role in promoting wildfires. Warm fuels burn more readily than cooler fuels because less energy is required to raise their temperature to the point of ignition. With temperatures soaring in northern Russia, it was easier for previously active fires to continue burning and for lightning to spark new ones . . . weather occurs within the broader context of the climate, and there's a high level of agreement among scientists that global warming has made it more likely that heat waves and wildfires of this magnitude will occur. ("Heat Intensifies" 2013)

"Even the Tundra Is Burning"

By 2016, fires were so widespread in Siberia that satellites were photographing them from a million miles away. Even the tundra was burning. One report commented:

> Currently, as we outline below, there are worrying reports of the tundra burning in the Arctic Yamal Peninsula, as well as other damaging fires, for example a 3,000-hectare blaze at the Lena Pillars Nature Park—a UNESCO World Heritage Site—which was finally extinguished in recent days in Yakutia, also known as Sakha Republic. Ecologists say the fires pose a direct threat to the role of Siberian pristine boreal forests in absorbing climate-warming emissions. ("Siberia's Wildfires" 2016)

The same report said that as weather becomes drier and warmer, wildfires increase. Annually, Russian forests absorb about 500 million tons of carbon from the atmosphere, equal to the emissions released by 534 coal-burning power plants. Yet "forest fire danger and carbon emissions will double or triple by the end of the century," according to expert Anatoly Shvidenko, who served on the UN's Intergovernmental Panel for Climate Change (IPCC) ("Siberia's Wildfires" 2016).

Based on such reports, Greenpeace asserted that the Russian government was vastly under-reporting the extent of wildfires and consequent damage to forests across large areas of Siberia. Independent observers said that satellite images indicated Siberian wildfires were 10 times more widespread than the Russian government acknowledged. The thinning of the forests is most acute in northern Siberia, where fires can ravage plant life and shallow roots, making it impossible for trees to regrow for centuries—a process known as green desertification. "'Already 7 million hectares have burned, that's more than average,' said Grigory Kuksin, in charge of the wildfire prevention programme at Greenpeace Russia. 'Everywhere it's the same scenario, where a small fire is ignored and then goes out of control'" ("Siberia's Wildfires" 2016). As temperatures rise, drying peat moss becomes fuel for more fires, feeding more carbon dioxide and methane into the air, all part of a cycle that is raising temperatures worldwide.

The decline of boreal forests also endangers more than 1 million indigenous people who live in them, including the Dene and Cree of Canada, the Sami (Laplanders) of Norway, Sweden, and Finland, the Ainu of northern Japan, and the Nenets, Yakut, Udege, and Altaisk of Siberia (Jardine 1994). Some of the forests' larger animals, such as the Siberian tiger, already are near extinction. The Greenpeace report concluded that rapid logging of the boreal forests is intensifying pressure on animal life and accelerating the release of even more carbon dioxide and other greenhouse gases into the atmosphere.

Further Reading

"Fires in Northwestern Siberia." 2016. NASA Earth Observatory, July 29. Accessed September 29, 2018. https://earthobservatory.nasa.gov/NaturalHazards/view.php?id=88430.

"Heat Intensifies Siberian Wildfires." 2013. NASA Earth Observatory, August 2. Accessed September 29, 2018. http://earthobservatory.nasa.gov/IOTD/view.php?id=78305&src =eoa-iotd.

Jardine, Kevin. 1994. "The Carbon Bomb: Climate Change and the Fate of the Northern Boreal Forests." Greenpeace International. Accessed September 29, 2018. http://dieoff .org/page129.htm.

Ranson, J. 2012. "Siberia 2012: A Slow and Smoky Arrival. Notes from the Field." NASA Earth Observatory, July 9. Accessed November 6, 2018. http://earthobservatory.nasa .gov/blogs/fromthefield/2012/07/09/siberia-2012-a-slow-and-smoky-arrival/.

"Siberia's Wildfires Seen from 1 Million Miles Away: Even the Tundra Is Burning." 2016. *Siberian Times*, July 23. Accessed September 29, 2018. http://siberiantimes.com/ecol ogy/casestudy/news/n0682-siberias-wildfires-seen-from-1-million-miles-away-even -the-tundra-is-burning/.

"State of Emergency Declared Due to Fires in Eastern Regions." 2012. *The Moscow Times,* June 18. Accessed September 29, 2018. https://themoscowtimes.com/news /state-of-emergency-declared-due-to-fires-in-eastern-regions-15500.

Wiedinmyer, C. 2011. "The Fire Inventory from NCAR (FINN): A High Resolution Global Model to Estimate the Emissions from Open Burning." *Geoscience Model Development* 4: 625–641.

"Wildfires in Siberia." 2012a. NASA Earth Observatory, July 12. Accessed September 29, 2018. http://earthobservatory.nasa.gov/IOTD/view.php?id=78515&src=eoa-iot.

"Wildfires in Siberia." 2012b. NASA Earth Observatory, September 13. Accessed September 29, 2018. http://earthobservatory.nasa.gov/IOTD/view.php?id=79161&src=eoa-iotd.

Chapter 3: Fossil Fuels and Mining

OVERVIEW

The universal desire of Earth's growing population for comfort, convenience, and profit requires acquisition of resources (usually by mining the Earth, often producing pollution), as well as generation of energy for transportation and industrial production, most of which is supplied by fossil fuels that add greenhouse gases to the atmosphere, raising temperatures and destabilizing climate. Two hundred years ago, the proportion of carbon dioxide, which comprises less than 1 percent of the atmosphere, had cycled between roughly 180 parts per million and 280 ppm for almost a million years. As coal was combusted to fire the first steam engines, that proportion began to rise. By 2015, it had reached 400 ppm, very likely the highest level since the Pliocene, 2 to 3 million years ago, when Earth had very little long-lasting ice, and sea levels were at least 50 feet higher.

Energy is the crux of the world's fossil-fuel dilemma. Aviation in 2016 accounted for about 15 percent of fossil fuels consumed in transport; surface transport accounts for 80 percent. During the 10 years before 2008, passenger air traffic increased by 60 percent. Passenger vehicles in the United States account for 40 percent of the country's oil consumption and 10 percent of the world's (Kolbert 2007, 88, 90).

Today energy is being generated from fossil fuels (especially coal, the most carbon-intensive fuel per unit of energy provided) at record rates as China, India, and other countries industrialize and populations rise. Ninety percent of Earth's remaining fossil-fuel reserves are in the form of coal, the most dangerous fossil fuel from a greenhouse point of view because most coals produce roughly 70 percent more carbon dioxide per unit of energy generated than natural gas and about 30 percent more than oil. Coal is also the most plentiful fossil fuel, especially in places with large populations, such as China, which controls 43 percent of remaining reserves. During the 1980s, China passed the Soviet Union as the world's largest coal producer. China also built 114,000 megawatts of coal-fired power in 2006 and 95,000 more in 2007. Coal poses environmental problems other than carbon emissions. The mining of coal also produces methane; its combustion produces sulfur dioxide and nitrous oxides, as well as carbon dioxide. Transport of coal also usually requires more energy than any other fossil fuel.

Realizing its dangers, many energy producers have begun to move away from coal, substituting natural gas, wind, and solar. By 2015, coal production and consumption were falling in the United States, much of Europe, and even in China. Financially, many coal-mining companies were in trouble, as were coal-transporting railroads. Wind power has reached price parity with coal for electrical generation

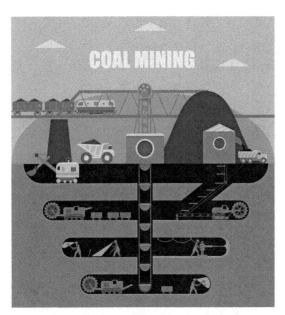

Currently, coal mining is largely done by machines, with a minimum of human labor, as illustrated by this cross section. (Piscari/Dreamstime.com)

in some cases, and capacity is growing quickly. Mid-America Energy, for example, which provides electricity to about half of Iowa, reached 50 percent of electricity from wind in 2016. Mid-America has plans to generate 89 percent from wind by 2020.

Even so, more people are traveling by air, which is especially carbon-intensive. One journey across the Atlantic Ocean from the United States to Europe on a jet aircraft will emit as much greenhouse gases *per person* as an average automobile commuter creates in an entire year. At the same time, efforts are underway to make air transport more energy efficient.

Alternative fuels do not provide the thrust required to lift commercial aircraft and keep them aloft; no one will ever build a solar- or wind-powered passenger airplane. Some improvement in fuel efficiency has been achieved, however. Hydrogen-powered and electric cars are of little help in reducing greenhouse-gas emissions as long as their power sources stem from coal, oil, and gas. Fuel efficiency has improved in this area as well, but not enough to counter the rising number of cars and trucks worldwide.

Even as it installs more solar power than any other nation, China was burning more coal than the United States, Europe, and Japan combined. In 2010, China overtook the United States as the world's largest consumer of energy, having used only half as much ten years previously. China has accelerated as a consumer of energy (mainly low-energy coal) more quickly than any other country in world history. For the first time, in 2009, the Chinese purchased more cars than residents of the United States, 12.8 million to 10.3 million. China's car sales increased 42 percent in 2009—a year when most of the world was in recession—compared to 2008, 72 percent of them SUVs.

China is beginning to embrace a need for an energy paradigm shift to renewable sources of energy, as is most of the world. The major question for energy planners (and everyone else) is: will this paradigm change fast enough to spare the Earth a climate catastrophe?

In addition to China, this chapter will examine mining and fossil-fuel extraction in Canada, Indonesia, Nigeria, the Philippines, Russia, the South Pacific Islands, and the United States.

Coal Ash as a Major Pollutant

One of the largest sources of toxic waste in the world is coal ash (residue) resulting from combustion to produce electricity. One electrical generating site, the Gallatin Fossil Plant in Tennessee, produces more than 200,000 tons of coal-ash residue in an average year, which leaks into local drinking-water supplies. More than 100 million tons of coal ash are produced in the United States during an average year, adding lead, arsenic, mercury, among other heavy metals into the environment (Schlossberg 2017).

A few years ago, coal provided half the United States' electricity. By 2017, facing competition from natural gas, wind, and solar power, coal's share was down to a one-third share, but coal-ash residue was still a major pollutant. The same is true worldwide, especially in China and India, where hundreds of millions of people are acquiring appliances, such as air conditioners, which use large amounts of power.

Tatiana Schlossberg of the *New York Times* described pollution by coal-ash residue:

> From 1970 to 1978, coal ash from one of the Gallatin slush ponds seeped into the groundwater and made its way to the Cumberland River beyond, the center's lawyers and several scientists say. The slow-motion coal slurry spill totaled 27 billion gallons, more than 100 times the size of the Deepwater Horizon oil spill. In 2008, an ash pond dike at its Kingston Fossil Plant in eastern Tennessee collapsed, releasing just over a billion gallons of coal-ash water into the Emory River, which flows into two other rivers, including the Tennessee. The slurry released in that spill, which has been called the largest environmental disaster of its kind, buried 300 acres of land in toxic sludge. (Schlossberg 2017)

Further Reading

Schlossberg, Tatiana. 2017. "Two Tennessee Cases Bring Coal's Hidden Hazard to Light." *New York Times*, April 15. Accessed October 11, 2018. https://www.nytimes.com/2017/04/15/climate/tennessee-coal-ash-disposal-lawsuits.html.

Further Reading

Kolbert, Elizabeth. 2007. "Running on Fumes: Does the Car of the Future Have a Future?" *New Yorker*, November 5: 87–90.

CANADA

Tar Sands and Moonscapes

Exploitation of tar sands at the surface (some of it also is mined underground) requires a form of strip mining that scars the earth in ways that will not be quickly nor easily repaired. The opponents assert that oil sands are a relatively new form of fossil fuel—the last thing the Earth needs when carbon-dioxide levels in the atmosphere have risen to more than 400 parts per million, more than 40 percent above peak preindustrial levels, with damage to climate, as well as rising seas and oceanic acidity. On a local level, many people worry about oil spills in fragile areas such as the Nebraska Sand Hills, where water is scarce and the Ogallala aquifer could be contaminated. This aquifer supplies 78 percent of public water and 83 percent of irrigation water in Nebraska, almost a third of the irrigation water used in the United States. The planned Keystone XL pipeline will carry tar-sands product from Alberta southward to the Gulf Coast through this area.

The oil industry and supportive politicians argue that consumers will receive a secure source of vital energy from tar sands. Oil companies produce profits by removing carbon from the Earth and selling it as products that inject carbon dioxide into the atmosphere as a by-product of both manufacture and consumption. Tar-sands advocates argue that Alberta collects a carbon tax (it is very small) and that carbon emissions from oil-sands mining have dropped about 25 percent since 1990, as refining become more efficient. Oil sands are still "dirtier" than conventional oil, however, because complex manufacturing is required to make products useful as fuel.

"Tar Sands and Tipping Points"

James Hansen, author of *Storms of My Grandchildren,* asserts that full usage of the Alberta tar sands may push the world over a climatic tipping point. The reserves contain twice as much carbon dioxide as humanity has thus far consumed by global oil use in its entire history. "If we were to fully exploit this new oil source and continue to burn our conventional oil, gas, and coal supplies, concentrations of carbon dioxide in the atmosphere eventually would reach levels higher than in the Pliocene era, more than 2.5 million years ago, when sea level was at least 50 feet higher than it is now," Hansen wrote. Disintegration of ice sheets will destroy many coastal cities within a century or two if development of the tar sands continues unchecked. "Global temperatures would become intolerable. Twenty to 50 percent of the planet's species would be driven to extinction. Civilization would be at risk. . . . If this sounds apocalyptic, it is," Hansen wrote (Hansen, 2012).

Hansen ran the numbers: The level of carbon dioxide in the atmosphere has risen from 280 parts per million to about 410 ppm over the last 160 years, as of 2017. The tar sands contain enough carbon (240 gigatons) to add 120 ppm. Tar shale contains at least an additional 300 gigatons of carbon. "If we turn to these dirtiest of fuels, instead of finding ways to phase out our addiction to fossil fuels, there is no hope of keeping carbon concentrations below 500 ppm—a level that would, as Earth's history shows, leave our children a climate system that is out of their control," Hansen wrote (2012).

According to NASA, the Alberta fields, which were first mined in 1967, are "the world's largest oil sands deposit, with a capacity to produce 174.5 billion barrels of oil—2.5 million barrels of oil per day for 186 years" ("Athabasca Oil Sands" 2011). (The United States as a whole consumes 15 to 20 million barrels of oil per day.) Environmental activist Bill McKibben has called tar-sands mining and the Keystone XL the "fuse to the biggest carbon bomb on the planet" (Tollefson 2013).

"Saying that the tar sands are not necessarily worse than coal is like saying that drinking arsenic is not necessarily worse than drinking cyanide," said geophysicist Raymond Pierrehumbert of the University of Chicago. He wrote that fully developing the tar sands could by itself, "even if we suddenly stopped burning coal," warm the planet an additional 3.6°F by century's end—an amount that climate scientists warn could be catastrophic (Koch 2014).

Environmental Debits of Tar-Sands Mining

Tar sands are a mixture of clay and sand with bitumen, a thick, low-grade form of petroleum similar to asphalt. According to Thomas Homer-Dixon, who teaches global governance at the Balsillie School of International Affairs:

> Tar sands production is one of the world's most environmentally damaging activities. It wrecks vast areas of boreal forest through surface mining and subsurface production. It sucks up huge quantities of water from local rivers, turns it into toxic waste and dumps the contaminated water into tailing ponds that now cover nearly 70 square miles. . . . Bitumen is junk energy. A joule, or unit of energy, invested in extracting and processing bitumen returns only four to six joules in the form of crude oil. In contrast, conventional oil production in North America returns about 15 joules. Because almost all of the input energy in tar-sands production comes from fossil fuels, the process generates significantly more carbon dioxide than conventional oil production. (Homer-Dixon 2013)

According to NASA, oil-sands refining produces the equivalent of 86 to 103 kilograms of carbon dioxide for every barrel of crude oil produced. By comparison, 27 to 58 kilograms of carbon dioxide are emitted in the conventional production of a barrel of crude oil.

By 2012, the mining and refining of tar sands in Canada consumed as much natural gas as that country uses for home heating (Kolbert 2007, 49). The gas is used to produce synthetic oil and by-products, such as gasoline. Tar sands require about 15 to 40 percent more energy in manufacture compared to conventional crude oil; oil shales require about twice as much. Converting tar sands into oil costs about as much as oil at $30 a barrel, however. With oil pushing $100 a barrel in late 2014, tar sands were becoming very profitable for many fossil-fuel companies (Kolbert 2007, 49, 50).

Ryan Lizza, writing in the *New Yorker*, described the mining of tar sands:

> Oil sand has the texture of soft asphalt; twenty percent of it lies close to the surface, and the area is effectively strip-mined. The bitumen-rich sand is removed, mixed with water into a slurry, and spun in centrifuges until the oil is separated, leaving behind vast black tailings ponds that are hazardous to wildlife. The mining operations sprawl ruinously for miles. The remaining eighty percent of the oil sands lie hundreds of feet down beneath a layer of hard rock. Steam is injected deep belowground until the oil naturally separates and is drawn out. The extra energy required to extract the sand makes it a more carbon-intensive fossil fuel—averaging seventeen percent more . . . than conventional oil. (Lizza 2013, 42)

A Toxic Wasteland

Carol Berry, of the Indian Country Today Media Network, described the environmental damage of tar-sands mining:

> If you can imagine the bleak landscape of the moon, you can envision the desolate, 54,000-square-mile tar sands of northern Alberta. . . . "It's literally a toxic wasteland—bare ground and black ponds and lakes—tailings ponds—with an awful smell," said Warner Nazile . . . [an] activist from British Columbia and member of the Wet'suwet'en First Nation. The mining is despoiling an area roughly the size of England. . . . University of Alberta scientists "found indications that contamination from the tailings ponds was polluting a huge aquifer that ultimately flows into the Arctic Ocean," Nazile said. Two aboriginal communities downstream from the oil sands have experienced higher-than-average rates of cancer and other health problems, he added. (Berry 2012)

"For a vast stretch of western Canada's boreal forest, the fight over extracting bitumen has already been lost. The question is, how much more will we lose?" wrote Andrew Nikiforuk, a Canadian journalist and author of *The Energy of Slaves: Oil and the New Servitude*, in the *New York Times* (2014). After intensive tar-sands mining accelerated after 2000, almost 2 million acres of this forest have been cleared or degraded, according to Global Forest Watch. Canadian economist Jeff Rubin said, "When you're schlepping oil from sand, you're probably in the bottom of the ninth inning in the hydrocarbon economy" (Nikiforuk 2014).

After the forests have been removed, "the landscape is reduced to a treeless wasteland" (Nikiforuk 2014). The energy intensity of the bitumen harvest begins with drilling deep into frozen ground that is melted with water that has been heated to steam and pumped to the surface. Some of the bitumen that the Cree once heated to repair leaks in their canoes lies near the surface, from which it is removed by electric shovels the size of large buildings, then transported to mills that remove the sands in trucks that carry 400 tons at a time. Imagine their gas mileage. The toxic sludge that comes out of the mills is dumped into lakes, forming mercury contamination.

Further Reading

"Athabasca Oil Sands." 2011. NASA Earth Observatory, November 30. Accessed September 30, 2018. http://earthobservatory.nasa.gov/IOTD/view.php?id=76559&src=eoa-iotd.

Berry, Carol. 2012. "Alberta Oil Sands Up Close: Gunshot Sounds, Dead Birds, a Moonscape." Indian Country Today Media Network, February 2.

Berry, Carol. 2013. "Tribal Members Sign Treaty Calling for an End to Alberta Oil Sands Development and Keystone XL." Indian Country Today Media Network, January 31.

Hansen, James. 2012. "Game Over for the Climate." *New York Times,* May 9. Accessed September 30, 2018. http://www.nytimes.com/2012/05/10/opinion/game-over-for-the-climate.html.

Homer-Dixon, Thomas. 2013. "The Tar Sands Disaster." *New York Times,* March 31. Accessed September 30, 2018. http://www.nytimes.com/2013/04/01/opinion/the-tar-sands-disaster.html.

Koch, Wendy. 2014. "Would Keystone Pipeline Unload 'Carbon Bomb' or Job Boom?" *USA Today*, March 10. Accessed September 30, 2018. http://www.usatoday.com/story /news/nation/2014/03/01/keystonexls-myths-debunked/5651099/.

Kolbert, Elizabeth. 2007. "Unconventional Crude: Canada's Synthetic-Fuels Boom." *New Yorker*, November 12: 46–51.

Lizza, Ryan. 2013. "The President and the Pipeline." *New Yorker*, September 16: 38–51.

Nikiforuk, Andrew. 2014. "A Forest Threatened by Keystone XL." *New York Times*, November 18. Accessed September 30, 2018. http://www.nytimes.com/2014/11/18/opinion /a-forest-threatened-by-keystone-xl.html.

Tollefson, Jeff. 2013. "Climate Science: A Line in the Sands." *Nature* 500 (August 8): 136–137. Accessed November 6, 2018. http://www.nature.com/news/climate-science-a -line-in-the-sands-1.13515.

CHINA

Slowly Working Away from Coal

By 2009, China was burning more coal than the United States, Europe, and Japan combined. In 2010, China overtook the United States as the world's largest user of energy, having used only half as much ten years previously. China has accelerated as a consumer of energy (mainly low-energy coal) more quickly than any other country in world history.

De-emphasizing Coal after a Rapid Buildup

In 2014, however, China's coal consumption fell 2.9 percent, according to its government. Glen Peters, a scientist at the Center for International Climate and Environmental Research (Oslo), estimated that "the drop in consumption, together with slowed growth in cement production, reduced China's annual emissions of carbon dioxide, the main greenhouse gas from human activity, 0.8 percent. This was the first fall in China's emissions after more than 15 years of fast growth" (Wong and Buckley 2015).

By 2015, China had reduced its consumption of coal by 3.7 percent below 2014 levels, according to statistics released by the Chinese government. China also leads the world in the deployment of renewable energy; in 2015, the country invested about US$110 billion—twice as much as the United States, according to the World Resources Institute (WRI) (Tollefson 2016, 425). The de-emphasis on coal was felt worldwide as Peabody Coal, the largest miner in the United States, skirted bankruptcy, and Union Pacific, a major coal-hauling railroad, laid off several hundred employees.

China's statistics are sometimes less than reliable, however. In 2015, China admitted to having underreported its coal usage (and thus carbon-dioxide emissions) by 17 percent—a billion tons of CO_2 annually. World emissions of CO_2 stabilized in 2014 at 32 billion tons. China's CO_2 emissions have risen from about 3 billion tons annually in 2000 to 8 billion in 2014. China had been underreporting its coal consumption since the year 2000, with the disparity increasing over time. By 2015, the billion-ton shortfall was as much coal as the United States burned in total (Buckley 2015, A-1).

While China has also become a world leader in manufacture, export, and deployment of wind- and solar-power technology and equipment, the largest proportion of its industrial development (as well as rapidly growing domestic electricity supply) has remained relatively dirty (and abundant) coal. China has been transforming gigatons of dirty coal with human ingenuity and labor to fill the world's retail stores with products and its own urban skies with some of the most noxious air pollution on the planet.

China's consumption of coal increased 185 percent between 2000 and 2010, according to the International Energy Agency. During the same decade, coal use in the rest of the world rose 17.5 percent, while consumption in the United States declined 1.6 percent. In 2010, coal produced 83 percent of China's electricity. The world as a whole burned 7.3 billion metric tons of coal during 2010. China burned nearly half of that (Bradsher 2011, B-8). Until 2015, when China curtailed construction, another large coal-fired power plant opened somewhere in the country every week to 10 days with enough capacity to serve all the households in Dallas or San Diego (Bradsher and Barboza 2006). Many of them use old, polluting technology, and they will be operating for an average of 75 years or until other sources allow authorities to shut them down.

Within a few years, China developed a major appetite for worldwide supplies of coal. China was importing coal from ports in Canada, Australia, Indonesia, Colombia, and South Africa; once an exporter of coal when its carbon-dioxide emissions were lower, China quickly became a large net importer.

Air Pollution in Urban China

Mainly because of pollution from coal-generated power, cities in China have some of the worst air quality in the world, past or present. The U.S. Embassy in Beijing monitors air quality on its roof on a scale that measures particles smaller than 2.5 microns in diameter, the deadliest form of air pollution, as it penetrates deep into people's lungs and bloodstreams, causing permanent DNA mutations, heart attacks, lung cancer, and premature death ("Beijing Bans" 2014). On January 12, 2013, that scale reached 728. Some days, readings of small particulate matter in Beijing have reached 1,000 micrograms per cubic meter, more than three times the level that the U.S. Environmental Protection Agency regards as hazardous. During its worst "killer" smogs in about 1950, industrial London never reached these levels.

At the end of January 2013, the Chinese government declared a national emergency in an attempt to reduce the thick smog that had shrouded its capital many days. More than 100 factories were shut down (but only for a few days), and one-third of the government's cars and trucks were removed from the roads, all

Climate Change Linked to Polluting Stagnant Air

The fact that severe air pollution is provoked by the burning of fossil fuels (especially coal) combined with stagnant air is so well known that it is common sense. The fact that global warming (caused by emissions of fossil fuels) may be a major contributor to air stagnation in some parts of the world is new. One such region is eastern China. The link is changing atmospheric wind patterns that slow "ventilation"—refreshing winds—over Chinese urban areas. "Everyone used to think that controlling smog hinged on reducing regional pollution," said Liao Hong, a professor at Nanjing University of Information Science and Technology and coauthor of a climate change study establishing the relationship. "Now it's clear that it will require a global effort" (Hernández 2017). Yuhang Wang, an atmospheric scientist at Georgia Institute of Technology in Atlanta and a coauthor of the paper with Liao, said: "In the long run, emission reductions of both pollutants and greenhouse gases are needed to mitigate the winter haze problem" (Hernández 2017).

Further Reading

Hernández, Javier C. 2017. "Climate Change May Be Intensifying China's Smog Crisis." *New York Times*, March 24. Accessed October 11, 2018. https://www.nytimes.com/2017/03/24/world/asia/china-air-pollution-smog-climate-change.html.

to combat a mid-winter spell of stagnant air from fossil-fuel-burning factories and vehicles that had reached hazardous levels.

Air pollution is a major political issue in China, the subject of copious political pressure that has been recognized by the Communist Party on national and local levels. The Environment News Service reported on August 6, 2014 that the Beijing Municipal Environmental Protection Bureau had announced that the districts of Dongcheng, Xicheng, Chaoyang, Haidian, Fengtai and Shijingshan would "stop using coal and close coal-burning power plants and other coal-fired facilities over the next six-and-a-half years" ("Beijing Bans" 2014). In many instances, coal will be replaced by natural gas, a cleaner-burning fossil fuel. Beijing still must tackle pollution from cars, the source of 31 percent of the city's air pollution. Industry makes up about 18 percent, and dust about 14 percent ("Beijing Bans" 2014).

End Coal Use by 2020?

Officials also curtailed the burning of oil, coal, burnable trash waste, and some biomass fuels. In 2014, Beijing, China's capital and second-largest city (with about 20 million people), announced plans to ban the use of coal by the end of 2020, a very ambitious goal given the country's continued dependence on it. Curtailing coal use was neither quick nor easy. By 2013, Beijing was recording its worst air quality in history. The *New York Times* reported, "The surge in pollution, which is happening across northern China, has angered residents and led the state news media to report more openly on air-quality problems" (Wong 2013).

Beijing was facing the results of a "growth-at-any-cost" policy, which has reaped widespread pollution and other environmental damage, as well as soaring greenhouse-gas emissions. In addition, Beijing's atmosphere collects the effluvia of coal-burning factories, power plants, and vehicles during stagnant weather, especially in winter, and seals them under an atmospheric inversion in which warm air sits on top of a colder layer near the ground. Like Mexico City and Los Angeles, Beijing is located in a valley nearly surrounded by mountains and prone to thermal inversions that hold pollution close to the ground. With 20 million people in 2007, a number that is steadily increasing, metropolitan Beijing has been undergoing one of the world's largest construction booms, adding residential and office space that requires more electricity, heating, and cooling, nearly all of it provided by coal. Automotive pollution pours into the city as well, as it adds 400,000 cars and trucks a year. While Shanghai has held new vehicle registration to 100,000 a year with license fees as high as $7,000 per vehicle, Beijing has allowed nearly untrammeled growth.

Even as China continued to build coal-fired and nuclear-power plants, in 2009, it passed the United States as the world's largest market for wind energy. The country was building the world's largest wind farms (at 10,000 to 20,000 megawatts each). China already was the world's largest producer of solar panels.

On August 9, 2010, China announced closure of 2,078 industrial plants that use large amounts of energy (most of it low-grade coal) as part of a drive for energy efficiency. The closures include 762 cement plants (very heavy users of energy), 279 that produce paper, 175 steel factories, and 84 leather plants (Bradsher 2010).

China's Auto Mileage Standards

China also has enacted tough automobile mileage standards. The booming Chinese economy and rising consumption of 1.4 billion Chinese is swamping these efforts, however. Total use of fossil fuels and production of greenhouse gases continue to rise rapidly. Apartment and office buildings continue to rise in cities, and rural people are purchasing appliances at record rates. In 2009, car sales in China surpassed those in the United States with a rise of 48 percent in one year (Bradsher 2010). Use of air-conditioning is spreading rapidly. "We really have an arduous task" even to reach China's existing energy-efficiency goals, said Gao Shixian, an energy official at the National Development and Reform Commission, in a speech at the Clean Energy Expo China in late June in Beijing (Bradsher 2010).

China has eliminated motor-vehicle fuel subsidies and set the fuel-efficiency standard for new vehicles at 36.7 miles per gallon seven years before the United States will reach that requirement. China also has set high efficiency requirements for new coal-fired electric plants. (The United States set no such standards until

2015.) "Regardless of whether the United States passes its own legislation, China will take positive measures because this is a requirement for our own economy to conserve resources," said Xie Zhenhua, vice chairman of the National Development and Reform Commission (Mufson 2009). China has set a renewable-energy target of 15 percent by 2020, in part by increasing its nuclear energy output 400 percent.

Further Reading

"Beijing Bans Coal Burning to Clear the Air." 2014. Environment News Service, August 6. Accessed September 30, 2018. http://www.ens-newswire.com/2014/08/06/beijing-bans-coal-burning-to-clear-the-air/.

Bradsher, Keith. 2010. "China Fears Warming Effects of Consumer Wants." *New York Times*, July 4, 2010. http://www.nytimes.com/2010/07/05/business/global/05warm.html.

Bradsher, Keith. 2011. "Cleaner China Coal May Still Feed Global Warming." *New York Times*, June 17: B-8.

Bradsher, Keith, and David Barboza. 2006. "Pollution from Chinese Coal Casts a Global Shadow." *New York Times*, June 11. Accessed September 30, 2018. http://www.nytimes.com/2006/06/11/business/worldbusiness/11chinacoal.html.

Buckley, Chris. 2015. "China Burns Much More Coal Than Reported, Complicating Climate Talks." *New York Times*, November 4: A-1.

Mufson, Steven. 2009. "China Steps Up, Slowly but Surely, to Address Emissions Issue." *Washington Post*, October 24. http://www.washingtonpost.com/wp-dyn/content/article/2009/10/23/AR2009102304075.html.

Rosenthal, Elisabeth. 2010. "Nations That Debate Coal Use Export It to Feed China's Need." *New York Times,* November 21. Accessed November 6, 2018. https://www.nytimes.com/2010/11/22/science/earth/22fossil.html.

Tollefson, Jeff. 2016. "China's Carbon Emissions Could Peak Sooner Than Forecast." *Nature* 531 (March 24): 425. Accessed September 30, 2018. http://www.nature.com/news/china-s-carbon-emissions-could-peak-sooner-than-forecast-1.19597.

Wong, Edward. 2013. "Beijing Takes Steps to Fight Pollution as Problem Worsens." *New York Times*, January 31. https://archive.nytimes.com/www.nytimes.com/2013/01/31/world/asia/beijing-takes-emergency-steps-to-fight-smog.html.

Wong, Edward, and Chris Buckley. 2015. "Chinese Premier Vows Tougher Regulation on Air Pollution." *New York Times*, March 16. Accessed September 30, 2018. http://www.nytimes.com/2015/03/16/world/asia/chinese-premier-li-keqiang-vows-tougher-regulation-on-air-pollution.html.

INDONESIA

Irian Jaya: Tidal Waves of Waste

Papua New Guinea (formerly known as Irian Jaya) is located on one of the most environmentally exploited large islands in the world. In West Papua, one of the world's largest gold and copper mines continues to grow, spewing waste that turns forests into moonscapes. Several hundred local indigenous people have died following protests of the mine's environmental and safety record during a quarter-century of strenuous opposition to the destruction, by strip mining, of mountains they regard as sacred.

The World's Largest Gold Mine

New Orleans–based Freeport-McMoRan's Grasberg mine, in West Papua's Jayawijaya district, operates in conjunction with the Rio Tinto Company (formerly Rio Tinto Zinc). Grasberg, the largest gold mine (and the third-largest copper mine) in the world, is situated on 16,500-foot high, snow-capped Mount Jaya, a few hundred miles south of the equator in Papua New Guinea, in an area considered sacred by indigenous people in the area. The Grasberg mine contains gold, silver, and copper valued at $50 billion (Bryce 1996). According to one observer, "Freeport's Grasberg mine is essentially grinding the Indonesian mountain into dust, skimming off the precious metals, and dumping the remainder into the Ajkwa River" (Bryce 1996).

The Mineral Policy Institute has called for an end to Rio Tinto's environmentally destructive mining activities at the Freeport mine, which it described as having "the world's worst record of human-rights violations and environmental destruction" ("Rio Tinto's Shame" 2000). The Freeport mine uses the alpine Lake Wanagon, which is considered sacred by the indigenous Amungme people, to dispose of waste rock from its massive gold and copper mining operation near the Grasberg gold mine.

In Freeport's five-square-mile strip mine, between 80 and 100 giant trucks haul 600,000 tons of rock daily from a pit almost 3,000 feet deep. Twelve miles of conveyor belts carry ore to a milling plant that uses more than a billion gallons of water a month. Most of the machinery in the mine was dismantled and hauled up rock walls in pieces on an aerial tramway (Roberts 1996). Mount Jaya, at 16,500 feet high, towers above the mine and contains three of the world's eight remaining equatorial glaciers. Gold production from the mine averages between 1 million and 1.5 million ounces a year; copper production averages 1 billion pounds a year. The mine employs about 17,000 people, 89 percent of them non-Papuans (Roberts 1996, 14).

Death and Devastation

Mine tailings are dumped into a tributary of the Ajkwa River, after which they flow down steep mountainsides into rain forests at lower elevations, producing a desolate landscape. The scene was described by one observer: "Dead and dying trees are everywhere, their broken branches protruding from tracts of gray sludge. . . . Vegetation is being smothered by accumulated sludge that is several yards deep in places. . . . By the company's own calculations, 51 square miles of rain forest is expected to be destroyed before the century is out" (Roberts 1996, 16).

An estimated 3 billion tons of rock will have been processed by the time the mine is exhausted in about 2040. According to the Mineral Policy Institute, "This waste is acidic and contains heavy metals. The water from Lake Wanagon flows into the Ajkwa River system that flows down to the Arafura Sea. In addition, the mine dumps 300,000 tons of waste tailings into the Ajkwa River every day" ("Rio Tinto's Shame" 2000).

In 1977, local indigenous peoples affiliated with the Free Papua Movement issued their own critique of its environmental record by blowing up one of its ore pipelines. According to Al Gedicks, reaction of the Indonesian military was swift and emphatic:

> The Indonesian military responded by sending United States-supplied OV-10 Bronco attack jets to strafe and bomb villagers. The retaliation was code-named Operation Tumpas ("annihilation"). Papuans claim that thousands of men, women, and children were killed in this action; the government admits to 900. Reports of the use of these counterinsurgency aircraft did not appear in the world press until a year later. (Gedicks 2001, 95)

Local Protests Swell

Local protests of the Grasberg mine have continued for many years. In 1996, after an indigenous man was hit and injured by a car driven by a Freeport employee, 6,000 tribal people laid siege to the mine's offices. When Freeport chief executive officer Jim Bob Moffett arrived at a local airstrip March 12, 1996, a group of similar size gathered at the airport to demand that the mine be shut down. Moffett was quoted in the September/October 1997 issue of *Mother Jones* as saying that the environmental impact of the mine is the equivalent of "me pissing in the Arafura Sea" (Ziman 1998).

During March 1997, several thousand villagers rioted in the towns of Timika and Tembagapura, located near the mine. Four people were killed and more than a dozen injured as protesters damaged Freeport's equipment. The Australian Council for Overseas Aid (ACFOA) and the Catholic Church of Jayapura reported that Freeport turned a blind eye while the Indonesian military killed and tortured dozens of native people in the area surrounding the mining concession. "Villagers were beaten with rattan, sticks, and rifle butts and kicked with boots," one tribal leader told Catholic Church officials. "Some were tortured until they died" (Ziman 1998).

Even as Freeport adamantly denied responsibility for alleged human-rights violations, the company and the Indonesian military responded to local indigenous protests by spending $35 million to assemble barracks and other facilities to house and support 6,000 troops, "more than one soldier for each adult Amungme" (Gedicks 2001, 106–107). The company asserted that the ACFOA had backtracked on its original claim that Freeport was involved in the killings (Bryce 1996). Indonesian military troops routinely guarded the area around the mine, and Freeport provided them with food, shelter, and transportation (Bryce 1996). The Indonesian government maintained a 9 percent share in the mine, enough to earn several hundreds of millions of dollars a year in royalties, taxes, and benefits, making Freeport Papua New Guinea's largest single taxpayer.

Waves of Waste

During May 2000, the Grasberg mine's waste-rock disposal dam collapsed, killing four workers and, according to one account,

> sending several 40-meter-high "tidal waves" of waste roaring down the Wanagon river toward Banti village. Incredibly, there was no loss of life at Banti despite most people being asleep when the waves arrived, passing just meters below homes, killing livestock and destroying the village graveyard. Adding insult to injury, 30 minutes after the flood reached Banti, an early-warning system installed by Freeport rang the alarm. ("Rio Tinto's Shame" 2000).

One witness at the site reported in *The Jakarta Post* that a "150-foot high wave had . . . destroyed pig sties, vegetable gardens, and a burial ground . . . about seven miles downstream of the basin" (Gedicks 2001, 30). One report described the resulting tidal wave of waste as "a mini-tsunami" (Freeport Faces 2000). The spill occurred, coincidentally, one day before the annual shareholders' meeting for Freeport McMoRan Copper & Gold.

Within days of the spill, on May 8 and again on May 18, protests against Freeport shut down the company's offices in Jakarta and prevented about 1,000 Freeport employees from entering their workplaces. In addition to protesting the environmental devastation and deaths caused by the spill, the protesters demanded that Freeport Indonesia provide a larger proportion of its earnings to support local people in the impoverished province surrounding the mine.

Soon after the accident, about 600 Amungme people from Banti, Tsinga, and Arwanop blockaded the Freeport mine's access road, preventing workers' buses from entering the mine. Roughly 100 police confronted the blockade but failed to break it until representatives met personally with Hermani Soeprapto, Freeport's General Manager, and addressed grievances to the company.

The dam at Lake Wanagon has failed three times (June 20, 1998, March 20, 2000, and May 2000) due to the company's dumping of overburden. After the third breach, dumping was halted pending an investigation. The investigation, conducted by Freeport and the Institute of Technology of Bandung (Indonesia), cleared the company to continue operations in January 2001.

Construction of a dormitory town at Tembagapura in association with Freeport Indonesia's mining operation at Mount Carstensz led to eviction of indigenous Amungme, who were barred from entering the town, which houses as many as 20,000 workers and family members ("Resource Boom," n.d.). Freeport moved the 1,000 inhabitants of a village, Lower-Waa, to the coastal lowlands, where, in one month, 88 of them died of malaria ("Resource Boom," n.d.).

"Freeport has taken over and occupied our land," said Tom Beanal, leader of LEMASA, an acronym for the Amungme Tribal Council, the community organization of the indigenous Amungme people. "Even the sacred mountains we think of as our mother have been arbitrarily torn up by them, and they have not felt the least bit guilty. Our environment has been ruined, and our forests and rivers polluted by waste" (Ziman 1998). "They take our land and our grandparents' land," said Beanal. "They ruined the mountains. They ruined our environment. . . . We can't drink our water anymore" (Bryce 1996).

Further Reading

Bryce, Robert. 1996. "Spinning Gold." *Mother Jones*, September/October. Accessed September 30, 2018. http://www.etan.org/news/kissinger/spinning.htm.

"Freeport Faces Investigation Due to Recent Disaster and Past Mismanagement." 2000. *Drillbits and Tailings* 5, no. 9 (May 31).

Gedicks, Al. 2001. *Resource Rebels: Native Challenges to Mining and Oil Corporations*. Boston: South End Press.

"Resource Boom or Grand Theft?" n.d. Australia West Papua Association, Sydney. Accessed September 30, 2018. http://www.cs.utexas.edu/users/cline/papua/deforestation.htm.

"Rio Tinto's Shame File: Indonesian Landowners' Discontent Represented at Rio Tinto AGM." 2000. Mineral Policy Institute, May 22.

Roberts, Greg. 1996. "Mining Big Money in Irian Jaya." *Sydney Morning Herald*, April 6, in *World Press Review*, July: 14–16.

Shulman, Susan. 2016. "The $100bn [Billion] Gold Mine and the West Papuans Who Say They Are Counting the Cost." *The Guardian*, November 21. Accessed September 30, 2018. https://www.theguardian.com/global-development/2016/nov/02/100-bn-dollar -gold-mine-west-papuans-say-they-are-counting-the-cost-indonesia.

Ziman, Jenna E. 1998. "Freeport McMoran: Mining Corporate Greed." *Z Magazine*, January. https://zcomm.org/zmagazine/freeport-mcmoran-mining-corporate-greed-by-jenna -e-ziman/.

NIGERIA

The Ogoni: Oil, Blood, and the Death of a Homeland

The 500,000 indigenous Ogoni of the Niger Delta in southern Nigeria have watched as their traditional fishing and farming livelihood has been laid waste by several multinational companies' extraction of oil, with the complicity of the national government, which has allowed large parts of the Ogoni's homeland to be ruined.

The Ogoni's land has been contaminated not only by oil wells and pipelines, but also by gas flares that burn 24 hours a day, producing intense heat and chemical gas fogs that pollute nearby homes as they render farm fields barren and unproductive. The constant flaring of natural gas also contributes measurably to global warming. Several Ogoni who protested the ruination of their homeland and the impoverishment of their people have been convicted of false charges and executed.

The constant flaring of oil and gas—24 hours a day, seven days a week, for as many as 40 years—leaves many people suffering from blindness, skin problems,

A boy stands near an abandoned oil well head leaking crude oil in Kegbara Dere, Ogoni Territory, Nigeria. (Lionel Healing/AFP/Getty Images)

and asthma among the young and headaches and other maladies among the elderly. In some villages, where flares are maintained in the middle of settlements, children never see darkness. Acid rain provoked by the flares is so strong that it sometimes disintegrates tin roofs as it reduces crop yields and contaminates the water and soil.

Shell Oil has extracted oil from the Niger Delta since 1958 as part of a joint venture with the Nigerian National Petroleum Corporation, Elf, and Agip. Shell is by far the largest foreign oil company in Nigeria, accounting for 50 percent of Nigeria's oil production. Nigeria generated roughly 12 percent of Shell's oil production worldwide in the late 1990s ("Shell: 100 Years" 1997).

The Toll of Oil Spills

High-pressure pipelines have been laid above ground through villages and farmlands, a major reason why the area suffered an average of 190 oil spills per year between 1989 and 1996, involving on average 319,200 gallons of oil ("Shell: 100 Years" 1997). According to one observer on the scene, "Rivers, lakes, and ponds are polluted with oil, and much of the land is now impossible to farm. Canals, or 'slots,' have permanently damaged fragile ecosystems and led to polluted drinking water and deaths from cholera. Gas flaring and the construction of flow stations near communities have led to severe respiratory and other health problems" ("Shell: 100 Years" 1997).

By 2000, oil accounted for more than 90 percent Nigeria's export earnings and roughly 80 percent of government revenue, about $20 million per day ("Nigeria: Godforsaken" 2002). More than 90 percent of Nigeria's oil is extracted from the Niger Delta. During the last 40 years, oil worth $30 billion has been extracted from the Ogoni's homeland (Wiwa 2000). By 2000, the Ogoni's homeland was home to 100 oil wells, two refineries, a petrochemical complex, and a fertilizer complex, while most of the Ogoni people do not have electricity or running water. It is a land where five physicians serve 500,000 people (Wiwa 2000).

Obsolete, leaking, rusty oil pipelines have become a major source of contaminating oil spills for the Ogoni. In 1992, a major oil blowout in the village of Botem lasted a week before it was stopped, creating a biological dead zone in the watercourses that supplied drinking water for local residents. Oil spills caused by obsolete pipelines are routinely blamed on sabotage, which allows companies to ignore repairs under Nigerian law. During October 1998, an explosion and leak flooded a large part of the village of Jesse, killing more than 700 people; two years later, two pipeline explosions in southern Nigeria killed 300 people (Gedicks 2001, 45).

Protests and Repression

The Ogoni's protests of such conditions have been met with brutal repression by Nigerian police. During 1990, people in the village of Umuechem protested oil pollution of their homeland, after which they were attacked by the notorious Mobile Police (known locally as the "Kill and Go"), who bombarded the village, killing more than 100 people and looting many homes. Survivors were forced to leave their homes (Gedicks 2001, 46). The Movement for the Survival of the Ogoni People (MOSOP), organized during 1990, adopted an Ogoni Bill of Rights demanding local control of political and environmental affairs, blaming Shell Oil for "full responsibility for the genocide of the Ogoni" (Gedicks 2001, 46).

Big oil spills have turned large areas of the Ogoni's homeland into wastelands. In mid-2001, for example, a United Nations Internet page described Yaata, an Ogoni village where dying vegetation in various shades of ochre stretch as far as the eye can see, poisoned by soil turned soggy and a dark, greasy hue since crude oil began seeping through. On April 29, at the Royal/Dutch Shell Yorla oil field, a "quake-like tremor sent shockwaves onto Yaata and surrounding villages" ("Nigeria: Focus" 2001). Within minutes, before people could guess the cause, jets of crude oil were already shooting up 100 meters, raining on the surroundings. The oil plume was quickly followed by strong fumes of natural gas, as the people of the village ran for their lives.

As the third millennium dawned, an old story continued in Ogoniland: poorly maintained oil-drilling infrastructure continued to leak and blow up, and anyone who spoke out against the devastation risked death. The August 31, 2001, issue of *Drillbits and Tailings* was dedicated to Vincent Ifelodun Bolarin Oyinbo (also known as Bola), who was reported to have died on July 19, 2001, in Lagos, Nigeria, at the age of 36 of a heart ailment, even though relatives said he never had any such history. He was, however, among 100 peaceful protesters on the Parabe offshore platform who were attacked and arrested by the Nigerian military during the protest in helicopters operated by Chevron personnel. He had been held for 12 days and tortured by the Nigerian military (*Drillbits* 2001).

"Devastation of Unimaginable Proportion"

As of August 21, 2001, community leaders from Gokana, Ogoniland, reported that fires caused by ruptured pipelines owned and operated by Shell Oil had been burning for two months with no response from authorities. "Agency reports yesterday [August 20, 2001] that the community faced being ravaged by 'devastation of unimaginable proportion' unless urgent steps were taken to put off the 'scores of fires ignited by pipeline excavators,'" reported *The Guardian* from Lagos ("Shell Oil" 2001). Other news correspondents reported that farmland has been lost to the fires.

Meanwhile, to quell any popular expression of notions that something might be dreadfully wrong in the oil fields, the Nigerian government set up a special committee to ensure total security for oil-producing areas. The authorities demanded that "a recent siege" in the oil-producing areas by "restive youths, communal agitators, and economic saboteurs" must end. The new committee signaled increasing vigilance on the part of the military against any sign of unhappiness among the Ogoni. Chief Ekaette explained that the recent "terrorism" had made the assured security of oil installations an urgent imperative. Felix Ekure, Delta State chairman of the National Youth Council of Nigeria (NYCN), warned that unless the youths of the Niger Delta are included in the development of the region, "The country may know no peace" ("Hotspots: Nigeria" 2001).

Oil-related disasters in Ogoniland assumed a nearly daily regularity. They are reported on the back pages of local newspapers. The Lagos *Vanguard*, for example, carried a report on July 18, 2001 describing how three children had died in Akwa Ibom by drowning in uncapped oil wells belonging to Shell Producing Nigeria Ltd. Addressing the World Conference of Mayors in Eket, Governor Victor Attah said, "Shell callously left uncapped wells in which three young children have so far drowned." Narrating the "evil side" of oil exploitation in the area by Exxon-Mobil, Addax, and Elf oil companies, he said that "pollution, environmental degradation, terminal diseases, and birth defects had affected many people in oil-producing areas" (Ashton-Jones 2001).

The same newspaper, on the same day and on the same page, carried the headline "Oil Spill: Strange Illness Hits Rivers Community." In the Ogbodo Isiokpo community in the Ikwerre area of the Rivers region, where the June 25 oil spill of Shell Petroleum Development Company occurred, residents were said to have reported "strange ailments" that had claimed four lives. The community said the spill had "spread quite extensively on the only stream that provided [the] source of drinking water for the area" (Ashton-Jones 2001).

In the midst of all this, Royal Dutch-Shell told a reporter for the *Wall Street Journal* that its "more urgent concern is to protect Ogoni lives and avert disaster"

(Moore 2002, A-10). The company also said it plans to spend $7.5 billion to extract 300 million barrels of remaining oil reserves in the region. This particular account portrayed the oil company as a victim of "local hostility," as well as enterprising thieves who can sell a purloined 25-foot section of oil pipeline for $87—more than an average Nigerian construction worker earns in a month (Moore 2002, A-10). Children were said in this account to "flock to the theft sites, collecting leftover oil with plastic bottles to sell to those who use it as medicine or to frighten away evil spirits" (Moore 2002, A-10). Shell executives are portrayed here as lamenting local hostility that keeps them from helping clean up the mess. "We have pleaded with the Ogoni people to let us come and make those wells safe," said Hubert Nwokolo, Shell's general manager of community development in Nigeria. "What worries me is one day we'll have a blowout and then they'll say, 'Shell, they planned it, they want to kill us all'" (Moore 2002, A-10).

Further Reading

Ashton-Jones, Nick. 2001. "Causes of Terrorism? Shell Oil in Nigeria, 1993 to 2001." October. *Drillbits & Tailings* 6, no. 7 (August 31, 2001).

Gedicks, Al. 2001. *Resource Rebels: Native Challenges to Mining and Oil Corporations*. Boston: South End Press.

"Hotspots: Nigeria." 2001. Catherine Baldi, ed. *Drillbits and Tailings* 6, no. 9 (November 30).

Johansen, Bruce E. "Nigeria: The Ogoni." *Indigenous Peoples and Environmental Issues*. Accessed January 4, 2019. https://ratical.org/ratville/IPEIE/Ogoni.html.

Moore, Sarah. 2002. "For Shell, Nigerian Debacle Isn't the End of the Line: Danger Lurks in Ogoniland for People and Firm, but the Place Beckons." *Wall Street Journal*, January 10: A-10.

"More Blood Is Spilled for Oil in the Niger Delta." 1999. *Drillbits and Tailings* 4, no. 20 (December 11).

"Nigeria: Focus on Ogoni Oil Spill." 2001. Integrated Regional Information Networks, United Nations Office for the Coordination of Humanitarian Affairs, June 12. Accessed September 30, 2018. http://www.irinnews.org/report/22109/nigeria-irin-focus-ogoni -oil-spill.

"Nigeria: Godforsaken by Oil." 2002. Ricardo Carrere, ed. World Rainforest Movement *Bulletin*, March. Accessed November 6, 2018. https://wrm.org.uy/articles-from-the-wrm -bulletin/section1/nigeria-godforsaken-by-oil/.

Robinson, Deborah. 1996. *Ogoni: The Struggle Continues*. 2nd ed. Geneva, Switzerland: World Council of Churches.

Saro-Wiwa, Ken. 1992. *Genocide in Nigeria: The Ogoni Tragedy*. Port Harcourt, Nigeria: Saros International Publishers.

"Shell: 100 Years Is Enough!" Corporate Watch. 1997.

"Shell Accused of Abuses in Nigeria's Ogoniland." 2017. BBC News, November 28. Accessed September 30, 2018. https://www.bbc.co.uk/news/world-africa-42151722.

"Shell Oil Spills Continue to Ravage Communities and the Environment in Nigeria." 2001. *Drillbits and Tailings* 6, no. 7 (August 31).

Wiwa, Owens. 2000. "Like Oil and Water: The Ogoni in Nigeria." August. Doctors for Global Health.

PHILIPPINES

Toxic Spills from a Copper Mine

The Philippines, a chain of 7,000 islands, hosts a wide variety of indigenous peoples, many of whom are seeing their traditional ways of life forced to the margins of survival by intrusions of various extractive enterprises. On Luzon, the largest of the islands, toxic tailings spills from copper mines have caused floods that extensively damaged several villages. Hydroelectric dams have threatened to flood native peoples' lands on Luzon and other islands, as growing urban areas seek new sources of power. The Cordillera Central mountain range of Luzon Island contains some of the world's largest gold reserves, with a familiar range of environmental problems afflicting native peoples, who, as in other areas, find themselves suffering environmental problems but enjoying none of the wealth that is being generated by mining of their lands. "We are sitting on gold, but where is this gold?" asked an Ibaloi elder in Benguet, as he surveyed poverty in a land of abundance ("The New Gold Rush" 1996).

Mining has been controversial on Luzon since the early 1990s, when the people of Itogon, in the Cordillera region of the island, first protested the open-pit mining of Benguet Corporation. On March 24, 1996, Marcopper Mining Corporation's copper mine in Boac, Marinduque, in the southern Tagalog region of Luzon (about 100 miles south of Manila), released toxic mine tailings from the Tapian Pit (holding 23 million metric tons of mine wastes) at the rate of 5 to 10 cubic meters per second into the Makulapnit and Boac rivers. Before the leak was staunched, roughly 1.5 million cubic meters of mining sludge had coursed into the two rivers (Tauli-Corpuz 1996). The site was mined between 1969 and 1996 by Placer Dome Inc. (then known as Placer Development, Ltd.) and the Marcopper Mining Corporation. In 1997, Placer Dome abandoned the mining project, selling its 39 percent share in Marcopper to a local mining company, holding it responsible for the unfinished cleanup.

Several Villages Buried

The waste spill quickly caused flash floods that isolated five villages, including 4,400 people, along the Boac River. According to first-person accounts, one village, Barangay Hinapulan, "was buried under six feet of muddy floodwater and 400 families had to flee to higher ground. Their sources of drinking water were contaminated while fish, freshwater shrimp, and pigs were killed" (Tauli-Corpuz 1996). Twenty other villages were evacuated. A 27-kilometer stretch of the Boac River was declared dead by government officials. Within three weeks, the Philippine Department of Health issued a report that said some area residents had unhealthy levels of zinc and copper in their blood. According to a report from the scene, Indigenous residents of the affected area "also complained of skin irritations and respiratory problems which could have been caused by the poisonous vapors emitted by hydrogen sulfide and nitrous oxide from the mine wastes" (Tauli-Corpuz 1996).

Could It Happen Again?

The threat of a major toxic spill from the former Marcopper mine has not ended. According to a report in *Drillbits and Tailings,* Placer Dome commissioned reports to assess the safety of the mine's infrastructure that said the Tapian Pit, which leaked in 1996, and the Maguila-guila siltation dam, which burst in 1993, "will both cave in and spill tons of toxic wastes in already devastated Marinduque" ("A Disaster" 2002).

A U.S. Geological Survey report noted that the

> high rate of sediment transport from Marcopper will continue to have adverse effects on the aquatic ecosystem, and on the ability of the river system to handle large flood events . . . the fine-grained, metal-rich, and potentially acid-generating nature of the sediments from Marcopper is likely to have been a substantial change from the natural condition of the Mogpog prior to mining. For example, fine-grained sediment from the mine site may fill in the pore spaces of the originally coarser riverbed sediments, thereby adversely affecting the habitat fish and aquatic invertebrates living on the river bottom. (Plumley et al. 2000)

The U.S. Geological Survey report also criticized a plan to use submarine tailings disposal (STD) to remove waste by flushing it into the ocean. The Geological Survey said that STD would create "a highly acidic, metal-enriched, and environmentally detrimental plume" ("Philippine Province" 2002).

The 1996 spill could repeat itself, given conditions in the area. The people of Marinduque, who have opposed the mine for 30 years, find themselves in the crosshairs of a disaster waiting to happen. According to Placer Dome's own experts, only the timing remains an issue. In addition, during 2001, engineers for the U.S. Geological Survey said that the waste dams were "virtually certain" to collapse again, placing 100,000 people downstream at risk ("Philippine Province" 2002). In December 2001, Placer Dome Technical Services pulled out of the Philippines, leaving behind toxic mine tailings in the Boac River, the threat of five dangerously unstable mine structures, and the incomplete compensation of Marinduqueos affected by the 1996 spill ("A Disaster" 2002). The company asserted that its departure absolved it of liability for past or potential environmental damage to indigenous lands and peoples.

During January 2002, Philippine president Gloria Macapagal led an official delegation to address Canada's prime minister with a demand that Placer Dome clean up the damaged river, repair leaking dams at the mountainous mine site, and provide compensation to people in Marinduque province whose livelihoods were nearly obliterated by the Marcopper disaster (Jaimet 2002; Nikiforuk 2002).

The Cordillera's Riches

The Cordillera of Luzon Island in the Philippines is endowed with rich mineral resources. Luzon Island's Cordillera (particularly the area near Benguet, which contains much of the Philippines' mineral wealth) is one of the richest mining lodes in the world. Measured in the amount of minerals taken from a given piece of land, the Philippines is the world's second richest producer of gold (404 pounds per square kilometer), the world's third richest in copper (0.75 pounds per square kilometer), and the world's sixth richest in chromate (0.57 pounds per square kilometer) ("The New Gold Rush" 1996). Despite the area's mineral wealth, however, most of the indigenous people continue to endure intense poverty.

Following approval of the Mining Act of 1995, foreign investment in Philippine mining surged. The Cordillera Peoples Alliance opposed the new act, arguing that it served only to surrender the Philippines' sovereignty to foreign corporations. "The new Act underlines the [Fidel] Ramos government's brazen disregard for indigenous peoples' rights and welfare, as the mineral resources being offered to foreign corporations are mostly located in the territories of indigenous peoples," said the Alliance ("The New Gold Rush" 1996).

The new mining law modified requirements of 60 percent Philippine ownership of mining operations. Under the new law, 100 percent foreign ownership was allowed for tracts up to 81,000 hectares after 25 years of operation, providing that at least US$50 million was invested ("The New Gold Rush 1996"). The new law doubled depreciation rates and granted a five-year tax holiday that could be extended for another five years. It also allowed repatriation of profit and capital in U.S. dollars, tax-free treatment of capital investments, and assignment of rights that guaranteed unhampered mining operations. Under this law, foreign investors also were granted control over water and timber rights, as well as mineral rights.

Along with local indigenous peoples, the National Council of Churches of the Philippines also condemned the new law, asserting that a large amount of new mining activity would provoke thousands of indigenous families to be evicted from mining areas. In the meantime, applications for new mine sites burgeoned, with Australian companies in the lead. Indigenous peoples on Luzon complained that the central government, in its rush to expand the Philippines' industrial base, was running roughshod over their inherent rights to ancestral lands, as well as their future prospects for economic prosperity and social development. Thus, they argued, the new mining law disrespected indigenous cultures and indigenous peoples' rights to pursue their cultural development and political integrity.

Mining Law Criticized

Indigenous peoples in the Cordillera reacted skeptically to the new mining law because they already had endured a century during which their lands and rights had been bargained away by the central government in favor of development by foreign companies. In Itogon, for example, nearly all of the rivers had dried up or been contaminated by the Benguet Corporation's open-pit mining operations. Clean drinking water had become scarce, and many rice fields had been abandoned because irrigation water was no longer available. Open-pit mining was encouraged by the new law because it speeded the pace of mining and profits, while paying little regard to environmental consequences. The new law also broadens companies' easement powers, speeding eviction of indigenous peoples whose homes stand in the way of expanding open-pit mines and infrastructure, such as roads, which are required for industrial-scale mining.

According to indigenous opponents of increased mining on Luzon, the Mining Act of 1995

> threatens to wipe out the very existence of the Cordillera indigenous peoples. For the Cordillera indigenous peoples, land is the source of life—from the land comes the materials needed for their production, and thus, their very sustenance. From the land emanates, and revolves, their collective culture and spiritual life. To destroy their land is to wipe out this web of life, and the very peoples themselves. ("The New Gold Rush" 1996)

Further Reading

"A Disaster Looms for Communities in Marinduque, Philippines." 2002. *Drillbits and Tailings* 7, no. 4 (April/May).

Jaimet, Kate. 2002. "Placer Dome Blamed in 'World-Calibre Disaster'; Canadian Mining Giant Says Philippine Government Blocking Cleanup." *Ottawa Citizen*, January 29.

"The New Gold Rush: The 1995 Philippine Mining Act Lures a New Wave of Profit-Hungry Gold Diggers." 1996. *KASAMA* 10, no. 2 (April-May-June). Accessed September 30, 2018. http://cpcabrisbane.org/Kasama/1996/V10n2/GoldRush.htm.

Nikiforuk, Andrew. 2002. "Still a Fine Mess; the Controversy over a Philippine Mine Cleanup Rages on." *Canadian Business*, February 22.

"Philippine Province Bans Mining for Twenty-Five Years." 2002. *Drillbits and Tailings* 7, no. 2 (February 28).

Plumley, Geoffrey S., Robert A. Morton, Terence P. Boyle, Jack H. Medlin, and José A. Centeno. 2000. "An Overview of Mining-Related Environmental and Human Health Issues, Marinduque Island, Philippines: Observations from a Joint U.S. Geological Survey—Armed Forces Institute of Pathology Reconnaissance Field Evaluation, May 12–19, 2000." U.S. Geological Survey Open-File Report 00-0397. Accessed November 7, 2018. https://pubs.usgs.gov/of/2000/ofr-00-0397/ofr-00-0397.pdf.

Tauli-Corpuz, Victoria. 1996. "The Marcopper Toxic Mine Disaster: Philippines' Biggest Industrial Accident." TWN: Third World Network. Accessed January 30, 2019. http://www.twn.my/title/toxic-ch.htm.

RUSSIA

The Nenets, Natural Gas, and Climate Change

The Nenets live on the Yamal Peninsula, which juts into the Arctic Ocean from northwest Siberia. They have developed a cold-weather economy that depends on reindeer herding. Weather records that have been kept since 1861 note accelerating warming since the middle 1970s. While winter temperatures used to reach minus 55°F and summers rarely broached 70°F, by 2016 and 2017, some summer days reached 100°F.

For decades, the Nenets resisted harsh assimilationist policies of the former Soviet Union that attempted to organize herders into collectives by redistributing reindeer and pasture lands as many people were coerced into resettlement.

Traditionally, reindeer have been the Nenets' major source of sustenance, similar to the buffalo for the Plains Indians in the United States before about 1880. Soviet authorities also tried to eradicate indigenous religions by killing shamans and destroying sacred sites. After the Soviets' fall, the Nenets continue to struggle against climate change and Russian exploitation of some of the world's richest deposits of natural gas. In 2017, about 7,500 Nenets lived on the Yamal Peninsula. Across four autonomous administrative districts called okrugs (Yamalo-Nenetsky, Nenetsky, Taymyrsky, and Khanty-Mansiysk), about 6,000 herders tend 250,000 reindeer, carrying on traditional lifeways amid a jungle of natural gas wells, pipelines, and associated infrastructure.

"Yamal" in the Nenets language translates to "end of the earth" or "back of the beyond." However, according to Andrei V. Golovnev and Gail Osherenko, writing in *Siberian Survival,* "The traditional lands and waters of the Khanty and the forest Nenets have been devastated by oil production and plundered by thousands of newcomers who helped themselves to wild mushrooms and berries, game and fish while the indigenous people were forced to survive in increasingly marginal circumstances" (Golovnev and Osherenko 1999, 105–106).

Outsiders began to arrive in large numbers during the 1930s, including about 50,000 peasants who had been removed from other parts of Stalin's Soviet Union by force and used as forced labor to develop industrial infrastructure on the Nenets' land. The Nenets, Khanty, and other indigenous peoples rebelled against forced collectivization and other impositions of Soviet authority. Their leaders were arrested and meetings broken up by the Red Army as reindeer herds were confiscated.

Oil and Gas Discovered

Oil and natural gas were discovered in the Nenets' homeland during 1964; substantial production began a year later. Since the 1960s, the Nenets' territory, especially the Yamal Peninsula, has been contaminated in many places by oil production and other chemical agents, with damage to birds, fish, and mammals, including the Nenets' reindeer herds. Pastures have been despoiled by oil development and transportation, which also has caused destabilization of the permafrost in some areas. Many more immigrants from Russia were drawn to the Nenets' homelands by the exploitation of natural gas reserves during the 1970s.

The Soviet state offered high salaries and bonuses to workers who moved to the Nenets' country to develop oil and gas resources and to build cities. A million tons of oil was produced across Northwestern Siberia during 1965, 28.5 million tons in 1970, 143.2 million in 1975, and 307.9 million in 1980. Gas production, 3.3 million cubic meters in 1965, rose to 9.5 million on 1970, 38 million in 1975, and 160 million in 1980. By the 1980s, thousands of miles of pipelines, roads, and railroads laced the taiga and tundra of Northwestern Siberia. One million tons of oil spilled from poorly maintained pipelines in an average year by the 1980s. By 1988, 42,460 square miles of reindeer pastures had become unusable because of oil and gas transport and spills in the Nenets and Khanty okrugs (administrative districts).

The Death Toll

At least 30 Nenets men, women, and children died within a few years from causes related to oil exploration, and many reindeer were killed by automobile accidents on heavily traveled roads. In 1990, an oil truck struck and killed an old man's reindeer, provoking families from five clans to raise a tent on a bridge that blocked traffic in both directions. The protest became a popular cause across Russia, one of many ecological confrontations that helped speed the dissolution of the Soviet Union. The response was so great because reindeer were being killed intentionally by outsiders, "decapitated and their carcasses left to rot" (Golovnev and Osherenko 1999, 106).

The Russian city of Nadym, a company town of 50,000, became known as the "gas capital of Russia," as it was built up by GAZPROM, the Russian natural-gas monopoly, to exploit 367 trillion cubic feet of gas on the Yamal Peninsula (Raygorodetsky 2017, 118). By the 1990s, Nadym housed thousands of workers who built a road and railway along the eastern side of the Yamal Peninsula, preparing the area for construction of a gas pipeline (Gasperini 1997). A hot-oil pipeline and tanker terminal were proposed, but thick wintertime ice and the shallowness of Ob Bay limited the size of tankers that could dock there (Golovnev and Osherenko 1999, 11–13).

By 1990, the Nenets, having been inundated by 500,000 outsiders, comprised only 6 percent of the population in their homeland (Golovnev and Osherenko 1999, 77). Amid the development of the gas fields on the Nenets' homelands, an estimated 2,500 indigenous nomadic families live in the area. "We wouldn't have it any other way, we've always lived just with our deer," said Anatoli Vanuito, who expressed concern about ongoing oil and gas development in the Yamal by GAZPROM. "In recent years there have been fewer fish, and the reindeer get sick more often now," he lamented. "If there are no fish and the deer are sick, that's it for us. We have no other way to live" (Gasperini 1997).

For several years, Nenets herders have said that their reindeer are becoming smaller and more prone to illness. "The tundra is the kind of environment where a single vehicle's track will remain for decades," said Bruce Forbes, an American ecologist who spent three years studying the Yamal ecosystem (Gasperini 1997).

Damage from Gas Exploitation and Climate Change

Nenets' survival comprises a constant battle against the elements, natural and manmade. Their children are sent to boarding schools and subject to systematic "Russification," an attempt to strip them of traditional language and culture. Russian industry entwines them in a forest of gas wells and pipelines as climate change decimates their reindeer herds.

Each year, the Nenets undertake an 800-mile seasonal migration with their reindeer herds. During the winter of 2013 and 2014, however, warming weather brought rain to southern Yamal, which then suddenly froze, encasing much of the reindeer's usual pasture under thick ice. The reindeer, which are accustomed to pawing snow away from their food sources, were unable to penetrate the ice. "Tens of thousands of starved," wrote Gleb Raygorodetsky in the *National Geographic*. "In the summer of 2016, the survivors were still recovering" (Raygorodetsky 2017, 114). By the summer of 2017, temperatures rose to as high as 94°F close to the Arctic, with severe drought. Several Nenets and many more reindeer also are dying of anthrax, "a direct result of thawing permafrost, which allowed animal carcasses buried during an outbreak in the 1940s to reemerge, still bearing infectious microbes" (Raygorodetsky 2017, 115).

Russia's government plans call for rapid acceleration of natural-gas development in the Yamal area to an extent that by 2030, it may comprise more than a third of the country's total production, along with construction of roads, a new airport, several pipelines, and railroads, as well as housing and commercial infrastructure, all of which will hinder the Nenets' ability to practice their traditional style of life. At the same time, thawing permafrost is undermining the stability of GASPROM's infrastructure. As Raygorodetsky wrote:

> Some effects of thawing permafrost are hard to prepare for. . . . In the summer of 2014, a gaping crater 130 feet wide by 115 feet deep suddenly formed in the tundra 19 miles southeast of Bovanenkovo. Experts blame an eruption of methane gas that had been trapped under frozen ground. They worry that if a similar eruption were to occur under a gas field itself, it could cause considerable damage. In the summer of 2017, two more eruptions were reported. One occurred near the camp of a [reindeer] herder. (Raygorodetsky 2017, 118)

Reindeer Deaths, Warming, and Anthrax

Andrei Listishenko, who has headed Yamal's veterinary service for 15 years, said that herders reported that their reindeer have become groggy and sluggish. "They were falling behind and had stopped responding to commands. . . . Many of them were dying" (Sneider 2018). Extreme heat is much more stressful on the reindeer than extreme cold. Visiting the area, Listishenko "saw reindeer clustered on the ground, shaking and panting. Patches of their fur seemed to have fallen off, leaving them splotchy; the animals were emaciated, their ribs visible." The pain was visible to Listishenko as the reindeer's heads drooped, they became feverish, breathing rapidly, and then died in convulsions. Dozens of dead reindeer became hundreds, then a thousand (Sneider 2018, 42).

Further investigation revealed that heat was only one factor in the reindeer's demise. They also were dying of anthrax, spread by the heat, liberated by melting permafrost, which the Nenet call *zhivaya zemlya,* "living earth" bringing bacteria and other pathogens, including anthrax, back to life (Sneider 2018, 49).

Further Reading

Gasperini, William. 1997. "In a Land 'Back of the Beyond' Reindeer Rule the Nomads' Life." *Christian Science Monitor,* April 25. https://www.csmonitor.com/1997/0425/042597. feat.feat.1.html.

Golovnev, Andrei V., and Gail Osherenko. 1999. *Siberian Survival: The Nenets and Their Story*. Ithaca, NY: Cornell University Press.

Raygorodetsky, Gleb. 2017. "Life on the Edge." *National Geographic*, October: 108–125.

Sneider, Noah. 2018. "Cursed Fields: What the Tundra Has in Store for Russia's Reindeer Herders." *Harper's Magazine*, April: 40–51.

SOUTH PACIFIC ISLANDS

Strip-Mining Paradise

North American newspaper readers of the early 21st century usually encounter the islands of the South Pacific in the travel sections as fantasies of paradise—palms waving on sandy beaches portending release and relief from the cares of industrial life. While such enclaves do exist, some of the South Pacific's scenery might shock the uninformed. Large parts of entire small islands have been shaved down to mud and stubs of dead trees, having been strip-mined for various metals and minerals.

The Mataiva, Nauru, and Banaba Islands, for example, have become open-pit phosphate mines. The Kanaky Mine of New Caledonia is the fourth-largest producer of nickel in the world. Indigenous people on these islands often have been relocated to other islands that do not contain salable minerals. A number of the indigenous inhabitants of the Solomon Islands, for example, have been removed to make way for gold mining. In some cases, native peoples have stayed at home, to be drafted into the industrial mining workforce.

New Caledonia's Nickel Mines

The Kanaky mine contains nearly a third of the world's nickel deposits. More than 150 million tons of nickel ore have been extracted from the island since 1864, with widespread environmental impacts. Indigenous opposition to nickel mining and smelting has become widespread on the island.

The Thio Mine on New Caledonia sits atop the world's largest single nickel deposit, which has been mined for more than a century. The Thio River and its tributaries have been polluted by massive generation of mining waste, as has the Kouaoua River basin, which flows into red-stained seas and offshore shoals.

Nickel is processed in New Caledonia at the Doniambo smelter in Noumea, which is owned by Société Le Nickel (SLN). Airborne nickel is a class-one carcinogen, considered so toxic that the World Health Organization has ruled that no safe level exists in the air. Kanaky has the world's highest level of asthma mortality and the highest levels of lung cancer in the Central and South Pacific. Pollution control is minimal, with local emissions from smelters permitted at levels more than 1,000 times those deemed safe under international standards ("Mining in the South Pacific," n.d.).

Early in 2002, indigenous peoples of New Caledonia attempted to rally against the opening of yet another nickel mine. *Drillbits and Tailings* reported that Canada's Inco Corporation had announced plans to mine the company's Goro property in the southern reaches of the island, which is said to contain some of the world's richest as-yet-undeveloped nickel and cobalt deposits. According to *Drillbits and Tailings*, "Inco chief executive officer Scott Hand wants to build a US$1.4 billion commercial nickel-cobalt project on the site" ("Inco Threatens" 2002). An organization of clan chiefs called the Sénat Coutumier, considered the voice of New Caledonia's Kanak people, opposes the project. Requested materials arrived in French (New Caledonia is a French colony), so most of the chiefs could not read them. Many indigenous people are concerned that mine wastes may be submerged in the ocean around the island, damaging the world's second-largest coral reef (after the Great Barrier Reef of Australia).

Phosphate Mining in the South Pacific

Phosphate has been mined from Mataiva, Nauru, and Banaba islands in the South Pacific for more than a century to provide feedstock for the agricultural industries in Germany, Britain, Australia, and New Zealand. Nauru first was mined under German control and later (after 1906) by joint British, Australian, and New Zealand "trusteeship," "leaving two-thirds of the island a barren and lifeless wasteland of jagged rock pinnacles" ("Mining in the South Pacific," n.d.). Indigenous islanders were forced to take legal action to force the trustees' governments to pay compensation for ruining their homelands.

Mining began on Banaba in 1900 after the Pacific Islands Company signed an agreement with two senior landowners. According to a local source, "Like many such agreements, the local community leaders did not and could not sign on behalf of all the islanders and apparently did not understand the implications of the document they were signing" ("Mining in the South Pacific," n.d.). Several indigenous families were resettled on Rabi Island during the 1940s to make way for expansion of phosphate mining on Banaba.

In 1968, the Banabans petitioned the United Nations regarding their grievances. They subsequently sued the Phosphate Commissioners and the British Government in 1971 for $120 million. Although the islanders lost in court on most arguments, the United Kingdom, New Zealand, and Australian governments eventually agreed to pay $10 million compensation in 1977, principally in the form of unpaid mining royalties. Ninety percent of Banaba Island's 1,500 acres had been strip-mined by the middle 1990s ("Mining in the South Pacific," n.d.).

On Mataiva Island, exploratory mining by Australian-owned GIE Raro Moana was halted in 1982 by local peoples' protests. Indigenous Mataiva islanders said they were never consulted regarding the proposed mine, which they said poisoned their lagoon, preventing them from eating fish, formerly a staple of their diet, for seven years. Mining is overwhelmingly opposed by the local population because of its environmental destructiveness ("Mining in the South Pacific," n.d.).

Gold Mining on the Solomon Islands

Environmental conflicts over gold mining and industrial-scale logging have brought violence, including two deaths, to the Solomon Islands, which lie to the northeast of Australia and east of Papua New Guinea. A land area of nearly 30,000 square kilometers is spread over 992 islands, with a combined population of about 340,000 people. Roughly two-thirds of the Solomon Islands is covered with tropical rainforest, and many of the people still live in villages whose residents depend on the forests for survival.

Since 1994, Gold Ridge (Ross Mining NL, Australia) has been mining gold on the Solomon Islands, relocating indigenous villagers. Local observers have noted, "Controversy has surrounded the potential impact of tailings on the local water supply and the supply of electricity, with opposition to the mine generating its own power rather than buying from the Solomon Island Electricity Authority, as originally agreed" ("Mining in the South Pacific," n.d.). Some members of the local community, such as David Thuguvada, a Guadalcanal community leader, have expressed concern regarding the impacts of mining, saying, "The villagers on the plains below Gold Ridge have been excluded from negotiations over the environmental agreement, and we are deeply concerned" ("Mining in the South Pacific," n.d.).

Local indigenous landowners, represented by the Australian legal firm Slater and Gordon (which also represented land owners at the Ok Tedi Mine in Papua New Guinea), filed a constitutional challenge to the mine in the Solomons' High Court. The suit sought to have the compensation agreement set aside on the grounds that it is unreasonable because of its probable environmental impacts.

Solomon Island Logging

While gold mining has become a subject of contention on some of the Solomon Islands, logging has sparked controversy on others. Maving Brothers Ltd., a Malaysian company, already has logged half the islands' remaining forests, and would like to clear-cut the rest, as indigenous peoples protest. Community tension over government-supported logging of disputed lands on Pavuvu Island has led to one murder and another death that has been characterized as suspicious. Martin Apa, a well-known local opponent of logging, was killed during early November 1995, in Yandina, Pavuvu's main port. A post-mortem found that his neck had been broken and pierced by a sharp object.

The *Solomon Star* reported November 10, 1995 that US$2.2 million was paid in bribes from the logging company Integrated Forestry Industry Ltd. (a subsidiary of Malaysian company Kumpulan Emas) to ministers and other government employees. Disclosure of these bribes provoked calls for the resignation or discharge of the ministers during a public rally and petition by churches, unions, and environmental advocates ("Solomon Islands" 1995). Journalist Duran Angiki, who reported the story for the *Solomon Star*, was fired following pressure from the government and logging companies. Even though Angiki was fired, seven Solomon Island government ministers were charged on December 4, 1995 with having received bribes from Maving Brothers.

Despite the indictments, logging in the Solomon Islands continued into the last years of the 1990s, exceeding sustainable levels, according to the 1994 Annual Report of the Central Bank of Solomon Islands ("Solomon Islands" 1995). Greenpeace asserted that "logging practices by the mainly Malaysian and South Korean companies are uncontrolled and destructive and supply the Japanese and Korean log market. With more than 60 percent of government revenue derived from log export levies, forest depletion means a looming disaster for the economy" ("Solomon Islands" 1995).

During the early 1990s, some Pavuvu Island landowners began to counter industrial-scale logging with their own program of sustainable development by working with Greenpeace to set up a village-based eco-forestry project. About 90 percent of the land in Solomon Islands is customarily owned by groups of families, providing an economic base for small-scale, sustainable forestry.

Is "Eco-Timber" the Answer?

Eco-timber is the local, indigenous answer to industrial-scale logging since the 1980s, when "Australian and Asian logging companies have swept through the Solomon Islands, leaving a trail of disintegrating communities, flattened and degraded forests, and silted coral reefs from runoff of exposed fragile soils" (Singh 2000). Some of the industrial-scale logging continues side by side with the indigenous peoples' more cautious approach.

Production and marketing of Solomon Islands "eco-timber" began in 1997. According to a report from the Environment News Service, the Solomon eco-timber "is managed in a way which causes minimal damage to the forests" and is harvested under local control in a manner and at a speed that will not strip the land. The export of eco-timber arose as a ray of economic and ecological hope "in a country torn by a tribal conflict, with nearly 100 people dead and the economy in tatters" (Singh 2000). A third of the eco-timber was being exported to New Zealand. In 2000, the first shipment of Solomon Islands eco-timber reached port in Sydney, Australia.

"Here is a community that has been trying to provide a future for their forests and children, only to be thwarted by aggressive logging. Martin Apa was a key supporter of eco-forestry, and we fear that he may have been targeted for the stand he took. Pavuvu landowners are being intimidated into agreeing to further unsustainable logging," said Grant Rosoman, a Greenpeace New Zealand Forests campaigner ("Solomon Islands" 1995).

Further Reading

"Inco Threatens Indigenous Kanuks and Environment of New Caledonia." 2002. Catherine Baldi, ed. *Drillbits and Tailings* 7, no. 3 (March 29).

"Kanaks March over New Caledonia Roll." 2016. Radio New Zealand, October 27. Accessed September 30, 2018. https://www.radionz.co.nz/international/pacific-news/316616 /kanaks-march-over-new-caledonia-roll.

"Mining in the South Pacific: On the Other Side of the Forests." n.d. Mineral Policy Institute.

Singh, Rowena. 2000. "Eco-Timber Export Brings Hope to Solomon Islands." Environment News Service, December 8. Accessed November 7, 2018. http://www.ens-newswire .com/ens/dec2000/2000-12-08-02.html.

Shannon Graves, a Nebraskan who will be directly affected by the proposed Keystone XL pipeline, stands at the wooden post where the pipeline would pass through her property in Polk, Nebraska. Graves is staunchly opposed to the pipeline. (Andrew Burton/Getty Images)

"Solomon Islands Murder and Corruption: Logging Takes Its Toll. Martin Apa Murdered; Greenpeace Calls for Investigation." 1995. Background Briefing by Greenpeace New Zealand, December.

UNITED STATES

The Keystone XL Pipeline's Oil and Water Don't Mix

The Keystone XL Pipeline, which has been proposed to transport Alberta oil refined from tar sands to the Texas coast for shipment by sea, has been resisted in Nebraska by many conservative farmers who assert that it may spoil their irrigation water. Why? First, the record of existing pipelines that convey Canadian tar sands is poor. Witness several documented spills. This is the main issue that has turned many otherwise conservative rural people who are in its path against this pipeline.

Second, the pipeline is designed as a conduit for two types of unconventional oil: tar (or oil) sands products from Alberta and fracking products from the Bakken Formation in North Dakota and Montana. Adding carbon dioxide from these sources over time will increase atmospheric levels (and eventually atmospheric warming) past sustainable levels.

Air-Conditioning's Climatic Irony

Air-conditioning engages us in dangerous climatic irony. As global temperatures rise (and as affluence spreads), more of it is used. Since most air-conditioning coolants produce greenhouse gases and increased use requires more electric power (and combustion of fossil fuels), its use aggravates climate change. Stan Cox, author of *Losing Our Cool: Uncomfortable Truths about Our Air-Conditioned World,* commented in the *Washington Post* that

> over the past half-century, American cities have taken on an unstable thermodynamic form, coming to resemble collections of boxes full of cool air crowded onto concrete heat islands. Turn off the AC, and office buildings would become uninhabitable, vehicles sitting in traffic would become torture chambers, apartment buildings would become death traps. (Cox, 2016)

In 1960, 12 percent of households in the United States had air-conditioning. That figure rose to almost 90 percent by 2016. Between 1950 and 2006, the size of the average U.S. house tripled, from 290 to 900 square feet per person. Between 1993 and 2005, consumption of electricity for air conditioning-doubled, according to the U.S. Energy Information Administration (Cox 2016). Use of air-conditioning, which produces waste heat, also aggravates the urban heat-island effect, exposing the poor and elderly to the worst of heat waves.

Further Reading

Cox, Stan. 2016. "Your Air Conditioner Is Making the Heat Wave Worse." *Washington Post,* July 22. Accessed October 11, 2018. https://www.washingtonpost.com/posteverything/wp/2016/07/22/your-air-conditioner-is-making-the-heat-wave-worse/.

A Signature Environmental Issue

So why did the Keystone XL Pipeline become a signature environmental issue in the Obama and Trump presidencies—a line in the sand over which several thousand people traveled to the White House fence from across North America to protest? The pipeline, designed to cross the U.S.-Canadian border, requires a presidential permit declaring that its construction is "in the national interest." More than 1,000 people were arrested during August 2011, including NASA climatologist and head of the NASA Goddard Institute for Space Studies in New York City (since retired) Dr. James Hansen.

The Keystone Pipeline became one of North America's most contentious environmental issues. The State Department received more than 1 million public comments by its April 22, 2013, deadline (also Earth Day), with a decisive balance opposing the project, which went against that agency's initial conclusion that the Keystone XL would have little or no impact on global warming. The U.S. Environmental Protection Agency also opposed the State Department's position, urging more study of greenhouse-gas emissions, the possible effects of spills, and the sensitive ecological nature of the proposed route.

Hansen joined 19 other scientists in a letter to President Obama dated August 3, 2011. He was quoted by the Environment News Service as saying, "We can say categorically that it's not only not in the national interest, it's also not in the planet's best interest." Several large U.S. environmental groups called the controversy over the Keystone XL a "watershed moment." In a letter quoted by ENS, they wrote:

Dear President Obama,

Many of the organizations we head do not engage in civil disobedience; some do. Regardless, speaking as individuals, we want to let you know that there is not an inch of daylight between our policy position on the Keystone Pipeline and those of the very civil protesters being arrested daily outside the White House. This is a terrible project—many of the country's leading climate scientists have explained why in their letter last month to you. It risks many of our national treasures to leaks and spills. And it reduces incentives to make the transition to job-creating clean fuels. ("NASA Scientist" 2011)

Signers included the Sierra Club, Natural Resources Defense Council, Greenpeace, National Wildlife Federation, Friends of the Earth, Rainforest Action Network, 350.org, the League of Conservation Voters, Environment America, and the Center for Biological Diversity.

Farmlands Turned to Wastelands

Many of these groups (Greenpeace being most active) also are advocating shutdown of the tar sands fields themselves and an end to "the industrialization of a vast area of Indigenous territories, forests, and wetlands in northern Alberta" ("NASA Scientist" 2011). "The tar sands are huge deposits of bitumen, a tar-like substance that is turned into oil through complex and energy-intensive processes that cause widespread environmental damage—polluting the Athabasca River, lacing the air with toxins, and turning farmland into wasteland," Greenpeace said. "Large areas of the boreal forest are being clear-cut to make way for development in the tar sands, the fastest growing source of greenhouse-gas emissions in Canada" ("NASA Scientist" 2011).

The 1,700-mile Keystone tar-sands pipeline is designed to carry as much as 800,000 barrels of tar-sands crude oil a day from Alberta to refineries in Oklahoma and Texas. It also is routed through the Bakken Formation of North Dakota and Montana, which is producing new oil from hydraulic fracturing ("fracking"). Both of these fields tap relatively new sources of oil that will increase in years to come, raising greenhouse-gas levels in the atmosphere. On a local level, many people worry about oil spills in fragile areas such as the Nebraska Sand Hills that could contaminate the Ogallala aquifer in an area where water is scarce. This aquifer supplies 78 percent of public water and 83 percent of irrigation water in Nebraska, almost a third of the irrigation water used in the United States.

In August 2011, the National Congress of American Indians (NCAI) declared opposition to the Keystone XL Pipeline. It was quoted by the Indian Country Today Media Network with a rationale that resonated for many other opponents. "Based on the relatively poor environmental record of the first Keystone pipeline, which includes numerous spills, U.S. regulators shut the pipeline down in late May 2011," said the NCAI resolution, which concluded that "it is probable that further environmental disasters will occur in Indian country if the new pipeline is allowed to be constructed." The Assembly of First Nations of Canada also has expressed similar opposition. The NCAI urged the United States to reduce dependence on oil from tar sands that makes such a pipeline necessary, with "work toward cleaner, sustainable energy sources" ("NCAI Condemns" 2011).

Opponents of the Keystone XL pointed to the less-than-sterling record of existing pipelines, especially recent leaks in communities that were bearing high cleanup costs, such as with the Enbridge Energy spill that ruptured a pipeline near Marshall, Michigan, in 2010, spilling more than 840,000 gallons of tar-sands crude oil. In March 2012, an Exxon pipeline sprang leaks in Mayflower, Arkansas, a town of 2,200 near Little Rock, provoking evacuation of two dozen homes. This pipeline also was carrying tar-sands crude oil, of which about 210,000 gallons spilled.

"All oil spills are pretty ugly and not easy to clean up," said Stephen K. Hamilton, a professor of aquatic ecology at Michigan State University who is advising the Environmental Protection Agency and the state on the cleanup in Marshall. "But this kind of an oil is even harder to clean up because of its tendency to stick to surfaces and its tendency to become submerged," he told Dan Frosch of the *New York Times* (Frosch 2013).

The Toll of Oil Spills

A 40-mile segment of the Kalamazoo River that traverses Marshall, the site of the Enbridge spill, was closed for two years, and several waterfront homes were abandoned. In the meantime, Enbridge initially underestimated the amount of oil that had spilled by a factor of one hundred. The Environmental Protection Agency ordered the company to dredge the river. Enbridge also bought out 154 homes in the area most intensely affected by the spill. The company also spent several million dollars to enhance or build roads and parks along the oil-soaked river.

It was reported that four months later in Mayfield, "the neighborhood of low-slung brick homes is largely deserted, a ghostly column of empty driveways and darkened windows, the silence broken only by the groan of heavy machinery pawing at the ground as remediation continues" (Frosch 2013). Exxon offered to buy the 22 vacated residences, but only at their diminished, post-spill value. The company also had spent $2 million on temporary housing for residents and more than $44 million on the cleanup by mid-2013.

The NCAI, which kept an intense watch on the State Department's handling of the Keystone XL debate, issued a strenuous critique of sections of the report on oil-spill prevention and remediation, as well as the report's lack of detail on American Indians' and Alaska Natives' water quality and supply. The report concluded:

> In total, if these concerns are not addressed sufficiently or mitigated to the fullest extent, it is in the best interest of the United States to reject the Keystone XL pipeline permit solely on the basis of the federal trust responsibility to tribal nations. The project as outlined in the DSEIS poses tremendous risks to the cultural and natural resources of tribal nations and is not in the best interest of the tribal nations and their citizens. ("Fill Gaps" 2013)

Opposition to the Keystone pipeline has forged an unusual "cowboy-Indian" alliance that unites Native Americans and white ranchers who seek to protect their water with scientists who make a case that additional fossil fuels will push global temperatures beyond sustainable levels. As of 2018, President Donald Trump had declared support of the pipeline, and it likely will be completed despite the protests.

Further Reading

"Fill Gaps in Keystone XL Draft Environment Report or Reject Pipeline, NCAI Tells Obama Administration." 2013. Indian Country Today Media Network, May 2.

Frosch, Dan. 2013. "Amid Pipeline Debate, Two Costly Cleanups Forever Change Towns." *New York Times*, August 10. Accessed November 26, 2018. https://www.nytimes .com/2013/08/11/us/amid-pipeline-debate-two-costly-cleanups-forever-change -towns.html.

Kolbert, Elizabeth. 2007. "Unconventional Crude: Canada's Synthetic-Fuels Boom." *New Yorker*, November 12: 46–51. Accessed November 26, 2018. https://www.newyorker .com/magazine/2007/11/12/unconventional-crude.

"NASA Scientist, Religious Leaders Arrested in Tar Sands Protest." 2011. Environment News Service, August 29. Accessed November 26, 2018. http://ens-newswire .com/2011/08/30/nasa-scientist-religious-leaders-arrested-in-tar-sands-protest/.

"NCAI Condemns Keystone XL Pipeline." 2011. Indian Country Today Media Network, August 18.

Woodard, Stephanie. 2011. "Planned Oil Pipeline Must Cross Pine Ridge's Water-Delivery System." Indian Country Today Media Network, September 21. Accessed November 7, 2018. https://newsmaven.io/indiancountrytoday/archive/planned-oil-pipeline-must -cross-pine-ridge-s-water-delivery-system-hlbIWz7KBUamNTPklwkx_g/.

Chapter 4: Endangered Species and Extinctions

OVERVIEW

Flora and Fauna: Coming Extinctions

We are now in the midst of one of the Earth's most intense, rapid, and pervasive mass extinctions, which has placed in harm's way many flora and fauna that humankind does not eat or keep as pets. The Earth has experienced mass extinctions before, but all of them have resulted from natural causes. Global warming is one product of humankind's increasing dominance of the Earth that is devastating the native habits of many animals and plants, driving many to extinction. Compared to past mass extinctions, which were driven by natural catastrophes such as meteor strikes or large-scale volcanism, the present-day human-driven wave of extinctions has been occurring with frightening speed. Given projected rises in temperature during decades to come, the flora and fauna of our home planet thus far have seen only their initial travails.

The *Proceedings of the National Academy of Sciences* published a worldwide survey in 2017 revealing that one-third of 27,600 land-based vertebrate species—including birds, amphibians, mammals and reptiles—face extinction. According to this survey, Earth is now experiencing a "biological annihilation" of its animal species because humans are becoming dominant much more quickly than previously thought, producing "a massive anthropogenic erosion of biodiversity" (Ceballos et al. 2017).

Human-caused climate change is often one of several factors driving destructive environmental change and provoking a high risk of mass extinctions at a speed heretofore unknown in human history, far more quickly than anyone had anticipated, according to a report from an interdisciplinary international workshop held in April 2011. "The findings are shocking," said Dr. Alex Rogers, scientific director of the International Programme on the State of the Ocean, which convened the workshop. "As we considered the cumulative effect of what humankind does to the ocean, the implications became far worse than we had individually realized" ("Mass Extinction" 2011).

Robert L. Peters describes how human-provoked climate change combines with humanity's subjugation of the Earth to shape the planet's biological future:

> Habitat destruction in conjunction with climate change sets the stage for an even larger wave of extinction than previously imagined, based on consideration of human

Experts assert that the Great Barrier Reef will be so degraded by warming seas that many of its corals will be dead within 20 years. (Dreamstime.com)

encroachment alone. Small, remnant populations of most species, surrounded by cities, roads, reservoirs, and farmland, would have little chance of reaching new habitat if climate change makes the old unsuitable. Few animals or plants would be able to cross Los Angeles on the way to the promised land. (Peters 1989, 91)

Global warming could play a role in the destruction or fundamental alteration of a third of Earth's plant and animal habitats within a century, bringing extinction to thousands of species, according to a study by Great Britain's World Wide Fund for Nature, an affiliate of the World Wildlife Fund. The study said that the most vulnerable plant and animal species will be in the Arctic and mountainous areas, where as many as 20 percent could be driven to extinction. In the north of Canada, Russia, and Scandinavia, where warming was predicted to be most rapid, up to 70 percent of habitat could be lost, according to this study.

The report, *Global Warming and Terrestrial Biodiversity Decline*, was written by Jay Malcolm, professor of forestry at Toronto University, and Adam Markham, the executive director of Clean Air/Cool Planet. Pests and weedy plant species would fare best, the report said. According to the report, "If past fastest rates of migration are a good proxy for what can be attained in a warming world, then radical reductions in greenhouse gas emissions are urgently required to reduce the threat of biodiversity loss" (Clover 2000, 9). To adapt and survive the expected rate of warming during the 21st century, plants may need to move 10 times more quickly than they did when recolonizing previously glaciated land at the end of the last ice age, the report said. Few plant species can adjust habitat at a rate of one kilometer per year, the speed that will be required for survival in many parts of the world.

In the first study of its kind, researchers in a range of habitats, including northern Britain, the wet tropics of northeastern Australia, and the Mexican desert, said early in 2004 that given "mid-range" climate change scenarios for 2050, they anticipate that 15 to 37 percent of the species in their sample of regions (covering 20 percent of Earth's surface) will be "committed to extinction" (Thomas et al. 2004, 145). The intensity of extinctions is expected to vary with the severity of warming. The study used United Nations projections that world average temperatures will rise 2.5° to 10.4°F by 2100. "We're not talking about the occasional extinction— we're talking about 1.25 million species. It's a massive number," the authors of this study wrote (Gugliotta 2004, A-1).

This study, described in *Nature,* marked the first time that scientists have produced a global analysis with concrete estimates of the effect of climate change on many various animal and plant habitats (Thomas et al. 2004, 145–148). Chris D. Thomas led a 19-member international team that surveyed habitat decline for

---

Extinctions in the Oceans

Marine animals will be affected by a warming habitat. A report by the World Wildlife Fund said: "Scientific evidence strongly suggests that global climate change already is affecting a broad spectrum of marine species and ecosystems, from tropical coral reefs to polar ice-edge communities" (Mathews-Amos and Berntson 1999). The level of the world's seas and oceans have been rising slowly for much of the 20th century, a millimeter or two a year, enough to produce noticeable erosion on 70 percent of the world's sandy beaches, including 90 percent of sandy beaches in the United States (Edgerton 1991, 18). In some areas, such as the United States Gulf Coast between New Orleans and Houston, the withdrawal of underground water and oil is causing some areas to sink as the oceans rise.

A team of scientists from several nations described ocean extinctions during the last 23 million years (Kerr 2010). Marine mammals have suffered a much higher rate of extinction than others, most notably those with low birth rates and limited ranges. "Mapping the geographic distribution of these genera identifies coastal biogeographic provinces where fauna with high intrinsic risk are strongly affected by human activity or climate change," the marine scientists reported, adding: "Such regions are disproportionately in the tropics, raising the possibility that these ecosystems may be particularly vulnerable to future extinctions. Intrinsic risk provides a pre-human baseline for considering current threats to marine biodiversity" (Kerr 2010).

Further Reading

Edgerton, Lynne T., and the Natural Resources Defense Council. 1991. *The Rising Tide: Global Warming and World Sea Levels.* Washington, DC: Island Press.

Kerr, Richard A. 2010. "Ocean Acidification Unprecedented, Unsettling." *Science* 328 (June 18): 1500–1501.

Mathews-Amos, Amy, and Ewann A. Berntson. *Turning Up the Heat: How Global Warming Threatens Life in the Sea.* Redmond, WA: Marine Conservation Biology Institute, 1999.

1,103 plant and animal species in Europe; Queensland, Australia; Mexico's Chihuahua Desert; the Brazilian Amazon; and the Cape Floristic Region at South Africa's southern tip (Gugliotta 2004, A-1).

According to the researchers, climate change during the past 30 years already has produced many shifts in the distribution and abundance of plants and animals. Climate change thus has become a major driver of biodiversity change. The survey team used one of ecology's few ironclad laws: the species-area relationship, first postulated by Charles Darwin in his *Origin of Species* (1859), which holds that a smaller habitable area will host a smaller number of viable species. They then projected habitat changes based on various warming scenarios. With a temperature rise of 0.8° to 1.7°C, they anticipated an 18 percent extinction rate; at 1.8° to 2.0°C, they projected 24 percent, and at more than 2.0°C., 35 percent (Pounds and Puschendorf 2004, 108).

The authors of this study considered a range of possibilities based on the ability of each species to move to a more congenial habitat to escape warming. If all species were able to move or "disperse," the study said, only 15 percent would be irrevocably headed for extinction by 2050. If no species were able to move, the extinction rate could rise as high as 37 percent (Gugliotta 2004, A-1). The scientists concluded: "These estimates show the importance of rapid implementation of technologies to decrease greenhouse-gas emissions and strategies for carbon sequestration" (Thomas et al. 2004, 145).

This chapter will examine ice melt and its effects on penguins in Antarctica, polar bears in Canada, and walrus in Russia. It will also look at warming temperatures and species in Australia, as well as the golden toad in Costa Rica and India's endangered elephants. Finally, this chapter will examine the preservation of China's panda bears and the plight of the honeybee in the United States.

Further Reading

Ceballos, Gerardo, Paul R. Ehrlich, and Rodolfo Dirzo. 2017. "Biological Annihilation via the Ongoing Sixth Mass Extinction Signaled by Vertebrate Population Losses and Declines." *Proceedings of the National Academy of Sciences* 114, no. 30 (July 10). Accessed October 1, 2018. http://www.doi.org/10.1073/pnas.1704949114.

Clover, Charles. 2009. "Thousands of Species 'Threatened by Warming.'" *London Daily Telegraph*, August 31: 9.

Gugliotta, Guy. 2004. "Warming May Threaten 37 Percent of Species by 2050." *Washington Post*, January 8: A-1. Accessed November 26, 2018. https://www.washingtonpost.com /archive/politics/2004/01/08/warming-may-threaten-37-of-species-by-2050/2bd 2872f-68c3-40c1-ab82-86a3b3be7ffc/.

"Mass Extinction of Ocean Species Soon to Be 'Inevitable.'" 2011. Environment News Service, June 21. Accessed October 1, 2018. http://ens-newswire.com/2011/06/21 /mass-extinction-of-ocean-species-soon-to-be-inevitable/.

Peters, Robert L. 1989. "Effects of Global Warming on Biological Diversity." In *The Challenge of Global Warming*, edited by Edwin Abrahamson, 82–95. Washington, DC: Island Press.

Pounds, J. Alan, and Robert Puschendorf. 2004. "Clouded Futures." *Nature* 427 (January 8): 107–108.

Quammen, David. 1998. "Planet of Weeds: Tallying the Losses of Earth's Animals and Plants." *Harpers*, October: 57–69.

Thomas, Chris D., Alison Cameron, Rhys E. Green, Michel Bakkenes, Linda J. Beaumont, Yvonne C. Collingham, . . . Stephen A. Williams. 2004. "Extinction Risk from Climate Change." *Nature* 427 (January 8): 145–148.

ANTARCTICA

Penguins Under Assault

One prominent example of a species driven to decimation by climate change are the Adélie penguins, eighteen-inch-tall birds that look as if they are wearing tuxedoes that have become scarcer near the Palmer Station on the Antarctic Peninsula because they eat krill that arrives with sea ice. The Adélie penguins that have not died have been moving their range southward toward areas that heretofore had been too cold and barren for them.

Adélie populations within two miles of the Palmer Station have declined by about 50 percent in 25 years, including a 10 percent decline within two years in the late 1990s (Petit 2000, 68). By the late 1980s, Penguin biologist William R. Fraser was among the first to associate retreating sea ice on the Antarctic Peninsula with the decline of Adélie Penguins. As ice-loving species such as the Adélies declined, ice-avoiding ones, including chinstrap penguins and fur seals, flourished. "It reeked of habitat change due to a warming climate," Fraser said (Montaigne 2010, 160).

Penguins' Varied Responses to Climate Change

Magellanic penguins' lives are being complicated by weather extremes that are increasing mortality for those not already dying from predators and starvation. Increasing intensity of storms is also a threat. "Rainfall is killing a lot of penguins, and so is heat," said P. Dee Boersma, of the University of Washington, describing the penguins' largest breeding colony near Punta Tombo, which comprises about 200,000 breeding pairs on Argentina's southeast coast. "And those are two new causes" (Fountain 2014). Even without heavy rain and heat, two-thirds of hatchlings die. With both, mortality rises, in large part because soaking rain can cause hypothermia for a newly hatched penguin. The size of the colony declined 24 percent from 1987 to 2013 (Fountain 2014).

Different types of penguins are suited to varying ecological niches that are shaped by climate, and as coastal Antarctica has warmed, they are dying or moving. Gentoo penguins, which are habituated to relatively mild weather, have been replacing Adélie penguins on the Antarctic Peninsula as it warms rapidly and sea ice retreats. The Adélies feed on krill that lives on the undersides of ice sheets that are eroding. Warmer weather also produces more snowfall, which interferes with the Adélies' ability to breed and incubate their eggs. Further south, where climate has become more amenable to the Adélies, they are increasing.

A warming climate poses several problems for the Adélies. For example, warmer ocean waters west of the Antarctic Peninsula have opened ocean surface as ice has retreated, allowing more evaporation and snowfall that inhibits their ability to create and maintain nests. The penguins evolved in a polar desert and are a snow-intolerant species. Ice by 2015 was present near major Adélie rookeries three months less per year on average compared to 1979. Less ice provides less krill to feed chicks, which die in larger numbers from malnutrition. All of these factors combine to destroy Adélie penguins on the west Antarctic Peninsula (Montaigne 2010, 216–17).

Different species of penguins feed in diverse ways that have been variously affected as ice retreats. Adélie penguins usually browse the undersides of sea ice and have gone hungry as this ice recedes or breaks up. Their numbers have been declining. By contrast, chinstrap and gentoo penguins, which usually forage along open shorelines, have found a greater range as ice recedes, encouraging their reproduction (Pollack 2009, 212–213).

Rapid Warming of the Antarctic Peninsula

On average, sea ice formed 54 days later than in the autumn and retreated 31 days earlier in the spring in 2009, compared with 1979. Eighty-seven percent of glaciers on the Antarctic Peninsula are retreating as well (Montaigne 2009, 78, 82). Chinstraps, which prefer open water, are not declining on the peninsula, whereas Adélies, denizens of the retreating ice, are disappearing from the area. About 2.5 million pairs of Adélies still breed in Antarctica, but their range is changing. On Avian Island, for example, they have doubled in 35 years, but scientists expect that they will decline here as well when sea ice begins to retreat with future warming (Montaigne 2009, 82).

At Palmer Station, winter temperatures rose 11°F during the 60 years ending in 2010, making it one of the most rapidly warming places on Earth; average annual temperatures rose 5°F in 50 years, five times the worldwide average. Great Britain's Rothera Station on the western shore of the Antarctic Peninsula may be the most rapidly warming single location on planet Earth. Average temperatures there rose 20°F during the last quarter of the 20th century (Bowen 2005, 33).

"It has become increasingly clear," wrote Fen Montaigne in *Fraser's Penguins*, "that humanity is every bit as responsible for the decline of the Adélie penguins along the northwestern Antarctic Peninsula as Nathaniel Palmer and his mates were for the near extirpation of Antarctica's fur seals nearly two centuries ago" (Montaigne 2010, 7–8). About 2.5 million breeding pairs of Adélie Penguins still existed in Antarctica in about 2010, even as shrinking ice drove them to near extinction near the Palmer Station. They will continue to exist, but perhaps not thrive, as the range of oceanic ice shrinks over time.

Some of the penguin mortality that has been attributed to climate change may also be aggravated by scientists' use of leg bands to track them. The banding of penguins also reduces some penguins' life span and ability to feed their young because the bands create drag when they swim, inhibiting gathering of krill. Rory P. Wilson, a zoologist at Swansea University, Wales, said that "if we stopped marking, we wouldn't have any way of measuring how birds are being affected by the climate, or anything else" (Bhanoo 2011, D-3; Saraux et al. 2011, 203–206; Wilson 2011, 164–185). Scientists may replace tags with microchips, although these cannot be traced by sight.

Penguins Decline off Namibia and South Africa

A century ago, at least 1.5 million African penguins waddled and swam on and along the coasts of Namibia and South Africa. In 2001, the population was about 120,000; by 2007, however, only roughly 20,000 survived. Populations have been slowly declining since then. With few exceptions, this species of penguin occupies only a few islands near South Africa's tip, including Robben Island, where Nelson Mandela (and others) were imprisoned during the apartheid regime. The penguin population has been declining rapidly as the anchovies and sardines they eat migrate southward out of warming water.

The African penguin, the continent's only native type, is less than two feet tall, "with large eyes set in a field of red and a distinctive black stripe circumnavigating its belly. Its calls range from a delicate trill to a loud nighttime bray, which leads some to call it the 'jackass penguin'" (Wines 2007). The cause of the penguins' decline "has to be either fishing or climate or a mixture of both," Rob Crawford, a marine scientist with South Africa's department of marine and coastal management, said of the sardine shift, "but my hunch is that it's environmental" (Wines 2007).

The African penguins can find no land base in the open ocean further south, so they are limited to food that is available within waters 25 miles out. They have been waiting for the sardines and anchovies to return and are starving. Commercial fishing also has depleted sardine stocks. Scientists have concluded the sardines move with changes in the cold, nutrient-rich Benguela Current that flows from Antarctica along the southwest African coast. In addition, the sardines may move in a 50-year cycle (Wines 2007).

Warming Favors Invasive Species

Warming in Antarctica causes ecosystems to change. For example, Douglas Fox wrote in *Nature* that "crabs invading the Antarctic continental shelf could deal a crushing blow to a rare ecosystem" (Fox 2012, 17). Fox described seafloor life that had been present for 30 million years: "Mostly invertebrates: sea lilies waving in the currents; brittle stars with their skinny, saw-toothed arms; and sea pigs, a type of sea cucumber that lumbers along the seafloor on water-inflated legs" (Fox 2012).

In some areas, however, sea life was largely absent, having been wiped out by an invader moving in as water temperatures had risen: "a red-shelled crab, spidery and with a leg-span as wide as a chessboard" (Fox 2012). An estimated 1.5 million crabs by 2012 had invaded the depths of the Palmer Sea, obliterating local fauna. All of these changes are taking place because local waters have risen from between 1°C and minus 2°C to about plus 1.8°C, just enough to allow the rapidly reproducing crabs to become the dominant species within a decade or two.

Many of the local fauna are adapted to water at just above the freezing point and struggle to survive in anything warmer. "Many of the species here are exquisitely sensitive to increases in temperature. The brittle stars and other invertebrates have extremely slow metabolisms—an adaptation to the cold water—and only meager ability to absorb and transport oxygen," wrote Fox (2012, 172). "'So what do those guys do if it warms up and their metabolic rate speeds up?'" asked Lloyd Peck, a biologist at the British Antarctic Survey in Cambridge, who has monitored these creatures in aquarium warming experiments. "Their oxygen demand revs beyond what their gills can supply—and they slowly suffocate" (Fox 2012, 172).

"There are no hard-shell-crushing predators in Antarctica," said Craig Smith, a marine ecologist from the University of Hawaii at Manoa. "When these come in they're going to wipe out a whole bunch of endemic species" (Fox 2012, 171). The invading crabs have been wiping out a seafloor ecology that is unique on Earth "It's a fascinating thing," said Richard Aronson, a marine biologist at Melbourne's Florida Institute of Technology. "A little scary, because it's a very obvious footprint of climate change" (Fox 2012, 171). The crabs are wiping out an ecosystem that may have been preserved in the chilly waters for millions of years. "All of this stuff has got a very Paleozoic flavor to it," said Aronson. "Westerly winds are strengthening, and the circumpolar current is intensifying, driven by atmospheric warming and a hole in the ozone layer over Antarctica," wrote Fox. "These changes are lifting warm, dense, salty water from 4,000 meters down in the Southern Ocean up over the lip of the continental shelf" (Fox 2012, 171).

Further Reading

Bhanoo, Sindya N. 2011. "Penguins Harmed by Tracking Bands, Study Finds." *New York Times*, January 18: D-3.

Bowen, Mark. 2005. *Thin Ice: Unlocking the Secrets of Climate in the World's Highest Mountains*. New York: Henry Holt.

Fountain, Henry. 2014. "For Already Vulnerable Penguins, Study Finds Climate Change Is Another Danger." *New York Times*, January 29. Accessed October 1, 2018. http://www.nytimes.com/2014/01/30/science/earth/climate-change-taking-toll-on-penguins-study-finds.html.

Fox, Douglas. 2012. "Polar Research: Trouble Bares Its Claws." *Nature* 492 (December): 170–172.

"Melting Antarctic Glacier Could Raise Sea Level." 1998. Reuters, July 24.

Montaigne, Fen. 2009. "The Ice Retreat: Global Warming and the Adélie Penguin." *New Yorker*, December 21: 72–82.

Montaigne, Fen. 2010. *Fraser's Penguins: A Journey to the Future of Antarctica.* New York: Henry Holt.

Petit, Charles W. 2000. "Polar Meltdown: Is the Heat Wave on the Antarctic Peninsula a Harbinger of Global Climate Change?" *U.S. News and World Report*, February 28: 64–74.

Pollack, Henry. 2009. *A World Without Ice.* London: Avery/Penguin.

Saraux, Claire, Céline Le Bohec, Joël M. Durant, Vincent A. Viblanc, Michel Gauthier-Clerc, David Beaune, . . . Yvon Le Maho. 2011. "Reliability of Flipper-Banded Penguins as Indicators of Climate Change." *Nature* 469 (January 13): 203–206.

"Sea Ice Penguin Theory Sinks." 2011. *Nature* 472 (April 21): 263.

Wilson, Rory P. 2011. "Animal Behaviour: The Price Tag." *Nature* 469 (January 13): 164–165.

Wines, Michael. 2007. "Dinner Disappears, and African Penguins Pay the Price." *New York Times*, June 4. Accessed October 1, 2018. http://www.nytimes.com/2007/06/04/world/africa/04robben.html.

AUSTRALIA

Death by Fire

Australia's Queensland state may lose half of its wet tropical highland rain forest, including many of its rarest animals, because of global warming. The state's emblem, the koala, is at risk because of rising carbon-dioxide levels, which could strip the gum leaf (the koala's principal food) of its nutritional value. The same goes for many other vulnerable species. A report released February 4, 2002 by Climate Action Network Australia "represents one of the most comprehensive pictures yet of the local ecological effects of global warming" (Ryan 2002, 1). More than half of Australia's eucalyptus species are unlikely to survive a 3°C average temperature rise, according to the report. In addition, higher carbon-dioxide levels are expected to reduce carbohydrates and nitrogen in leaves, undermining food supplies for animals such as koalas (Ryan 2002, 1).

Ian Hume, professor emeritus of biology at Sydney University, said that rising levels of carbon dioxide increase levels of "anti-nutrients" in the leaves, making the koalas' sole source of food toxic to them. Hume expects such problems to interfere with koala reproduction and reduce their populations substantially within 50 years. As it is, eucalyptus leaves contain very little energy; koalas cope by sleeping as much as 20 hours a day to conserve energy. Populations already have been reduced in some parts of Australia by habitat loss, including the spread of farms and suburbs ("Scientist: Warming" 2008, 7-A).

Heat and Extinctions

The study reviewed possible damage to Queensland's tropical areas, asserting that "90 Australian animal species, including a third of those on the endangered list, also [are] likely to suffer in the hotter, more extreme climate forecast this century"

(Ryan 2002, 1). David Hilbert, principal research scientist at the CSIRO Tropical Forest Research Center, said even a 1°C temperature rise would devastate half of the rain forests of north Queensland's wet tropics.

In the meantime, ring-tailed possums were falling dead out of trees in Australia's far north Queensland because the climate has become too hot for them. Green ring-tailed possums can't survive more than five hours at an air temperature above 30°C. The ring-tailed possums are not alone. Australia's Rainforest Cooperative Research Centre anticipates that half of all the unique mammals, reptiles, and birds in far north Queensland's rain forests could become extinct, given a 3.5°C rise in temperature, at the mid-point of the IPCC's range for the end of the 21st century. Most of these animals are found only in the Wet Tropics World Heritage Area near Cairns, at elevations above 600 meters. "It potentially takes just one day of six or seven hours of extreme heat . . . and you find dozens of dead possums on the ground," said Steve Williams, a tropical biologist at Australia's James Cook University (Braasch 2007, 83).

Australia's wildlife now has been ravaged by heat, drought, and fire for at least 20 years, nearly without respite. Record temperatures in 2015, 2016, and 2017 had precedents during the summers of 2002–2003, when wildfires pushed by raging hot, dry winds from Australia's interior seared parts of Canberra, charring hundreds of homes, killing four people, and forcing thousands to flee the area. "I have seen a lot of bushfire scenes in Australia . . . but this is by far the worst," Australia's prime minister, John Howard, said (Australia Assesses 2003, 6-A). Flames spread through undergrowth and exploded as they hit oil-filled eucalyptus trees. The 2002–2003 drought knocked 1 percent off Australia's gross domestic product and cost $6.8 billion in exports. It reduced the size of Australia's cattle herd by

---

Temperatures Skew Bearded Dragons' Sex Ratio

The temperature at which eggs are laid also affects the sex ratio of Australian bearded dragons. Warming temperatures favor females in this case, too. This propensity is so common in reptiles that scientists have given it a name: temperature override. Some amphibians and fish also display the same tendency. With bearded dragons, even a 1°C warming can skew the sex ratio significantly.

Clare E. Holleley and colleagues wrote in *Nature*: "Sex determination in animals is amazingly plastic. Vertebrates display contrasting strategies ranging from complete genetic control of sex (genotypic sex determination) to environmentally determined sex (for example, temperature-dependent sex determination)" (Holleley et al. 2015).

Further Reading

Holleley, Clare E., Denis O'Meally, Stephen D. Sarre, Jennifer A. Marshall Graves, Tariq Ezaz, Kazumi Matsubara, . . . Arthur Georges. 2015. "Sex Reversal Triggers the Rapid Transition from Genetic to Temperature-Dependent Sex." *Nature* 523 (July 2): 79–82.

5 percent and its sheep flock by 10 percent (Macken 2004, 61). Recovery was impeded during 2004 by continuing drought.

Heat Worst on Record

The Australia Green Party urged the country's federal government to expand an inquiry into the devastating bushfires to consider global warming. Bob Brown, a Green Party Australian senator, said that new research indicates record day-time temperatures and unprecedented rates of water evaporation made the current drought the worst on record. The extreme dryness of vegetation arising from global warming was what made recent fire season so devastating, he said ("Greens Want" 2003).

Australia's decade-long drought, provoked by climate change and natural variability, also has been aggravated by deforestation, according to Clive McAlpine of the University of Queensland in Brisbane and colleagues, who used a climate model that simulated Australian climatic conditions from the 1950s to 2003, comparing them to ground conditions before European immigration began in 1788. The drought is most intense in southeast regions of the continent, where less than 10 percent original vegetative cover remains, increasing the number of days with temperatures over 35°C by 300 percent (six to 18 a year), and the number of dry days by a similar ratio (five to 15 a year) ("Land Clearances" 2009).

By 2010, Australia's extended drought was being called "The Big Dry." The scientific journal *Nature* described work by Gavan McGrath at the University of Western Australia in Crawley and his colleagues:

> [They] analyzed satellite data from across the continent and found evidence of decreased water storage, rainfall, and plant growth throughout the country between 2002 and 2010. In the southeast, the drought correlated with an irregular Indian Ocean circulation, whereas in the northwest it was associated with a decreased frequency of tropical cyclones. The authors say that the northwest drought coincided with and probably exacerbated the one in the southeast. The findings suggest that distinct climatic factors such as decadal cyclone trends and changes in ocean circulation can combine to create a continental-scale drought. ("The Extended Reach" 2012).

By 2017, summers in Australia had become even hotter, with temperatures reaching as high as 45°C (117°F) and destruction of plants and animals intensifying, especially in the deserts.

The First Known Victim of Climate Change

The first animal known to have gone extinct because of global warming was Australian—a rodent called the Bramble Cay melomys, which was killed off during 2016 when rising seas covered Bramble Cay, a tiny atoll outcrop of the Great Barrier Reef near the Queensland coast, in the northeast Torres Strait's Cape York Peninsula. "The key factor responsible for the death of the Bramble Cay melomys is almost certainly high tides and surging seawater, which has traveled inland across the island," said Luke Leung, a scientist from the University of Queensland. "The seawater has destroyed the animal's habitat and food source. This is the first documented extinction of a mammal because of climate change" (Innis 2016). Anthony D. Barnosky, a professor at the University of California, Berkeley, who studies climate change's effects on the natural world, called the disappearance of the melomys "a cogent example of how climate change provides the coup de grâce to already critically endangered species. . . . I think this is significant because it illustrates how the human-caused extinction process works in real time" (Innis 2016).

Ocean Extinctions

Australian scientists by 2017 had compiled a list of more than a hundred plants and animals that face extinction or out-migration in the country's southernmost waters around Tasmania. The scientists expect them to die away or move southward to cooler water before the 21st century ends. These include several types of seaweeds, some invertebrates, and a number of fish, as well as giant, ethereal kelp jungles that are unique to Tasmania's adjoining waters. Neville Barrett, a research fellow at the Institute for Marine and Antarctic Studies in Hobart, said that the waters around Tasmania were a global hot spot for warming. "I mentioned that there were 100 or more species in general of kelps and endemic fishes and things that will probably disappear over the coming century, certainly by the turn of the next century under the current bottom end of predictions of climate change," he said. "There's a whole lot of species on the southern end of Australia that are as far south as they can currently go and some of them are already pushed to their upper thermal limit, as far as summer temperatures will go," said Barrett (Mathiesen 2017).

Many of these species will be lost because no land mass (and no relatively shallow ocean water) exists between Tasmania and Antarctica, thousands of miles away, with a completely different climate. The species that are threatened near Tasmania have "nowhere else to go," said Barrett (Mathiesen 2017). One such species, the giant kelp, *Macrocystis pyrifera,* died out along Tasmania's east coast during 2016. Tasmania's east coast is warming two to three times faster than the worldwide average because the East Australian Current has been transporting unusually warm water southward along Tasmania's east coast.

Further Reading

"Australia Assesses Fire Damage in Capital." 2003. *Omaha World-Herald,* January 20.

Braasch, Gary. 2007. *Earth Under Fire: How Global Warming Is Changing the World.* Berkeley: University of California Press.

"The Extended Reach of Australian Drought." *Nature* 483 (March 1, 2012): 8.

"Greens Want Global Warming Examined in Bushfire Inquiry." 2003. Australian Associated Press, January 21. (LEXIS).

Innis, Michelle. 2016. "Australian Rodent Is First Mammal Made Extinct by Human-Driven Climate Change, Scientists Say." *New York Times,* June 14. Accessed October 1, 2018.

https://www.nytimes.com/2016/06/15/world/australia/climate-change-bramble-cay
-rodent.html.

"Land Clearances Turned up the Heat on Australian Climate." 2009. *New Scientist*, May 16. Accessed October 1, 2018. http://www.newscientist.com/article/mg
20227084.700-land-clearances-turned-up-the-heat-on-australian-climate.html.

Macken, Julie. 2004. "The Double-Whammy Drought." *Australian Financial Review*,
May 4: 61.

Mathiesen, Karl. 2017. "100-Plus Species Face Extinction as Warming Hits Australia's
Southern Waters." Climate Home News, February 21. Accessed October 1, 2018.
http://www.climatechangenews.com/2017/02/21/more-than-100-species-face-extinc
tion-as-warming-hits-australias-southern-waters/.

Ryan, Siobhain. 2002. "National Icons Feel the Heat." *Courier Mail* (Australia), February 4: 1.

"Scientist: Warming Threatens Koalas." 2008. Associated Press in *Omaha World-Herald*,
May 8: 7-A.

CANADA

Can Polar Bears Adapt?

Climate-change skeptics who assert that polar bears recently adapted from brown bears and could easily transform themselves back in a few years ignore recent explorations of their genome by scientists indicating that the species is 4 to 5 million years old, not the roughly 600,000 years previously thought. Webb Miller and colleagues, writing in the *Proceedings of the National Academy of Sciences*, said:

> Polar bears are superbly adapted to the extreme Arctic environment and have become emblematic of the threat to biodiversity from global climate change. Their divergence from the lower-latitude brown bear provides a textbook example of rapid evolution of distinct phenotypes. However, limited mitochondrial and nuclear DNA evidence conflicts in the timing of polar bear origin as well as placement of the species within versus sister to the brown bear lineage. We gathered extensive genomic sequence data from contemporary polar, brown, and American black bear samples. . . .
>
> [P]olar and brown bears [are] sister species . . . [in an] ancient admixture between the two species. . . . We also provide paleodemographic estimates that suggest bear evolution has tracked key climate events and that polar bears in particular experienced a prolonged and dramatic decline in [their] effective population size during the last c. 500,000 years. We demonstrate that brown bears and polar bears have had sufficiently independent evolutionary histories over the last 4–5 million years to leave imprints in the polar bear nuclear genome that likely are associated with ecological adaptation to the Arctic environment. (Miller et al. 2012)

Polar Bears Pushed onto Land

The loss of summer sea ice is pushing polar bears more onto land occupied by people in northern Canada and Alaska, causing inflated estimates of their numbers,

A polar bear with her cubs in the Canadian tundra, an area where their survival is at risk because of warming temperatures and melting ice. (Andrey Gudkov/Dreamstime.com)

said NASA scientist Claire Parkinson, who studies the bears ("Arctic Ice Melting" 2006). Steven Kazlowski and colleagues, writing in *The Last Polar Bear*, described how the Alaskan North Slope town of Barrow was inundated by land-bound polar bears during the early fall of 2002, after sea ice retreated too far offshore to be used as a hunting platform:

> The strange congregation of dozens of stranded bears . . . threw the community into fear and consternation. A 1,100-pound bear—a large male—was shot when it wouldn't leave the school. Visitors were warned not to walk outdoors. Biologists employed by the local government were constantly on the run to keep polar bears off the streets. Polar bears normally don't eat people, but they easily can and, when hungry enough, they occasionally have. Bears on land in the summer/fall have been more common all along the coast since the trend of warmer weather shortened the winters and diminished the sea ice. But no one had seen an invasion like this before. Biologists counted more than one hundred bears near town. (Kazlowski et al. 2008, 63)

During the summer of 2004, hunters found half a dozen polar bears that had drowned about 200 miles north of Barrow, on Alaska's northern coast. They had tried to swim for shore after the ice had receded about 400 miles. A polar bear can swim 100 miles—but not 400 ("Buncombe and Carrell," 2005). As many as 36 other polar bears may have died as they tried to swim to land after the Arctic ice shelf along the north shore of Alaska receded (Tidwell 2006, 57).

Without ice, polar bears can become hungry, miserable creatures, especially in unaccustomed warmth. During record heat in Iqaluit (on Baffin Island) during July 2001, two tourists were hospitalized after they were mauled by a bear in a park south of town. On July 20, a similar confrontation occurred in northern Labrador as a polar bear tried to claw its way into a tent occupied by a group of Dutch tourists. The tourists escaped injury, but the bear was shot to death. "The bears are looking for a cooler place," said Ben Kovic, Nunavut's chief wildlife manager (Johansen 2001, 18).

Until recently, polar bears had their own food sources and usually went about their business without trying to steal food from humans. Beset by late freezes and early thaws, hungry polar bears are coming into contact with people more frequently. In Churchill, Manitoba, polar bears waking from their winter's slumber have found the ice on Hudson's Bay has melted earlier than usual. Instead of making their way onto the ice in search of seals, the bears walk along the coast until they get to Churchill, where they block motor traffic and pillage the town dump. Churchill now has a holding tank for wayward polar bears that is larger than its jail for people.

TIME magazine described Churchill:

Polar bears that ordinarily emerge from their summer dens and walk north up Cape Churchill before proceeding directly onto the ice now arrive at their customary departure point and find open water. Unable to move forward, the bears turn left and continue walking right into town, arriving emaciated and hungry. To reduce unscheduled encounters between townspeople and the carnivores, natural-resource officer Wade Roberts and his deputies tranquilize the bears with a dart gun, temporarily house them in a concrete-and-steel bear "jail" and move them 10 miles north. In years with a late freeze—most years since the late 1970s—the number of bears captured in or near town sometimes doubles, to more than 100 (Linden 2000).

Polar Bears' Unfamiliar Neighbors

By 2007, several tour operators were charging $3,000 to $5,000 for two- to four-day "last chance" excursions to Churchill to witness the bears in their natural habitats—or in town, as the case may be. In the meantime, the amount of time during which polar bears had access to ice for hunting had been reduced by a month in about two decades. Residents of Churchill remarked at wearing flip-flops and shirtsleeves in mid-October, when Hudson's Bay had once been beginning to freeze. The average weight of female bears had fallen from 600 pounds to a few pounds over 400, inhibiting reproduction. The mortality for bear cubs under five years of age was up 50 percent, according to Robert Buchanan, president of Polar Bears International (Clark 2007, 2-D). The town dump was closed in 2005 after hungry bears made a habit of pillaging it for food scraps. A few emaciated bears even were boarding the back ends of garbage trucks before they reached the dump.

Polar bears face other problems related to warming. For example, some polar bear dens have collapsed because of lightning-sparked brush fires on the tundra that may be increasing because of global warming, according to Ian Stirling, adjunct professor in the Department of Biological Sciences at the University of Alberta, who said that the Canadian government should consider fighting fires in prime bear denning areas in the North ("Brush Fires" 2002, 18).

"The fires burn off all of the trees and bushes that are on the upper part of the banks holding the roofs together," he said ("Brush Fires" 2002, 18). "The fires melt the permafrost in the adjacent areas, and in particular the roofs, so there's nothing to hold the roofs together. They just collapse," he said. The bears then must find another denning site where they can give birth over the winter. "The fires are definitely affecting the ability of the bears to use some of the prime areas," said Stirling. He also said that, after a fire, vegetation may require 70 years to grow back to a density suitable for denning sites ("Brush Fires" 2002, 18). Yet another potential risk to polar bears is the increased chance that rain in the late winter will cause polar bear dens to collapse before females and cubs have departed. Scientists surveying polar bear habitat in Manitoba, Canada, have observed large snow banks used for dens that had collapsed under the weight of wet snow (Stirling and Derocher 1993, 244).

Polar Bear Cannibalism

The survival crisis facing polar bears took a horrifying turn for the worse after 2004, when scientists from the United States and Canada found evidence that lack of access to food sources (mainly ringed seals) caused by shrinking ice cover might be forcing some of them to engage in cannibalism in the southern Beaufort Sea north of Alaska and western Canada. The scientists found three examples of polar bears preying on each other between January and April 2004, including the killing of a female in a den shortly after she gave birth. The bears use the sea ice not only for feeding, but also for mating and giving birth. The predation study was published in an online version of the journal *Polar Biology* (Amstrup et al. 2006).

According to the study's principal author, Steven Amstrup of the U.S. Geological Survey Alaska Science Center, polar bears sometimes kill each other for population regulation, dominance, and reproductive advantage, but killing for food heretofore had been very unusual (Joling 2006). "During 24 years of research on polar bears in the southern Beaufort Sea region of northern Alaska and 34 years in northwestern Canada, we have not seen other incidents of polar bears stalking, killing, and eating other polar bears," the scientists said (Joling 2006). The Center for Biological Diversity of Joshua Tree, California, petitioned the federal government in February 2005 to list polar bears as threatened under the federal Endangered Species Act (Joling 2006). "It's very important new information," she said. "It shows in a really graphic way how severe the problem of global warming is for polar bears," said Kassie Siegal, lead author of that petition (Joling 2006).

According to Dan Joling's summary of the predation study for the Associated Press, researchers discovered the first kill in January 2004. A male bear had pounced on a den, killed a female, and dragged it 245 feet away, where it ate part of the carcass. Females are about half the size of males. "In the face of the den's outer wall were deep impressions of where the predatory bear had pounded its forepaws to collapse the den roof, just as polar bears collapse the snow over ringed seal lairs," the paper said, continuing: "From the tracks, it appeared that the predatory bear broke through the roof of the den, held the female in place while inflicting multiple bites to the head and neck. When the den collapsed, two cubs were buried, and suffocated, in the snow rubble" (Joling 2006).

Three months later, the researchers found a partially eaten carcass of an adult female that had been walking with a cub. The killer, a male, had not followed the cub, indicating to the scientists that he was motivated by a need for food, not a desire to mate. A few days after that, the remains of a year-old cub were found. Tracks indicated that the cub had been stalked and killed.

Further Reading

Amstrup, S. C., I. Stirling, T. S. Smith, C. Perham, and G. W. Thiemann. 2006. "Recent Observations of Intraspecific Predation and Cannibalism among Polar Bears in the Southern Beaufort Sea." *Polar Biology* 29, 997. Accessed October 1, 2018. http://www.doi.org/10.1007/s00300-006-0142-5.

"Arctic Ice Melting Rapidly, Study Says." 2006. *New York Times,* September 14.

"Brush Fires Collapsing Bear Dens." 2002. Canadian Press in *Calgary Sun*, November 2: 18.

Buncombe, Andrew, and Severin Carrell. 2005. "Melting Planet," *Counter Currents*, October 4. First published in *The Independent*. Accessed November 26, 2018. https://www.countercurrents.org/cc-carrell041005.htm.

Clark, Jayne. 2007. "Tours Bear Witness to Earth's Sentinel Species." *USA Today*, November 2: 1-D, 2-D.

Johansen, Bruce E. 2001. "Arctic Heat Wave." *The Progressive*, October: 18–20.

Joling, Dan. 2006. "Study: Polar Bears May Turn to Cannibalism." Associated Press, June 14. (LEXIS).

Kazlowski, Steven, with Theodore Roosevelt IV, Charles Wohlforth, Daniel Glick, Richard Nelson, Nick Jans, and Frances Beinecke. 2008. *The Last Polar Bear: Facing the Truth of a Warming World, A Photographic Journey*. Seattle: Braided River Books.

Linden, Eugene. 2000. "The Big Meltdown." *TIME,* August 27. Accessed November 26, 2018. http://content.time.com/time/magazine/article/0,9171,53418,00.html.

Miller, Webb, Stephen C. Schuster, Andreanna J. Welch, Aakrosh Ratan, Oscar C. Bedoya-Reina, Fangqing Zhao, . . . Charlotte Lindqvist. 2012. "Polar and Brown Bear Genomes Reveal Ancient Admixture and Demographic Footprints of Past Climate Change." *Proceedings of the National Academy of Sciences* 109, no. 36 (September 4): E2382–E2390. Accessed October 1, 2018. http://www.doi.org/10.1073/pnas.1210506109.

Stirling, Ian, and Andrew Derocher. 1993. "Possible Impacts of Climatic Warming on Polar Bears." *Arctic* 46, no. 3 (September): 240–245.

Tidwell, Mike. 2006. *The Ravaging Tide: Strange Weather, Future Katrinas, and the Coming Death of America's Coastal Cities.* New York: Free Press.

CHINA

Panda Bears Revived

China's iconic giant panda bears, regarded as a national symbol there, once were so imperiled that many conservationists feared that they would survive only in captivity. In recent years, the pandas have been revived as a wild species with the help of a worldwide support effort. The government of China also has created a huge natural reserve for their benefit.

The World Wildlife Fund (WWF), which adopted the panda as its trademark image and asserts that it is "the primary international conservation organization protecting pandas and their habitat," proclaimed, "In a welcome piece of good news for the world's threatened wildlife, the giant panda has just been downgraded from 'Endangered' to 'Vulnerable' on the global list of species at risk of extinction, demonstrating how an integrated approach can help save our planet's vanishing biodiversity" ("Giant Panda" 2016).

For many years, the WWF worked with Chinese organizations and government agencies to construct and maintain several large panda reserves (which have protections similar to the national parks of the United States and other countries), as well as "wildlife corridors" that connect isolated panda populations with each other. All panda advocates also worked with local people to respect panda habitats. By 2017, two-thirds of wild pandas in China were being sheltered in the 67 reserves. "It's a good day to be a panda," said Ginette Hemley, senior vice president for wildlife conservation at WWF. "We're thrilled" (Dell'Amore 2016).

Why Are Pandas Vulnerable?

As a species, the giant panda is vulnerable for several reasons. First, it is found in a restricted habitat (only in temperate-zone forests of China, which have been severely reduced by human incursion); secondly, because their diet usually is limited to certain types of bamboo; and third, because they breed rarely and only under the best of conditions. Many attempts to breed pandas in captivity (that is, in zoos around the world) have failed.

There are no exact figures for the numbers of panda cubs in the wild, but estimates listed the total number of giant pandas at 1,864 in 2014 and 2,060 in 2016, a rise of 17 percent in twelve years, according to the International Union for Conservation of Nature (IUCN). In addition, around 200 pandas live in captivity in zoos around the world. The giant panda had been listed as endangered since 1990. The IUCN's "Red List" includes 82,954 species, of which 23,928 are listed as threatened with extinction.

"For over 50 years, the giant panda has been the globe's most beloved conservation icon as well as the symbol of WWF. Knowing that the panda is now a step further from extinction is an exciting moment for everyone committed to conserving the world's wildlife and their habitats," said Marco Lambertini, WWF director general. "The recovery of the panda shows that when science, political will, and engagement of local communities come together, we can save wildlife and also improve biodiversity," added Lambertini ("Giant Panda" 2016). The WWF's panda logo was designed by its founding chairman, painter and naturalist Sir Peter Scott, in 1961, after Chi-Chi, a panda at the London Zoo. In 1981, the WWF became the first international organization to work in China concentrating on panda preservation.

Panda Characteristics and Diet

Giant pandas may be instantly identified by their identical coloring; their ears, eyes, muzzle, eyes, legs, and shoulders are black on an otherwise white body. They live in cool, usually wet mountainous areas where thick, downy fur is essential. According to one online animal fact guide, one of the interesting evolutionary traits of the panda is their protruding wrist bone that acts like a thumb. This helps the pandas hold bamboo while they munch on it with their strong molar teeth ("Animal Fact Guide," n.d.).

In addition to their preference for bamboo, giant pandas will eat eggs, small rodents, and fish (as well as plants other than bamboo) if necessary to survive. Reproduction is severely limited by the fact that female pandas are able to get pregnant for only two or three days each spring. Gestation requires roughly 100 to 160 days, after which a helpless, hairless three- to five-ounce cub at birth is completely dependent on its mother for three months. Cubs remain with their very protective mothers for several more months.

Giant pandas grow to a mature weight averaging 330 pounds as adults, surviving solely for 14 to 20 years on massive amounts of bamboo—26 to 84 pounds per day. They occupy an essential ecological niche as they spread seeds and fertilize plants with their feces. They roam in a solitary manner but can communicate with each other by vocalizing and marking territories with scents in their urine.

China's policies regarding preservation of panda populations recognize that

one of the main reasons that panda populations have declined is habitat destruction. As the human population in China continues to grow, pandas' habitat gets taken over by development, pushing them into smaller and less livable areas. Habitat destruction also leads to food shortages. Pandas [usually] feed on several varieties of bamboo that bloom at different times of the year. If one type of bamboo is destroyed by development, it can leave the pandas with nothing to eat during the time it normally blooms, increasing the risk of starvation. ("Animal Fact Guide," n.d.)

"Evidence from a series of wide-range national surveys indicate that the previous population decline has been arrested and the population has started to increase," said the IUCN's updated report. "The improved status confirms that the Chinese government's efforts to conserve this species are effective," it added. The rebound could be short-lived, the IUCN warned. Climate change may wipe out more than one-third of the panda's bamboo habitat by the end of the century. "And thus panda population is projected to decline, reversing the gains made during the last two decades," the report said. The IUCN also added: "To protect this iconic species, it is critical that the effective forest protection measures are continued and that emerging threats are addressed." John Robinson, a primatologist and chief conservation officer at the Wildlife Conservation Society, said, "When push comes to shove, the Chinese have done a really good job with pandas. So few species are actually down-listed, it really is a reflection of the success of conservation" ("Giant Pandas Rebound" 2016).

Debate Over the Pandas' Future

"This is a deserved status," said M. Sanjayan, a senior scientist at the nonprofit Conservation International, of the pandas' removal from the endangered list. "The Chinese government has put in 30 years of hard work in pandas—[they are] not going to let the panda go extinct." However, Marc Brody, senior adviser for conservation and sustainable development at China's Wolong Nature Reserve, said:

> It is too early to conclude that pandas are actually increasing in the wild—perhaps we are simply getting better at counting wild pandas. While the Chinese government deserves credit and support for recent progress in management of both captive and wild giant pandas . . . there is no justifiable reason to downgrade the listing from endangered to threatened. (Dell'Amore 2016)

Suitable panda habitat is decreasing due to fragmentation caused by highway construction, as well as increasing tourism development in Sichuan Province, along with other human economic activities (Dell'Amore 2016) "Whereas the decision to down-list the giant panda to 'vulnerable' is a positive sign confirming that the Chinese government's efforts to conserve this species are effective, it is critically important that these protective measures are continued and that emerging threats are addressed," the IUCN wrote in its giant panda assessment (Stack 2016).

China said it was less optimistic about the panda's progress, however. The State Forestry Administration disputed the conservation group's decision in a statement to the Associated Press, saying that pandas struggle to reproduce in the wild and live in small groups spread widely apart. "If we downgrade their conservation status or neglect or relax our conservation work, the populations and habitats of giant pandas could still suffer irreversible loss, and our achievements would be quickly lost," the forestry administration told the AP. "Therefore, we're not being alarmist by continuing to emphasize the panda species' endangered status" (Stack 2016).

Panda Reserves Expanded

In 2017, China was continuing to expand reserves for pandas, and one has been expanded to three times the size of Yellowstone National Park in the United States, spanning parts of Gansu, Shaanxi, and Sichuan provinces; the reserve was planned to unify the 67 exiting habitats. The 10,476-square-mile reserve required eviction of 170,000 people (Karimi 2017).

The national reserve also will help to protect about 8,000 endangered animals and plants. "It will be a haven for biodiversity and provide protection for the whole ecological system," Hou Rong, the director of the Chengdu Research Base for Giant Panda Breeding in Sichuan province, told Xinhua, the state news agency. "Many scientists and conservation experts support the building of a national park" ("China to Create" 2017). David Wildt, a senior scientist from the Smithsonian's conservation biology institute, said, "Past experience has told us how much a national park can do for a country's environment and ecology. I am delighted to see China's breakthrough in panda breeding and reintroduction programs. But it's time to test if these measures work out in the new system of national parks" ("China to Create" 2017).

Further Reading

"Animal Fact Guide: Giant Panda." n.d. Animal Fact Guide. Accessed October 1, 2018. http://www.animalfactguide.com/animal-facts/giant-panda/.

"China to Create 'Giant' Panda Reserve to Boost Wild Population." 2017. *The Guardian,* March 31. Accessed October 1, 2018. https://www.theguardian.com/world/2017/mar/31/china-to-create-giant-giant-panda-reserve-to-boost-wild-population.

Dell'Amore, Christine. 2016. "Giant Pandas, Symbol of Conservation, Are No Longer Endangered." National Geographic News, September 4. Accessed October 1, 2018. http://news.nationalgeographic.com/2016/09/pandas-vulnerable-endangered-species/.

"Giant Panda No Longer Endangered; Iconic Species Is One Step Further Away from Extinction." 2016. World Wildlife Fund, September 4. Accessed October 1, 2018. https://www.worldwildlife.org/stories/giant-panda-no-longer-endangered.

"Giant Pandas Rebound off Endangered List." 2016. BBC News, September 4. Accessed October 1, 2018. Accessed October 1, 2018. http://www.bbc.com/news/world-asia-china-37272718.

Karimi, Faith. 2017. "China Plans Massive Reserve for Giant Pandas." Cable News Network (CNN), April 1. Accessed October 1, 2018. http://www.cnn.com/2017/04/01/asia/china-panda-national-reserve/index.html.

Stack, Liam. 2016. "The Giant Panda Is No Longer Endangered. It's 'Vulnerable.'" *New York Times,* September 6. Accessed October 1, 2018. https://www.nytimes.com/2016/09/07/science/giant-panda-endangered-vulnerable.html

COSTA RICA

The Golden Toad Vanishes

Deforestation and a warming environment may help to explain the disappearance of *Bufo periglenes*, the golden toad of the Monteverde tropical rain forest, which was found nowhere else on Earth. No one knows how long the golden toad had lived in the cloud forest that runs along the Pacific coast of Costa Rica. By 1987, however, the level of mountain clouds had risen, reducing the frequency of mists during the dry season and probably playing a role in a massive population crash that affected most of the 50 species of frogs and toads in the forest. No fewer than twenty species became locally extinct (Moss 2001, 18).

Galápagos Iguanas Starve as Planet Warms

The marine iguanas of the Galápagos Islands in the equatorial Pacific Ocean, which played a key role in Charles Darwin's theory of natural selection 200 years ago, may not survive a warming planet. Their food supply, algae that grow on shore and in nearby waters, have been dying as water warms, and the raccoon-sized lizards have gone hungry, sometimes to the point of starvation. This species of iguanas, which live only on the Galápagos, are so finely tuned to their environment and singular food supply that a small change in their habitat has been killing them. Local corals, which produce the algae, have been bleaching as warming waters kill them. The rate of bleaching accelerated during a major El Niño, which warmed ocean waters in the equatorial Pacific more than 4°F above long-term averages during 2016.

Further Reading

Solomon, Christopher. 2017. "Life in the Balance: A Warming Planet Threatens the Galápagos Species That Inspired Darwin's Theory of Natural Selection." *National Geographic*, June: 52–69.

According to journalist Carol Yoon:

A flashy orange creature last seen in the late 1980s, the golden toad has become an international symbol of the world's disappearing amphibians. Seeing how even pristine and protected forests such as those at Monteverde can lose their crucial mists and clouds, researchers say it becomes less mysterious how a water-loving creature like the golden toad could vanish even from a forest where every tree still stands. (Yoon 2001, F-5)

Disease Is the Bullet; Climate Change Pulls the Trigger

By 2006, an increasing number of scientific studies pointed to warming temperatures as a crucial (but not sole) reason why several species of amphibians (frogs and toads) have been going extinct or face sharp population declines around the world. The primary reason for some of the extinctions is skin diseases, such as *chytridiomycosis*, but these may be aggravated by climate change (Alford 2011, 461).

These reports of extinctions include studies of the Monteverde harlequin frog (*Atelopus sp.*), which vanished along with the golden toad (*Bufo periglenes*) during the late 1980s from the mountains of Costa Rica. According to an analysis by J. Alan Pounds, resident scientist at the Tropical Science Center's Monteverde Cloud Forest Preserve in Costa Rica, and colleagues, an estimated 67 percent of roughly 110 species of *Atelopus* that are endemic to the American tropics are now extinct, having been afflicted by a pathogenic chytrid fungus (*Batrachochytrium dendrobatidis*). Pounds and colleagues concluded with more than 99 percent confidence that large-scale environmental warming is a key factor in the disappearances. They found that rising temperatures at many highland localities have encouraged growth of *Batrachochytrium*, encouraging outbreaks. "With climate change promoting infectious disease and eroding biodiversity, the urgency of reducing greenhouse-gas concentrations is now undeniable," they concluded (Pounds et al. 2006,161).

"Disease is the bullet killing frogs, but climate change is pulling the trigger," said Pounds, who has taken a lead role in these studies. "Global warming is wreaking havoc on amphibians and will cause staggering losses of biodiversity if we don't do something first" (Eilperin 2006, A-1). The chytrid fungus kills frogs by growing on their skin, then attacking their epidermis and teeth, as well as by releasing a toxin. Higher temperatures allow more water vapor into the air, which forms a cloud cover that leads to cooler days and warmer nights. These conditions favor the fungus, which grows and reproduces best at temperatures between 63°F and 77°F. "There's a coherent pattern of disappearances, all the way from Costa Rica to Peru," Pounds said. "Here's a case where we can show that global warming is affecting outbreaks of this disease" (Eilperin 2006, A-1).

Earth's Life-Support System Is in Trouble

Pounds and Karen L. Masters, both of whom have been active in studying amphibian die-offs in the Monteverde Cloud Forest Preserve of Costa Rica, criticized mono-causal interpretations that attribute frogs' and toads' problems to one factor, such as chytrid fungus in isolation. In a review of a book (James P. Collins and Martha L. Crump's *Extinction in Our Time: Amphibian Decline*, 2009) that draws just such a conclusion, they wrote that concentration on the fungus in isolation ignores many other factors, one of which is climate change. It is, they wrote, like blaming an automobile accident on excessive speed while ignoring the alcohol on a driver's breath. "Amphibians belong to a chorus of canaries telling us one thing: Earth's life-support system is on trouble," they believe (Pounds and Masters 2009, 39).

In 2004, the first worldwide survey of 5,743 amphibian species (frogs, toads, and salamanders) indicated that one in every three species was in danger of extinction, many of them likely victims of an infectious fungus possibly aggravated by drought and warming (Stokstad 2004, 391). Findings were contributed by more than 500 researchers from more than 60 countries. "The fact that one-third of amphibians are in a precipitous decline tells us that we are rapidly moving toward a potentially epidemic number of extinctions," said Achim Steiner, director-general of the World Conservation Union based in Geneva (Seabrook 2004, 1-C).

"Amphibians are one of nature's best indicators of the overall health of our environment," said Whitfield Gibbons, a herpetologist at the University of Georgia's Savannah River Ecology Laboratory (Seabrook 2004, 1-C; Stokstad 2004, 391). Amphibians are more threatened and are declining more rapidly than birds or mammals; "The lack of conservation remedies for these poorly understood declines means that hundreds of amphibians species now face extinction," wrote the scientists who conducted the worldwide survey (Stuart et al. 2004, 1783–1786).

In North and South America, the Caribbean, and Australia, a major culprit appears to be the highly infectious fungal disease *chytridiomycosis*. New research shows that prolonged drought may cause outbreaks of the disease in some regions, although some scientists attribute the disease's spread to global warming. Other threats include loss of habitat, acid rain, pesticides and herbicides, fertilizers, consumer demand for frog legs, and a depletion of the stratospheric ozone layer that exposes the frogs' skin to radiation (Seabrook 2004, 1-C.). Gibbons said that the loss of wetlands and other habitat to development, agriculture, and other reasons may be the leading cause of amphibian declines in Georgia and elsewhere in the southeastern United States. Pollution also may be playing a significant role, he said. "It's hard to find a pristine stream anymore," he said (Seabrook 2004, 1-C).

Begun as a colony of expatriate North American Quakers, Monteverde has been a major attraction for American biologists, many of whom have devoted their careers to the flora and fauna of the region. Carol Kaesuk Yoon described the area in the *New York Times:* "A quick look at the forest's inhabitants makes clear why—from the record-breaking diversity of orchids to creatures like the resplendent quetzal, a glittering green, kite-tailed bird truly worthy of its name, and the so-called singing mice that whistle and chirp like birds" (Yoon 2001, F-5). In addition to its value as a shelter for plants and animals, Monteverde also has become a center for ecotourism. Nearly 50,000 tourists visit each year to walk its many trails, "stopping for a bite to eat or perhaps picking up a golden toad mug or Monteverde T-shirt along the way" (Yoon 2001, F-5). Scientists say that if the cloud forest falters, much business will be lost along with the unique species.

Further Reading

Alford, Ross A. 2011. "Bleak Future for Amphibians." *Science* 480 (December 22 and 29): 461–462.

Eilperin, Juliet. 2006. "Warming Tied to Extinction of Frog Species." *The Washington Post*, January 12: A-1. Accessed October 1, 2018. http://www.washingtonpost.com/wp-dyn /content/article/2006/01/11/AR2006011102121_pf.html.

Moss, Stephen. 2001. "Casualties." *The Guardian*, April 26: 18.

Pounds, J. Alan. 2001. "Climate and Amphibian Decline." *Nature* 410 (April 5): 639–640.

Pounds, J. Alan, Martín R. Bustamante, Luis A. Coloma, Jamie A. Consuegra, Michael P. L. Fogden, Pru N. Foster, . . . Bruce E. Young. 2006. "Widespread Amphibian Extinctions from Epidemic Disease Driven by Global Warming." *Nature* 439 (January 12): 161–167.

Pounds, J. Alan, and Karen L. Masters. 2009. "Amphibian Mystery Misread." [Review, *Extinction in Our Times: Global Amphibian Decline* by James P. Collins and Martha L. Crump] *Nature* 462 (November 5, 2009): 38–39.

Seabrook, Charles. 2004. "Amphibian Populations Drop." *Atlanta Journal-Constitution*, October 15: 1-C.

Stokstad, Erik. 2004. "Global Survey Documents Puzzling Decline of Amphibians." *Science* 306 (October 15): 391.

Stuart, Simon N., Janice S. Chanson, Neil A. Cox, Bruce E. Young, Ana S. L. Rodrigues, Debra L. Fischman, and Robert W. Waller. 2004. "Status and Trends of Amphibian Declines and Extinctions Worldwide." *Science* 306 (December 3): 1783–1786.

Yoon, Carol Kaesuk. 2001. "Something Missing in Fragile Cloud Forest: The Clouds." *New York Times*, November 20: F-5.

INDIA

Elephants Endangered

With 1.2 billion human beings (and rising) competing with about 25,000 elephants (and declining), the largest land animals (which weigh as much as five tons each) find themselves on the defensive, despite their elevated status as mythological deities in the Hindu belief system. Despite their esteem as cultural and religious icons, Indian elephants find themselves under assault and within decades may all but cease to exist in the wild. The Indian elephant (*Elephas maximus indicus*) has been listed as endangered since 1986 by the International Union for the Conservation of Nature (IUCN) because its population had declined by more than 50 percent over three generations of about 50 years each.

India is home to between 50 and 60 percent of all of Asia's wild elephants and about 20 percent of its domesticated elephants. The elephant has had a vital role in India's religion and culture for many centuries.

> According to Hindu mythology, the gods (deva) and the demons (asura) churned the oceans in a search for the elixir of life so that they would become immortal. As they did so, nine jewels surfaced, one of which was the elephant. In Hinduism, the powerful deity honored before all sacred rituals is the elephant-headed Lord Ganesha, who is also called the Remover of Obstacles. ("Indian Elephant" 2017)

Elephants Hemmed in by Habitat Loss

Centuries ago, India's forests teemed with elephants. "Although no census or estimates of the wild population exist, it is said that in the early 17th century, the Moghul Emperor Jahangir had 113,000 captive elephants throughout his empire. Extrapolating from this figure, it is easy to imagine a wild population comfortably in excess of a million" ("Elephants in India" 2017).

Today, India maintains 25 elephant reserves covering about 21,000 square kilometers of the elephants' usual 58,000-square-kilometer range. Another 3,400 elephants live in captivity, two-thirds of them having private owners. The remainder are owned by Hindu temples, government forestry departments (which often use them in logging), zoos, and (until 2016, when their use was outlawed) by circuses.

Indian elephants are recycling machines. They feed as long as 19 hours a day, mainly on 300 pounds of grasses, as well as tree bark, roots, leaves, and stems, producing more than 200 pounds of dung, which spreads germinating seeds as they range over as much as 125 square miles. When available, bananas in the field, as well as rice and sugarcane, are game for roving elephants (and often a provocation for conflict with farmers). Several Indian elephants have been killed in collisions with railroad trains. Elephants need hydration constantly (they must drink at least twice a day), so they center their travels on water sources.

Given their wide range in the wild, India's elephants have been hemmed in by habitat loss. According to the World Wildlife Fund:

> In South Asia, an ever-increasing human population has led to many illegal encroach-ments in elephant habitat. Many infrastructure developments like roads and railway tracks also fragment habitat. Elephants become confined to "islands" as their ancient migratory routes are cut off. Unable to mix with other herds, they run the risk of inbreeding. Habitat loss also forces elephants into close quarters with humans. In their quest for food, a single elephant can devastate a small farmer's crop in a single feeding raid. This leaves elephants vulnerable to retaliatory killings, especially when people are injured or killed. ("Indian Elephant" 2017)

Elephants usually form extended families of six or seven related females headed by the eldest, the matriarch. These groups sometimes join others in herds. As in other parts of Africa and Asia, elephants that maintain complex family structures attempt to defend well-defined territories and mourn their dead.

Asian elephants are sometimes hunted (poached) for the ivory in their tusks, as well as their hides. Some of the ivory ends up as incidental items, such as Chinese chopsticks. Laws are stricter in India and elsewhere in South Asia, so poaching is more frequent against African elephants. Some elephant herds have declined to a point where inbreeding poses a problem, threatening the gene pool and raising incidences of juvenile mortality and failure to breed.

Use and Abuse in Circuses Banned

In 2016, India's Central Zoo Authority (CZA) banned the use of elephants in circus performances, following several years of pressure from animal-rights advocates. The CZA issued the order after animal advocates' allegations about extreme cruelty were substantiated by official inspections that included photos and video. Training elephants to perform unnatural tricks was found to be inherently abusive. More than 160 injured elephants were identified and freed from practices that if inflicted on human beings would constitute torture. The *Times of India* reported that more than 100 activists from 40 organizations had been involved in the probe.

The CZA's report outlined the dimensions of cruelty, according to a report in the *Times of India*:

> From cutting the flight feathers of birds to chaining elephants with spikes, circuses train animals under the harshest conditions to make them perform acts that are com-pletely unnatural to them. The entire process of breaking these innocent creatures during training includes burning them with hot iron, piercing their genitals, threat-ening them with fire, tying them for days on end without food or water, among other

gruesome acts. Displaying such abused creatures for entertainment is a reflection of circuses' callous attitude towards suffering of animals. ("India Bans" 2016)

In 2014, British Broadcasting Corporation (BBC) News reported:

> Wild elephants are being held in horrific conditions in Myanmar [Burma] by smugglers looking to resume a lucrative trade. . . . The animals, mainly calves, are being brutally treated as they are "tamed" for tourist camps in Thailand. . . . A resurgence in smuggling could seriously threaten the elephant's survival. (McGrath 2014)

Elephant calves were being sold for about US$33,000 each. "Wildlife in Myanmar is being completely hammered, by habitat loss but also by the trade," said Dr. Chris Shepherd from Traffic, an anti-smuggling organization (McGrath 2014).

"Domesticated elephants have been used to herd wild ones into pit traps, where the older family members are often shot. The younger elephants are then transported to the Myanmar border with Thailand, where they are 'broken in,'" according to the BBC report. "They are put in small log boxes and just beaten into submission," said Shepherd. "They are a bit like a light switch—you take a wild elephant and beat it long enough and suddenly the switch goes off and you have a tame elephant. The welfare implications are horrendous—it is a cruel business" (McGrath 2014). This is only one of several ways in which elephant populations have been depleted across Southeast Asia. Others include logging (which destroys elephant habitat), poaching for the ivory in their tusks, and capture for exhibition and personal use. "If you add up all these pressures, any off-take at all has a conservation impact. . . . It's a cruel business," said Shepherd (McGrath 2014).

On a less cruel note, Elephants in India's Jaipur state have been taught to paint large canvases with long brushes held in their trunks. The paintings are then touched up by some of India's most prominent human artists before they are sold at an arts festival to pay for their upkeep, as well as to fund advocacy for elephant welfare. "To enhance their artistic tendencies, the owners fed the elephants with favorite treats including sugarcane, bananas, and oranges. And the special diet seemed to work," reported India's CNS News. "The elephants seem quite happy to paint, moving their ears vigorously, which they do only when they enjoy something," said elephant owner Ayoob Khan (Mahaan 2008).

Wild Elephants Defend Their Turf

The declining number of wild elephants defend their territories and have been known to kill human interlopers. During February 2017, according to EleAid, a member of the Ranchi community was killed by a "musth" (rogue) elephant in a remote hamlet in the country's North Andaman region. The victim, Julius Lakra, was performing household chores near his home when the elephant appeared from the bushes and tossed the man into the air before crushing him to death ("Elephants in India" 2017).

Many of these deaths occur during dry weather when elephants range widely, desperate for water. In January 2017, "A gang of four poachers entered Thattekad bird sanctuary in Kerala, but they ended up getting trampled by elephants, with one dying and another in a critical condition. The poachers were looking for ivory when they were trampled" (Reid 2017).

A 26-year-old man known only as Tony was trampled to death by a herd of elephants, and his friend, a 30-year-old named Basil, was critically injured. Witnesses said that the men were caught in a herd of elephants and couldn't escape ("Elephants in India" 2017).

The government of India has been making elephant protection a priority since Project Elephant began in 1992 through the Ministry of Environment and Forests to protect their habitat and corridors, address issues of man-animal conflict, and enhance the welfare of captive elephants. The same initiative also emphasizes restoration of natural habitats and migratory routes of elephants and measures to protect wild elephants from poachers. Project Elephant also has been involved in the development of ecotourism that generates revenue for captive breeding programs.

Further Reading

"Elephants in India." 2017. EleAid. Accessed October 1, 2018. http://www.eleaid.com/country-profiles/elephants-india/.

"India Bans All Wild Animal Performances in Circuses." 2016. *Times of India*, December 14.

"Indian Elephant." 2017. World Wildlife Fund. Accessed October 1, 2018. http://www.worldwildlife.org/species/indian-elephant.

Mahaan, Deepak. 2008. "Elephants Paint for Their Own Survival in India." CNS News, July 7. Accessed October 1, 2018. http://www.cnsnews.com/news/article/elephants-paint-their-own-survival-india.

McGrath, Matt. 2014. "Cruel Trade in Asian Elephants Threatens Survival." British Broadcasting Corporation News, July 6. Accessed October 1, 2018. http://www.bbc.com/news/science-environment-28161472.

Reid, Claire. 2017. "Elephant Herd Tramples Poacher to Death in India." *LAD Bible*. January 18. Accessed November 27, 2018, http://www.ladbible.com/news/animals-elephant-herd-tramples-poacher-to-death-in-india-20170118.

RUSSIA

Winter Without Walrus

Hunting animals, including walrus, is still key to survival for Native people across the Arctic. Imported food is not only culturally out of character, but also so expensive that few Native people can afford it. Food imported by air to the one small grocery store in Gambell, in extreme northwest Alaska, within sight of Siberia's Chukchi Peninsula, is no alternative to country food. A single small chicken cost $25 in 2013. Wild swings in weather and ice conditions have devastated harvests of walrus. In recent years, ice, the walrus' primary habitat, has been too scarce—except in cases like the winter of 2012–2013, when weather

A walrus comes to shore on the Russian archipelago at Franz Josef Land. (Vladimir Melnik/ Dreamstime.com)

was so cold and ice so thick that hunters could not get out often enough to provide for their families.

With ice receding hundreds of miles offshore of Russia during the late summer of 2007, walrus gathered by the thousands onshore. Joel Garlich-Miller, a walrus expert with the U.S. Fish and Wildlife Service, said that walrus began to gather onshore late in July, a month earlier than usual. A month later, their numbers had reached record levels.

Walrus usually routinely feed as deep as 100 meters (328 feet) below the surface for clams, snails, and other bottom-dwelling creatures from the ice. In recent years, the ice has receded too far from shore to allow the usual feeding pattern. A walrus can dive 600 feet, but water under ice shelves in late summer is now several thousands of feet deep. The walrus have been forced to swim much farther to find food, using energy that could cause increased calf mortality. More calves are being orphaned. Russian researchers also reported many more walrus than usual on shore—tens of thousands in some areas along the Siberian coast—which would have stayed on the sea ice in earlier times (Joling 2007).

A Sobering Year, Tough on Walrus

Walrus are prone to stampedes once they are gathered in large groups. Thousands of Pacific walrus above the Arctic Circle were killed on the Russian side of the

Bering Strait, where more than 40,000 had hauled out on land at Point Schmidt as ice retreated further northward. A polar bear, a human hunter, or noise from an airplane flying close and low can send thousands of panicked walrus rushing to the water.

"It was a pretty sobering year, tough on walrus," said Garlich-Miller ("3,000 Walruses" 2007). Several thousand walrus died late in the summer of 2007 from internal injuries suffered in stampedes. The youngest and the weakest animals, many of them calves born the previous spring, were crushed. Biologist Anatoly Kochnev, of Russia's Pacific Institute of Fisheries and Oceanography, estimated 3,000 to 4,000 walrus out of a population of perhaps 200,000 died, two or three times the usual number on shoreline haul-outs, during stampedes on the Russian side of the Chukchi Sea, the body of water touching Alaska and Russia just north of the Bering Strait. Instead of spending the summer spread over sea ice, thousands of walrus were stranded on land in unprecedented numbers for up to three months ("Walrus Latest" 2007). Kochnev said the loss of 3,000–4,000 animals in one year could be disastrous (Joling 2007).

"We were surprised that this was happening so soon, and we were surprised at the magnitude of the report," said wildlife biologist Tony Fischbach of the U.S. Geological Survey (Joling 2007). Walrus lacking summer sea ice that they had used to dive for clams and snails may strip coastal areas of food and then starve in large numbers. "The ice is melting three weeks earlier in the spring than it did 20 years ago, and it's re-forming a month later in the fall," said Carleton Ray, of the University of Virginia, who has studied walrus since the 1950s (Angier 2008).

The walrus are imperiled by expanding human industry as well as eroding sea ice. In February 2008, for example, the U.S. Interior Department gave Royal Dutch Shell oil-drilling rights in the Chukchi Sea near the northwestern shore of Alaska, enabling the first exploitation of this prime walrus fishing ground.

Open Water and Walrus Carnage

According to an account in the *New York Times*, "Females and calves have been forced to abandon the ice in midsummer and follow the males to land. The voyage leaves them emaciated and easily panicked. With the slightest disturbance, the herd desperately heads back into the water, often trampling one another to death as they flee" (Angier 2008).

"The ones that take the brunt of it are the calves," said Chad Jay of the walrus research program at the Alaska Science Center in Anchorage. "Our Russian colleagues have observed thousands of calves killed in episodes of beachside mayhem" (Angier 2008).

Melting Arctic sea ice is separating walrus young from their mothers and sometimes killing them. During two months in 2004, a U.S. Coast Guard icebreaker found nine lone walrus calves in deep waters flowing from the Bering Sea to the shallower continental shelf of the Chukchi Sea near Russia. Researchers reported in the journal *Aquatic Mammals* that the pups, living with their mothers, probably fell into the sea following collapse of an ice shelf that broke up in water where the temperature had increased 6°F in two years.

"We were on a station for 24 hours, and the calves would be swimming around us crying," said Carin Ashjian, a biologist at Woods Hole Oceanographic Institution in Massachusetts and a member of the research team. "We couldn't rescue them" (Kaufman 2006). Walrus pups usually live with their mothers for two years. It was possible that the walrus young were separated from their mothers by a severe storm or that the mothers were killed by hunters from nearby native communities. They said that local hunting ethics frown on killing walrus mothers with young, and it was far more likely that retreating sea ice—rather than a storm—caused the young to become separated.

During 2009, for the third consecutive year, the walrus' range along the Chukchi Sea again was mostly open water in the summer. On September 14, 2009, 131 walrus were found dead, having been crushed during stampedes. "I think there is reason to be concerned," said University of Alaska marine biologist Brendan P. Kelly (Revkin 2009). While fatal stampedes occurred before recent ice melt, open water increases their frequency, as well as the number killed and injured.

By the autumn of 2013, at Cape Sertsekamen in Russia, an estimated 115,000 walrus hauled out, raising concerns "that such huge concentrations of walrus could wipe out the mussel and clam populations in some areas" (Sandford 2013). Conservationists worry that an oil spill close to a haul-out of that size could risk a large proportion of the Pacific walrus population: "The areas that have been earmarked for oil and gas exploration around Chukotka in the Russian Arctic are exactly the same places where mothers go with young calves. Drilling there will have unknown consequences," explained Anatoly Kochnev of Chukottinro, the Chukotka branch of Russia's Pacific Research Fishery Centre (Sandford 2013).

Summer haul-outs on both sides of the Bering Strait were becoming customary for thousands of walrus by 2015, as summer sea ice continued to retreat. One report described photos from Ryrkaypiy in Chukotka, Russia, which show an estimated 5,000 walrus hauled out in that spot, with thousands more hauled out across the strait.

> Villagers in both places are working to protect resting walrus herds from curious onlookers, as walrus hauled out in such large numbers on beaches are prone to being stampeded, killing smaller animals in the crush. During the late summer and early fall, the Pacific walrus of the Chukchi Sea north of Alaska and of Russia's Chukotka prefer to rest on sea ice over the shallow waters of the continental shelf. In those areas they can readily access food on the seabed. However, in most years since 2007, when Arctic sea ice extent plummeted to a record low, walrus have been forced ashore because there has been no sea ice cover over their preferred shallow feeding areas. ("Epic Walrus" 2015)

With arctic ice melting at current rates, "Certainly we look like we're on a death spiral right now," said Mark Serreze, senior research scientist at the National Snow and Ice Data Center at the University of Colorado. "Losing that summer sea ice-over by 2030, within some of our lifetimes, is a reasonable expectation," he said. "Ultimately it's beyond my scope. I can't make ice cubes out there" ("Walrus Latest" 2007).

Further Reading

Angier, Natalie. 2008. "Ice Dwellers Are Finding Less Ice to Dwell On." *New York Times*, May 20. Accessed October 1, 2018. http://www.nytimes.com/2008/05/20/science/20count.html.

"Epic Walrus Gathering on Again." 2015. Phys.Org, August 31. Accessed October 1, 2018. https://phys.org/news/2015-08-epic-walrus.html.

Joling, Dan. 2007. "Thousands of Pacific Walrus Die; Global Warming Blamed." Associated Press, December 14. (LEXIS).

Joling, Dan. 2007. "Melting Ice Pack Displaces Alaska Walrus." Associated Press in the *Washington Post*, October 4. http://www.washingtonpost.com/wp-dyn/content/article/2007/10/06/AR2007100601206.html.

Kaufman, Marc. 2006. "Warming Arctic Is Taking a Toll; Peril to Walrus Young Seen as Result of Melting Ice Shelf." *Washington Post*, April 15: A-7. Accessed October 1, 2018. http://www.washingtonpost.com/wp-dyn/content/article/2006/04/14/AR2006041401368_pf.html.

Revkin, Andrew C. 2009. "Walrus Suffer Substantial Losses as Sea Ice Erodes." *New York Times*, October 3. Accessed October 1, 2018. http://www.nytimes.com/2009/10/03/science/earth/03walrus.html.

Sandford, Daniel. 2013. "Russia's Arctic: Laptev Walrus." BBC News, August 26. Accessed October 1, 2018. http://www.bbc.com/news/world-europe-23793078.

"3,000 Walruses Die in Stampedes Tied to Climate." 2007. *Associated Press* on NBC News, December 14. Accessed November 27, 2018, http://www.nbcnews.com/id/22260892/ns/us_news-environment/t/walruses-die-stampedes-tied-climate/#.W_2W5ttKiUk.

"Walrus Latest to Be Threatened by Warming." 2007. CBS News, December 24. Accessed October 1, 2018. http://www.cbsnews.com/news/walrus-latest-to-be-threatened-by-warming/.

UNITED STATES

Bees Declining? Enter the Drones

Across the United States, bees have been disappearing, and concern has been rising that their decline may imperil the many plants that they pollinate, including many staple fruits and vegetables that most people eat. The plague on bees has been spreading to wild as well as domesticated species. In 2016, the rusty patched bumblebee was the first wild species of this type to be listed as endangered under the administration of President Barack Obama. Within his first two weeks in office, President Donald J. Trump rescinded the designation.

"Perhaps the biggest foreboding danger of all facing humans is the loss of the global honeybee population," wrote Joachim Hagopian for Global Research (2017). He continued, "The consequence of a dying bee population impacts man at the highest levels on our food chain, posing an enormously grave threat to human survival" (Hagopian 2017). All in all, 90 percent of flowering plants and one-third of food crops are sustained by pollination, according to the U.S. Department of Agriculture's Natural Resources Conservation Service (Khan 2017).

Since no other single animal species plays a more significant role in producing the fruits and vegetables that humans commonly take for granted and require near daily use to stay alive, the greatest modern scientist Albert Einstein once prophetically remarked, "Mankind will not survive the honeybees' disappearance for more than five years" (Hagopian 2017).

Scientists have been debating the causes of the bees' decline, and consensus has not settled on a single provocation. The bees may be victims of pesticides, of climate change, and habitat restrictions (including invasive species), or a combination. All the factors have human links, and now technology may be proposing a cure: robot bees—tiny drones that may do nature's duty. With drones, some technophiles ask, who needs bees?

The decline of bees first came to light in 2006. Within six years, honey production in California had declined by half; within a decade, honeybee population in Iowa was down by 70 percent (Hagopian 2017). Soon scientists had given the

---

Endocrine Disruptors Destroying Sperm

The number and vigor of men's sperm have fallen swiftly since about 1950, and tests on humans and animals link this decline to endocrine disruptors, chemicals in cosmetics, plastics, pesticides, and many other products. Sperm counts (their density in semen) have been falling consistently, and, as Nicholas Kristof wrote in the *New York Times*,

> Even when properly shaped, today's sperm are often pathetic swimmers, veering like drunks or paddling crazily in circles. Sperm counts also appear to have dropped sharply in the last 75 years, in ways that affect our ability to reproduce. "There's been a decrease not only in sperm numbers, but also in their quality and swimming capacity, their ability to deliver the goods," said Shanna Swan, an epidemiologist at the Icahn School of Medicine at Mount Sinai, who notes that researchers have also linked semen problems to shorter life expectancy. (Kristof 2017)

This trend is worldwide and has been accompanied by increases in testicular cancer, as well as undescended testicles (congenital malformation of the penis) and a condition called hypospadias, in which the urethra exits the side or base of the penis, not the tip. "There is still disagreement about the scale of the problem, and the data aren't always reliable," wrote Kristof (2017). "But some scientists are beginning to ask, at some point, will we face a crisis in human reproduction? Might we do to ourselves what we did to bald eagles in the 1950s and 1960s?"

Further Reading

Kristof, Nicholas. 2017. "Are Your Sperm in Trouble?" *New York Times*, March 11. Accessed October 11, 2018. https://www.nytimes.com/2017/03/11/opinion/sunday/are-your -sperm-in-trouble.html.

decline a name—Honeybee Colony Collapse Disorder (CCD)—as a robust debate evolved regarding its cause.

Causes of Colony Collapse Disorder

Commercial beekeepers follow an annual cycle in the United States. They transport hives to the Midwest during summer, where bees gather nectar and pollen, then to California and other states in the spring to pollinate many of the country's food crops. The U.S. Department of Agriculture has been funding bee-disease studies; it also spends money "to help reseed pastures in Michigan, Minnesota, Wisconsin, and the Dakotas with bee-appropriate plants. Increasing the availability of plants like alfalfa and clover will provide more foodstock for the thousands of commercial beekeepers who bring hives to those states each year" (Schlanger 2014). The more work scientists do on colony-collapse disorder, the more human influence they find; as many as 35 pesticides and fungicides may be polluting bees' pollen and hives, many in lethal doses.

With bee populations declining rapidly, the U.S. Department of Agriculture encouraged farmers to seed fields with alfalfa and clover to augment habitat, as part of a campaign to preserve the pollinators of food crops worth an estimated $40 billion each year. Honeybees pollinate a third of the United States' food, including more than 130 varieties of fruits and vegetables. This includes pollination by both wild and commercially raised bees that are moved between farms on trucks. Despite measures to preserve habitat, more than a third of bees in the United States had disappeared within a decade by 2016.

Scientists were tracing the plague to several possible factors, including "increased use of pesticides especially in the United States, shrinking habitats, multiple viruses, poor nutrition and genetics, and even cell phone towers, [as well as] the parasite called the Varroa destructor, a type of mite found to be highly resistant to the insecticides that U.S. beekeepers have used in attempts to control the mites from inside the beehives" (Hagopian 2017). Varroa was first detected in 1987, as Monsanto, Bayer, Dow, Bayer, and other large chemical companies sold genetically modified herbicides and insecticides as a remedy against parasites. Later, scientists found that the herbicides and insecticides were weakening the bees' natural genetic defenses against the parasite.

"It's a Quiet but Staggering Crisis"

The Center for Biological Diversity reported in 2017 that "of the 1,437 native bee species for which there was sufficient data to evaluate, about 749 of them were declining. Some 347 of the species, which play a vital role in plant pollination, are imperiled and at risk of extinction" (Cherulus 2017). Pesticides, coupled with urbanization (leading to habitat loss) and climate change, are the major reasons for bees' decline. "It's a quiet but staggering crisis unfolding right under our noses that illuminates the unacceptably high cost of our careless addiction to pesticides and monoculture farming," said Kelsey Kopec of the Center for Biological Diversity (Cherelus 2017).

In February 2017, the rusty patched bumble bee became the first species in the continental United States to be listed as endangered by U.S. federal authorities. "If we don't act to save these remarkable creatures, our world will be a less colorful and more lonesome place," Kopec said (Cherelus 2017). The rusty patched bumble bee's population has "plummeted by 87 percent" and is "balancing precariously on the brink of extinction" after a dramatic decline starting in the late 1990s, the U.S. Fish and Wildlife Service said. "Bees are dying, and we need to do something to save them," Rebecca Riley, senior attorney with the Natural Resources Defense Council, told NBC News (Silva 2017). Seven other types of bees once abundant in Hawaii also have been added to the endangered-species list.

Ishan Daftarder explained on Science ABC (a website) that "pesticides, known as neonicotinoids (a relatively new class of insecticides that affect the central nervous system of insects, resulting in paralysis and death), [have] had a major role in the bees' decline. When bees are exposed to neonicotinoids, they go into a shock and forget their way home (sort of like the insect version of Alzheimer's). Along with pesticides, parasites known as Varrao mites (also called Varrao destructors) are also responsible for their death. The Varrao can only reproduce in a bee colony. They are blood-sucking parasites that affect adult and young bees equally. The disease inflicted by these mites can result in bees losing legs or wings, essentially killing them" (Daftarder, n.d.).

Greenpeace developed a "Save the Bees" campaign against the use of these chemicals, including neonicotinoids, which have been causing poisoning of entire colonies as bees return to their hives with contaminated nectar and pollen. Toxicity destroys the bees' central nervous systems to the point where they can no longer fly. At the same time, the practice of feeding many domesticated bees with high-fructose corn syrup instead of their own honey was undermining their defenses against parasites and viruses.

Calling Out the Drones

By 2017, the robo-bees had been described as laboratory creations in the journal *Chem* but had not been field tested. Pollinators include bats, butterflies, and some birds, as well as bees, and drone technology may be adapted to mimic nature for all of them, as they also are declining due to the same lethal ecological factors that threaten bees.

The drones may spread pollen on horsehairs attached to their mechanical bellies as they are guided by techno-farmers of the future at computer keyboards through fields and orchards—that is, perhaps, before climate becomes too hot during the growing season to pollinate at all, given that human ingenuity has not yet come up with invented pollen. Until then, "gardeners might not just hear the buzz of bees among their flowers, but the whirr of robots, too. Scientists in Japan say they've managed to turn an unassuming drone into a remote-controlled pollinator by attaching horsehairs coated with a special, sticky gel to its underbelly" (Khan 2017).

Some proposals for robo-bees may not work. The authors of a study discussed by Amina Khan in the *Los Angeles Times* wrote that one

> artificial pollination technique requires the physical transfer of pollen with an artist's brush or cotton swab from male to female flowers. Unfortunately, this requires much time and effort. Another approach uses a spray machine, such as a gun barrel and pneumatic ejector. However, this machine pollination has a low pollination success rate because it is likely to cause severe denaturing of pollens and flower pistils as a result of strong mechanical contact as the pollens bursts out of the machine. (Kahn 2017)

In other words, the bees can pollinate without destroying the flowers. As of this writing, robo-bees are too clumsy to mimic them correctly. One other problem is powering the drones. Models tested thus far require a power source with a wire, once more a problem. Wires are destructive of flowers and often become ensnared in foliage.

"It's very tough work," said Eijiro Miyako, a chemist at the National Institute of Advanced Industrial Science and Technology in Japan (Khan 2017). Miyako has developed a gel that may be used someday to pick up pollen and transfer it between male and female plants. Not even Miyako believes that drones will replace bees. "In combination is the best way," he said (Khan 2017). The robots need work not only at mimicking pollination, but with improved artificial intelligence and guidance. The day of treating pollination as if it is a video game has not yet arrived. Biology still trumps technology.

Further Reading

Cherelus, Gina. 2017. "Hundreds of North American Bee Species Face Extinction: Study." *Science News*, March 1. Accessed October 1, 2018. http://www.reuters.com/article/us-usa-bees-idUSKBN1685NG.

Daftarder, Ishan. n.d. "Why Bee Extinction Would Mean the End of Humanity." Science ABC. Accessed October 1, 2018. https://www.scienceabc.com/nature/bee-extinction-means-end-humanity.html.

Hagopian, Joachim. 2017. "Death and Extinction of the Bees." Global Research, January 4. Accessed October 1, 2018. http://www.globalresearch.ca/death-and-extinction-of-the-bees/5375684.

Khan, Amina. 2017. "As Bee Populations Dwindle, Robot Bees May Pick up Some of Their Pollination Slack." *Los Angeles Times*, February 9. Accessed October 1, 2018. http://www.latimes.com/science/sciencenow/la-sci-sn-robot-bees-20170209-story.html

Schlanger, Zoë. 2014. "As Bee Populations Vanish, USDA Tries to Keep Them Fed." *Newsweek*, February 26. Accessed October 1, 2018. http://www.newsweek.com/bee-populations-vanish-usda-tries-keep-them-fed-230271.

Silva, Daniella. 2017. "Bumble Bee Species Declared Endangered in the U.S. for First Time." NBC News, January 12. Accessed October 1, 2018. http://www.nbcnews.com/science/environment/bumble-bee-species-declared-endangered-u-s-first-time-n706321.

Chapter 5: Indigenous Peoples

OVERVIEW

The toxic legacy of Native North American land is pervasive but largely invisible to most of us. Many toxic sites are located in out-of-the-way rural areas largely forgotten by the majority of Americans. Today, one-fourth of Environmental Protection Agency "Superfund" sites in the United States are on Indian reservations (Hansen 2014). From the (now-closed) uranium mines of Navajo Country to the PCB-laced turtles on the Akwesasne Mohawks' homeland in Upstate New York, Native American peoples are resisting energy and resource colonization to survive.

Examples abound, sea to shining sea. To cite only one of many: zinc and lead were mined within the jurisdiction of the Quapaw in Picher, Oklahoma, until 1967, when, according to a retrospective by Terri Hansen for the Indian Country Today Media Network:

> Mining companies abandoned 14,000 mine shafts, 70 million tons of lead-laced tailings, 36 million tons of mill sand and sludge, as well as contaminated water, leaving residents with high lead levels in blood and tissues. Cancers skyrocketed, and 34 percent of elementary-school students suffered learning disabilities. (Hansen 2013)

This is not an exceptional case. A catalogue of such environmental atrocities would fill a very thick volume. What follows is merely a sampler from across North America, beginning in the Arctic, which, in today's world, is nowhere nearly as pristine as many people believe. In Nunavut, the semi-sovereign Inuit nation in northern Canada, it's been snowing poison, as PCBs, dioxins, and other chemicals produced by southern industries are swept northward by prevailing winds. Mothers have been warned not to breastfeed their infants. At Akwesasne, in northernmost New York State and nearby Canada, the fish, laced with the same chemicals, are often inedible, and mothers also have been told not to breastfeed babies (and to avoid eating anything from the St. Lawrence River). The Huicholes of Mexico live on a land laced with toxic chemicals. In all of these locations, Native peoples have organized to restore a livable environment. The Inuit, in particular, have taken an active role in efforts to outlaw use of these toxins in an international agreement, the Stockholm Convention.

Following negotiation of the Stockholm Convention, which outlaws most of the "Dirty Dozen" persistent organic pollutants (POPs, to toxicologists), activist Sheila Watt-Cloutier evoked tears from some delegates with her note of gratitude on behalf of Inuit people. The treaty, she said, had "brought us an important step closer to fulfilling the basic human right of every person to live in a world free

The political-environmental indigenous movement Idle No More stages a rally in 2013 to
support treaties made between the Canadian government and the First Nations people.
(Lostafichuk/Dreamstime.com)

of toxic contamination. For Inuit and indigenous peoples, this means not only a
healthy and secure environment, but also the survival of a people. For that I am
grateful. *Nakurmiik.* Thank you" (Cone 2005, 202).

It's a long road back, however. Once they become part of the food chain, POPs
are very difficult to dislodge. Following the negotiation of international law to
outlaw these pollutants came the decades-long battle to enforce the ban and to rid
the Arctic food chain of their effects. Persistent organic pollutants tend to become
locked in the ecosystem, and as of this writing, Inuit still consume traditional foods
and breastfeed their babies with caution.

In this chapter, we'll examine the environmental issues that indigenous people
in the United States and around the world are facing. We'll look at the health effects
that mining of uranium has had on Canada's Dene Nation and the Navajos of the
United States. We'll also look at toxic chemicals affecting Canada's Nunavut Nation,
and how oil extraction is harming indigenous groups in Russia. The chapter will

also examine those who have been displaced by hydroelectric dams in India, as well as the plight of the Maya in Mexico.

Further Reading

Cone, Marla. 2005. *Silent Snow: The Slow Poisoning of the Arctic.* New York: Grove Press.

Hansen, Terri. 2013. "7 Major Environmental Disasters in Indian Country." Indian Country Today Media Network, October 9. https://newsmaven.io/indiancountrytoday/archive/7-major-environmental-disasters-in-indian-country-NfNsgQ1d90Slhh TW5xwrmA/.

Hansen, Terri. 2014. "Kill the Land, Kill the People: There Are 532 Superfund Sites in Indian Country!" Indian Country Today Media Network, June 17. https://newsmaven.io/indiancountrytoday/archive/kill-the-land-kill-the-people-there-are-532-superfund-sites-in-indian-country-LpCDfEqzlkGEnzyFxHYnJA/.

CANADA: DENE NATION

Decimated by Uranium Mining

At the dawn of the nuclear age in the middle 1940s, Paul Baton and more than 30 other Dene hunters and trappers who were recruited to mine uranium called it "the money rock" in their homeland, in Canada's Northwest Territories. Paid $3 a day by their employers, the Dene hauled burlap sacks of the grimy ore from one of the world's first uranium mines at Port Radium across the Northwest Territories to Fort McMurray. Since then, at least 14 Dene who worked at the mine between 1942 and 1960 have died of lung, colon, and kidney cancers, according to documents obtained through the Northwest Territories Cancer Registry (Nikiforuk 1998, A-1). The Port Radium mine supplied the uranium to fuel some of the first atomic bombs. Within half a century, uranium mining in northern Canada had left behind more than 120 million tons of radioactive waste, enough to cover the Trans-Canada Highway two meters deep across Canada. By the year 2000, production of uranium waste from Saskatchewan alone occurred at the rate of more than 1 million tons annually (LaDuke 2001).

Cancer Deaths Accelerate

The experiences of the Dene were similar to those of Navajo uranium miners in Arizona and New Mexico at the same time. Since 1975, a 30-year latency period after uranium mining began in the area, hospitalizations for cancer, birth defects, and circulatory illnesses in northern Saskatchewan had increased between 123 and 600 percent. In other areas impacted by uranium mining, cancers and birth defects have increased, in some cases to as much as eight times the Canadian national average (LaDuke 1995).

"Before the mine, you never heard of cancer," said Baton, 83. "Now, lots of people have died of cancer" (Nikiforuk 1998, A-1). The Dene were never told of uranium's dangers. Declassified documents have revealed that the U.S. government, which bought the uranium, and the Canadian federal government, at the time the world's largest supplier of uranium outside the United States, withheld health and safety information from miners and their families. A 1991 federal aboriginal health survey found that the Deline Dene community reported twice as much illness as any other Canadian aboriginal community.

While mining, many Dene slept on the ore, ate fish from water contaminated by radioactive tailings, and breathed radioactive dust while on the barges, docks, and portages where they hauled the ore. More than a dozen men carried sacks of ore weighing more than 45 kilograms six days a week, 12 hours a day, four months a year, as "children played with the dusty ore at river docks and portage landings. And their women sewed tents from used uranium sacks" (Nikiforuk 1998, A-1). In exchange for their labors in the uranium mines, the Dene received a few sacks of flour, lard, and baking powder.

While many of the Dene blame uranium mining and its waste products for their increased cancer rates, some Canadian officials compiled statistics indicating only marginally increased mortality from uranium exposure. André Corriveau, the Northwest Territories' chief medical officer of health, noted that high cancer rates among the Dene do not differ significantly from the overall territorial profile. He said that the death rate was skewed upward by high rates of smoking (Nikiforuk 1998, A-1). The Dene, in the meantime, maintained that the fact that almost half the workers in the Port Radium mine (14 of 30) died of lung cancer cannot be explained by smoking alone.

"The Incurable Disease"

Until his death in 1940, Louis Ayah, known as "Grandfather," one of the North's great aboriginal spiritual leaders, repeatedly warned his people that the waters in Great Bear Lake would turn a foul yellow. According to him, the yellow poison would flow toward the village, recalls Madelaine Bayha, one of a dozen scarfed and skirted "uranium widows" in the village (Nikiforuk 1998, A-1).

The first Dene to die of cancer, or what elders still call "the incurable disease," was Old Man Ferdinand in 1960. He had worked at the mine site as a logger, guide, and stevedore for nearly a decade. "It was Christmastime, and he wanted to shake hands with all the people as they came back from hunting," recalled René Fumoleau, then an Oblate missionary working in Deline. After saying goodbye to the last family that came in, Ferdinand declared: "'Well, I guess I shook hands with everyone now,' and he died three hours later" (Nikiforuk 1998, A-1).

According to Nikiforuk's account, others died during the next decade. Victor Dolphus's arm came off when he tried to start an outboard motor. Joe Kenny, a boat pilot, died of colon cancer. His son, Napoleon, a deckhand, died of stomach cancer. The premature death of so many men left so many widows that it disrupted the transmission of culture between generations. "In Dene society, it is the grandfather who passes on the traditions, and now there are too many men with no uncles, fathers, or grandfathers to advise them," said Cindy Gilday, Joe Kenny's daughter and chair of Deline Uranium Committee (Nikiforuk 1998, A-1).

"It's the most vicious example of cultural genocide I have ever seen," Gilday said. "And it's in my own home" (Nikiforuk 1998, A-1). According to Nikiforuk, "Watching a uranium miner die of a radioactive damaged lung is a job only for the brave." He described Al King, an 82-year-old retired member of the Steelworkers union in Vancouver, British Columbia, who has held the hands of the dying. King described one retired Port Radium miner whose chest lesions were so bad that they had spread to his femur and caused it to burst. "They couldn't pump enough morphine into him to keep him from screaming before he died," said King (Nikiforuk 1998, A-1). The Dene town of Deline was described by residents as a village of widows, after many of its men died of cancer. "The widows," said one resident, "were left to raise their families with no breadwinners," depending on welfare and other young men for food sources (Gilday, n.d.).

On March 21 and 22, 1998, residents in the Dene town whose families had experienced early deaths due to cancers held workshops and were briefed by the Deline Uranium Committee. "For the first time they're getting independent information outside of government, completely truthful information," said committee chair Cindy Gilday of the research compiled by doctors, scientists, and lawyers from Vancouver to Montreal (Korstrom, 1998).

According to an article by Glen Korstrom:

> For three dollars a day, Dene workers hauled and ferried burlap sacks at what was the world's first uranium mine, and at least 14 workers have since died of lung, colon, and kidney cancers, according to documents obtained through the NWT [Northwest Territory] Cancer Registry. Declassified documents from the eventual buyer of the uranium, the U.S. government, show the danger from uranium was known when the mine was operating. But neither the U.S. nor Canadian governments shared that vital health information, so both aboriginal and non-aboriginal workers treated the uranium casually, often sleeping on the sacks or putting bits of ore in their pockets. A 1991 federal aboriginal health survey found Deline reporting twice as many illnesses as any other Canadian aboriginal community. (Korstrom, 1998)

Analysis was complicated by the higher-than-usual rate of cigarette smoking among the miners. "The rates of smoking in Nunavut are quite a bit higher actually," said NWT chief medical officer André Corriveau, who estimated that 70 percent of Native Canadian Arctic people smoked at that time. The nature of the cancers was eventually attributed to radiation from uranium mining. After the toxic nature of the mines was established, some suggestions were made that the Dene ought to leave their home villages. Gilday rejected that idea, however. "Where would they move? This is their home. This is tens of thousands of years of living on Great Bear Lake," she said of the 600 people who live on the world's fourth largest inland lake. "Their home has been poisoned by someone [else]," she said, "who is (supposed to be) a responsible party" (Korstrom 1998). "It's a huge lake, so even if there was a local source of contamination, it would be unlikely to have a measurable effect across the lake or everywhere," said Corriveau. "There may be local hot spots, especially for bottom-feeding fish" (Korstrom 1998).

The Conflict Continues

During and after 2014, many Dene continued to protest uranium mining on their lands, to the point of setting up blockades against industrial traffic aiming to open one of the largest such mines in North America. Organized as the Northern Trappers Alliance, the Dene also opposed exploitation of tar sands. Press reports said that the blockade, established November 22, 2014, was "forcibly dismantled" by officers of the Royal Canadian Mounted Police on December 1 (Toledano 2015). Eighty days later, wrote Michael Toledano in *Vice*, "The trappers remain[ed] camped on the side of the highway in weather that has routinely dipped below -40[°]C. They are constructing a permanent cabin on the site that will be a meeting place for Dene people and northern land defenders."

All of this is being built upon the ruins of mines that killed many Dene in earlier generations; "abandoned and decommissioned uranium mines already host millions of tons of radioactive dust (also known as tailings) that must be isolated from the surrounding environment for millennia, while no cleanup plans exist for the legacy of severe and widespread watershed contamination that is synonymous with Uranium City, Saskatchewan" (Toledano 2015). More than 85 percent of the people in this area (northern Saskatchewan) are aboriginal. They are fighting exploratory mining that could increase in scale in coming years, a new neighbor to the open-pit tar sands mines further south that have turned indigenous lands into "moonscapes."

"When they spew the pollution, it affects our water, lakes, fish—any kind of species. Our traditional life [has been] destroyed with these oil mines around us," said Kenneth, one of the trappers. "We're in the middle of these oil mines, and the government's still not listening." Candyce Paul, spokesperson for the Northern Trappers Alliance, said, "We know our water isn't as good as it used to be. You see more fish with lesions" (Toledano 2015).

Further Reading

Gilday, Cindy Kenny. n.d. "A Village of Widows." *Arctic Circle*. Accessed October 2, 2018. http://arcticcircle.uconn.edu/SEEJ/Mining/gilday.html.

Korstrom, Glen. 1998. "Deline Poisoned? Past Area Mining Linked to Cancer." Northern News Services, March 23. Accessed October 1, 2018. http://arcticcircle.uconn.edu/SEEJ/Mining/korstrom.html.

LaDuke, Winona. 1995. "The Indigenous Women's Network: Our Future, Our Responsibility." Statement of Winona LaDuke at the United Nations Fourth World Conference on Women, Beijing, China, August 31. Accessed on October 2, 2018. https://ratical.org/co-globalize/WinonaLaDuke/Beijing95.html.

LaDuke, Winona. 2001. "Insider Essays: Our Responsibility." Electnet/Newswire, October 2.

Nikiforuk, Andrew. 1998. "Echoes of the Atomic Age: Cancer Kills Fourteen Aboriginal Uranium Workers." *Calgary Herald*, March 14: A-1, A-4. Accessed October 2, 2018. http://www.ccnr.org/deline_deaths.html.

Toledano, Michael. 2015. "Indigenous Canadians Are Fighting the Uranium Mining Industry." *Vice*, February 11. Accessed October 2, 2018. https://www.vice.com/en_us /article/jmbwx8/a-dene-alliance-formed-to-resist-uranium-and-tar-sands-mining-in -saskatchewan-892.

CANADA: NUNAVUT (INUIT)

It's Snowing Poison

Many residents in the temperate zones hold fond stereotypes of a pristine Arctic largely devoid of the human pollution that is now so ubiquitous there. To a tourist with no interest in environmental toxicology, the Inuits' Arctic homeland may seem as pristine as ever during its long, snow-swept winters. Many Inuit hunt polar bears and seals on the pack ice that surrounds their Arctic-island homelands. To the naked eye, the Arctic still *looks* pristine. In Inuit country these days, however, it's what you *can't* see that may kill you. Such a scene may look pristine, until one realizes that the polar bears' and seals' body fats are laced with dioxins and PCBs.

"As we put our babies to our breasts, we are feeding them a noxious, toxic cocktail," said Sheila Watt-Cloutier, a grandmother who also has served as president of the Inuit Circumpolar Conference. "When women have to think twice about breastfeeding their babies, surely that must be a wake-up call to the world" (Johansen 2000, 27). Watt-Cloutier was raised in an Inuit community in remote northern Quebec. Unknown to her at that time, toxic chemicals were being absorbed by her body and by those of other Inuit in the Arctic.

During a career as activist and diplomat, Watt-Cloutier ranged between her home in Iqaluit (pronounced "Eehalooeet"), capital of the new semi-sovereign Nunavut Territory, and Ottawa, Montreal, New York City, and other points south, doing her best to alert the world to toxic poisoning and other perils faced by her people. The Inuit Circumpolar Conference (ICC) represents the interests of roughly 140,000 Inuit spread around the North Pole, from Nunavut (which means "our home" in the Inuktitut language) to Alaska and Russia. Nunavut itself, a territory four times the size of France, has a population of roughly 27,000, 85 percent of whom are Inuit.

The Arctic Is No Longer Pristine

In 1908, Roald Amundsen, one of the earliest European explorers of the Earth's polar regions, said that Arctic peoples "living absolutely isolated from civilization of any kind are undoubtedly the happiest, healthiest, most honorable, and most contented of peoples. . . . My sincerest wishes for our Nechilli Eskimo friends is that civilization may never reach them" (Amundsen 1908, 93). Scarcely more than a century later, the environmental threats facing the Inuit today are entirely outside their historical experience. Welcome to ground zero on the road to environmental apocalypse: a place with a people who never asked for any of the curses that industrial societies to the south have brought to them. Persistent organic pollutants, such as DDT, dioxins, and PCBs, are multiplying up the food chain of the Inuit as air and ocean currents transport the effluvia of southern industry into polar regions. Geographically, the Arctic could not be in a worse position for toxic pollution, as a ring of industry in Russia, Europe, and North America pours pollutants northward.

"At times," said Watt-Cloutier, "We feel like an endangered species. Our resilience and Inuit spirit and of course the wisdom of this great land that we work so hard to protect gives us back the energy to keep going" (Johansen 2000, 27).

Most the chemicals that now afflict the Inuit are synthetic compounds of chlorine; some of them are incredibly toxic. For example, one millionth of a gram of dioxin will kill a guinea pig (Cadbury 1997, 184). In addition to evidence documenting the spread of persistent organic pollutants through the flora and fauna of the Arctic, studies describe the saturation of the same area by high levels of mercury, lead, and nuclear radiation in fish and game ("PD 2000 Projects" 2001).

Chemicals Linked to Many Maladies

These Persistent Organic Pollutants (POPs) have been linked to cancer and birth defects, as well as other neurological, reproductive, and immune-system damage in people and animals. At high levels, these chemicals also damage the central nervous system. Many of them also act as endocrine disrupters, causing deformities in sex organs as well as long-term dysfunction of reproductive systems. "POPs also can interfere with the function of the brain and endocrine system by penetrating the placental barrier and scrambling the instructions of the body's naturally produced chemical messengers. The latter tell a fetus how to develop, from the womb through puberty; should interference occur, immune, nervous, and reproductive systems may not develop as programmed by the genes inherited by the embryo.

Pesticide residues in the Arctic today may include some used decades ago in the southern United States. Although the United States is several hundreds of kilometers from its borders, Nunavut receives up to 82 percent of its dioxins from industries there (Suzuki 2000). Some of the pollution in the Canadian Arctic arrives on prevailing winds from as far away as Western Europe and Japan.

The Arctic's cold climate also slows decomposition of these toxins, so they persist in the Arctic environment longer than at lower latitudes. The Arctic acts as a cold trap, collecting and maintaining a wide range of industrial pollutants, from PCBs to toxaphene, chlordane to mercury, according to the Canadian Polar Commission ("Communities Respond," n.d.). As a result, "Many Inuit have levels of PCBs, DDTs, and other persistent organic pollutants in their blood and fatty tissues that are five to ten times greater than the national average in Canada or the United States" ("Communities Respond," n.d.).

Effects Increase Exponentially

Dioxins, PCPs, and other toxins accumulate with each succeeding generation in breastfeeding mammals, including the Inuit and many of their food sources. The reproductive cycle of humans and other mammals compounds the toxic effects of these chemicals because their effects increase exponentially with each step up the food chain, a process that scientists call "biomagnification." Airborne toxic substances are absorbed by plankton and small fish, which are then eaten by dolphins, whales, and other large animals. The mammals' thick subcutaneous fat stores the hazardous substances, which are transmitted to offspring through breastfeeding.

Sea mammals are more vulnerable to this kind of toxicity than land animals, so the levels of chemicals in their bodies can reach exceptionally high levels. In ecosystems, POPs "bioaccumulate" in animals at the top of their respective food chains, in the bodies of large meat-eaters such as marine mammals, polar bears, raptors, and human beings. Dolphins, seals, and whales in the northern seas are being contaminated. Large land animals, such as caribou, also are affected. Humans who eat these animals form the top (most polluted) rung on the food chain. At the very top of this hierarchy of toxicology is the Inuit mother breastfeeding her infant.

Effects on Inuit Infants

Inuit infants have provided "a living test tube for immunologists" (Cone 1996, A-1). Due to their diet of contaminated sea animals and fish, Inuit women's breast milk contains six to seven times the PCB level of women in urban Quebec, according to Quebec government statistics. Their babies have experienced strikingly high rates of meningitis, bronchitis, pneumonia, and other infections compared with other Canadians. One Inuit child out of every four has chronic hearing loss due to infections. "In our studies, there was a marked increase in the incidence of infectious disease among breastfed babies exposed to a high concentration of contaminants," said Eric Dewailly, a Quebec Public Health Center researcher who was among the first scientists to detect high levels of persistent organic pollutants in the bodies of Inuit mothers and their breast milk (Cone 1996, A-1).

Dewailly, a Laval University scientist, accidentally discovered that Inuit infants were being heavily contaminated by PCBs. During the middle 1980s, Dewailly had first visited the Inuit seeking a pristine group to use as a baseline with which to compare women in southern Quebec who had PCBs in their breast milk. Instead, Dewailly found that Inuit mothers' PCB levels were several times higher than the Quebec mothers in his study group.

Dewailly and colleagues then investigated whether organochlorine exposure is associated with the incidence of infectious diseases in Inuit infants from Nunavut. They reported that serious ear infections were twice as common among Inuit babies whose mothers had higher-than-usual concentrations of toxic chemicals in their breast milk. More than 80 percent of the 118 babies studied in various Nunavut communities had at least one serious ear infection in the first year of life. The three most common contaminants that researchers found in Inuit mothers' breast milk were three pesticides (dieldrin, mirex, and DDE, a derivative of DDT) and two industrial chemicals—PCBs and hexachlorobenzene. The researchers could not pinpoint which specific chemicals were responsible for making the Inuit babies more vulnerable because chemicals' effects may intensify in combination (Dewailly et al. 1993, 1994, 2000).

Born with depleted white blood cells, Inuit children have suffered excessive bouts of diseases, including a 20-fold increase in life-threatening meningitis compared to other Canadian children. These children's immune systems sometimes fail to produce enough antibodies to resist even the usual childhood diseases. A study published in the *New England Journal of Medicine* confirmed that children exposed to low levels of PCBs in the womb grow up with low IQs, poor reading comprehension, difficulty paying attention, and memory problems (Jacobson and Jacobson 1996, 783–789).

Infants in the North are being developmentally harmed by exposure to several contaminants, among them lead, methylmercury, and polychlorinated biphenyls (PCBs). Indigenous infants' cognitive abilities also are penalized by environmental contamination beginning in the womb and continuing after birth. Physiological damage due to chemical exposure impedes both biological and psychological development. Watt-Cloutier and other Inuit have been waging a long, difficult worldwide battle to remove these debilitating toxins from their lives.

Further Reading

Amundsen, Roald. 1908. *The North West Passage*. London: E. P. Dutton & Co.

Boucher O., G. Muckle, J. L. Jacobson, R. C. Carter, M. Kaplan-Estrin, P. Ayotte, . . . S. W. Jacobson. 2014. "Domain-Specific Effects of Prenatal Exposure to PCBs, Mercury, and Lead on Infant Cognition: Results from the Environmental Contaminants and Child Development Study in Nunavik." *Environmental Health Perspectives*, January 17. Accessed October 2, 2018. http://dx.doi.org/10.1289/ehp.1206323.

Cadbury, Deborah. 1997. *Altering Eden: The Feminization of Nature*. New York: St. Martin's Press.

"Communities Respond to PCB Contamination." n.d. PCB Working Group, IPEN.

Cone, Marla. 1996. "Human Immune Systems May be Pollution Victims." *Los Angeles Times*, May 13: A-1.

Cone, Marla. 2005. *Silent Snow: The Slow Poisoning of the Arctic*. New York: Grove Press.

"Contaminants Have Variety of Effects on Arctic Baby IQs." 2014. Environmental Health News, February 7.

Dewailly, E., P. Ayotte, S. Bruneau, S. Gingras, M. Belles-Isles, and R. Roy. 2000. "Susceptibility to Infections and Immune Status in Inuit Infants Exposed to Organochlorines." *Environment Health Perspectives* 108, no. 3 (March): 205–211.

Dewailly, E., S. Bruneau, C. Laliberte, M. Belles-Iles., J.P. Weber, and R. Roy R. 1993. "Breast Milk Contamination by PCB and PCDD/Fs in Arctic Quebec. Preliminary Results on the Immune Status of Inuit Infants." *Organohalogen Compounds* 13: 403–406.

Dewailly, E., J. J. Ryan, C. Laliberté, S. Bruneau, J. P. Weber, S. Gingras, and G. Carrier. 1994. "Exposure of Remote Maritime Populations to Coplanar PCBs." *Environmental Health Perspectives* 102, suppl. 1 (January): 205–209.

Jacobson, Joseph L., and Sandra W. Jacobson. 1996. "Intellectual Impairment in Children Exposed to Polychlorinated Biphenyls in Utero." *New England Journal of Medicine* 335, no. 11 (September): 783–789. Accessed October 1, 2018. http://www.doi.org/10.1056/NEJM199609123351104.

Johansen, Bruce E. 2000. "Pristine No More: The Arctic, Where Mother's Milk Is Toxic." *The Progressive*, December: 27–29.

"PD 2000 Projects." 2001. Arctic Monitoring and Research—Project Directory, April 11.

Suzuki, David. 2000. "Science Matters: POP Agreement Needed to Eliminate Toxic Chemicals." David Suzuki Foundation, December 6.

GUYANA

Mercury Poisoning from Gold Mining

Logging and gold mining, two staples of industrial intrusion into indigenous lands around the world, pose major problems for surviving tribal peoples in Guyana. The Isseneru have been struggling for many years with the toll of mercury poisoning caused by gold mining in their territories, while the people of the Akawaio Nation are working for legal recognition of their rain forest lands before commercial interests log them to the ground. In both cases, the indigenous peoples of Guyana find themselves realizing nearly nothing from commercial exploitation that ruins their lands and their health. Increased gold mining and logging have been promoted by Guyana's government to help reduce its international debt and to meet the terms of "adjustments" required by the International Monetary Fund and the World Bank.

Health Toll of Mercury Exposure

Gold mining is exposing residents of Isseneru, indigenous peoples who live on the Essequibo River in Guyana's Upper Mazaruni, to unhealthy levels of mercury contamination. "We have obtained evidence that the residents of the village of Isseneru are significantly exposed to mercury and that they are closely associated with gold-mining activity," said Institute of Applied Science and Technology Chairman David Singh. "It was important to seek the means by which the mercury loading might be minimized through the use of appropriate technologies," Singh said ("Isseneru Villagers" 2001).

Long-term exposure to mercury vapor can damage the brain, nervous system, and kidneys. Such exposure also may disrupt the functioning of the thyroid gland. Severe brain damage also is possible in children born to mothers who have been exposed. Babies fathered by men who have been exposed to mercury vapor have a greater chance of dying in utero. Consumption of organic mercury in fish can permanently damage the brain and cause mental and physical defects, especially in children. According to Cultural Survival, short-term exposure to mercury vapor also "can permanently damage the lungs and can lead to death" ("Isseneru Villagers" 2001).

These indigenous peoples have been resisting invasion of their lands by miners since the late 1950s. In 1959, one-third of the Upper Mazaruni indigenous reservation was classified by the government as a mining district. In the beginning, the area was invaded by small numbers of independent miners. A few years later, however, the first of several multinational companies, most of them from Canada, began mining gold on an industrial scale in the area. The Canadian companies

Golden Star Resources and Vannessa Ventures have been most notable for their environmental impact.

Dead Fish and Hogs

Large-scale fish kills and deaths of several hogs were reported in August 1995 after a wastewater dam at the Omai gold mine (about 100 miles from the coast) broke and spilled 3.2 million liters of cyanide-laced gold-mining waste into the Essequibo River, "a frightening reminder of the consequences of uncontrolled industrial mining" ("Little Progress" 1997, 12).

Studies by the Pan American Health Organization revealed that all aquatic life in a four-kilometer-long creek that runs from the mine into the Essequibo was killed (Chatterjee 1998). According to a report in *Native Americas*, "Dead hogs and fish were seen floating down the river" ("World Bank" 1996, 4). The mine was closed following the accident, a blow to the Guyanese economy because it had been providing about 25 percent of the nation's export income ("World Bank" 1996, 4).

The Omai gold mine disaster did not end in August of 1995 for the Essequibans in Guyana. Resumption of operations and the continuous dumping of chemicals into the river was ignored except for the people who have had to live with it. Led by Gustav Jackson, a Guyanese geologist and an environmental scientist, a group of professionals from both Guyana and North America joined efforts to secure clean (cyanide-free) drinking water for the people of the Essequibo.

Spillage of mining waste occurred in a formerly pristine tropical rain forest. According to a report in the World Rainforest Movement *Bulletin*, the Upper Mazaruni River basin once comprised a luxurious forest that is home to Akawaio (Kapon) and Arenuca (Pemon) indigenous peoples. Their ancestral territory also encompasses parts of the Gran Sabana in Venezuela, as well as northern Roraima State in Brazil. These peoples' traditional economy long has been based on seasonal migrations between the lower and upper reaches of the Mazaruni and Kamarang rivers for hunting, fishing, and farming.

Mining Waste Ruins Rivers

According to the World Rainforest Movement *Bulletin,* the companies use "missile dredging," which it describes as the use of "enormous vacuum cleaners shaped like missiles, that are attached to river dredges to remove alluvial deposits" to extract gold-bearing ore ("Guyana: Transnational Mining" 2001). The missile dredges destroy riverbanks and nearby forests, meanwhile increasing sedimentation and killing large numbers of fish that once fed local people. Miners also use mercury to separate gold from overburden.

The World Rainforest Movement *Bulletin* described the environmental impacts of mining in the area: "The water is discolored and heavy with sediments, [and] piles of debris accumulate at the riverbanks, some of which have disappeared because of missile dredging. Environmental impact assessments required by law exist only on paper" ("Guyana: Transnational Mining" 2001). Noise caused by mechanized mining also chases away game animals, on which local indigenous peoples depend for food.

The invasion of miners also has brought an increased incidence of malaria to the area, along with alcoholism, prostitution, drug use, and violence. The miners have moved in without consent of local indigenous people, who have had no role in advising government on the future uses of their lands.

Gold mining in Guyana continues to expand. According to the WRM *Bulletin*:

> On November 2, 1998, the government of Guyana and Vancouver-based mining company, Vannessa Ventures Ltd., signed an agreement granting Vannessa more than 2 million hectares of land in which to conduct geophysical and geological surveys for gold and primary diamond sources over the next two years. This concession includes the heavily forested Kanuku mountain range, as well as the upper reaches of the Corentyne River on the border with Suriname in the eastern region of Guyana. The area is part of the ancestral territory of the Wai Wai, Wapisiana and Macusi indigenous peoples. They have vigorously objected to any mining or logging company operating on their lands and are demanding that their rights to their ancestral lands be legally recognized and respected. ("Indigenous Peoples Fight" 1998)

Native Claims Ignored

Toward that end, Guyana established a Native claims process in 1967, but more than 30 years later, only a quarter of the land that Native peoples assert they own has been recognized as such. In October 1997, community leaders of the Wai Wai, Wapisiana, and Macusi peoples formed the Touchau's Amerindian Council of Region 9 to defend their ancestral territories from miners and loggers. Six Akawaio and Pemon indigenous leaders from the Upper Mazaruni also, for the first time, filed a lawsuit in the High Court of Guyana asserting their land rights. "Our communities have been requesting title to these lands, which we know to be ours, since the Amerindian Lands Commission visited our communities in 1967. Since then we have attempted to discuss this matter on many occasions without result" read a statement presented to the High Court ("Indigenous Peoples Fight" 1998).

In addition to Canadian gold-mining companies, several Asian logging companies also operate on Native lands in Guyana. Like many other indigenous peoples the world over, the Akawaio Nation of Guyana (in South America) is working to gain recognition of its traditional land base before a "timber frenzy" strips it bare.

The Guyanese government in 1997 awarded the Canadian Buchanan Group a 1.5-million acre "exploratory lease" in the Middle Mazaruni region of Guyana, the traditional homeland of the Akawaio. Buchanan proposed a complete cultural and economic makeover for the Akawaio—log the forest that sustains their traditional economy and put the Native people to work in its mills, until the resource has been exhausted. "When the timber is gone," wrote Mark Westlund in *Native Americas*, "the company will move on, leaving the ecosystem devastated, and the Akawaio without their traditional way of life" (Westlund 1997, 11). The Guyanese government is accelerating the logging with tax breaks. The government charges royalties, taxes, and fees that are as low as a tenth of what is charged for similar concessions in Africa and Asia.

The Akawaio hope to stall logging by gaining new titles to their lands. Their titles as now written are inadequate because they allow the government to assert ownership of subsurface minerals, a situation that has allowed indigenous lands to be inundated by hundreds of small-scale miners, most of whom are moving from exhausted deposits (usually of gold) in Brazil.

Further Reading

Chatterjee, Pratap. "Gold, Greed & Genocide in the Americas California to the Amazon." *Abya Yala News: Journal of the South and Meso American Indian Rights Center (SAIIC)* 11, no. 1 (1998): 6–9. Accessed November 28, 2018. https://saiic.nativeweb.org /ayn/document/1527.

"Guyana: Transnational Mining Companies' Impacts on People and the Environment." 2001. World Rainforest Movement *Bulletin* 43 (February 13). Accessed November 28, 2018. https://wrm.org.uy/articles-from-the-wrm-bulletin/section1/guyana-transnational -mining-companies-impacts-on-people-and-the-environment/.

"Indigenous Peoples Fight for Territorial Rights in Guyana." 1998. World Rainforest Movement *Bulletin* 17 (November 27). Accessed November 28, 2018. https:// wrm.org.uy/articles-from-the-wrm-bulletin/section1/indigenous-peoples-fight -for-territorial-rights-in-guyana/.

"Isseneru Villagers Face Risk of Mercury Contamination." 2001. *Stabroek News*, July 19.

"Little Progress in the Recognition and Demarcation of Indigenous Lands, in Guyana." 1997. *Native Americas* 14, no. 1 (Spring): 11–12.

Westlund, Mark. 1997. "Akawaio Nation in Cross-Hairs of Timber Frenzy." *Native Americas* 14, no. 1 (Spring): 11.

"World Bank Quietly Insures Major Polluters." 1996. *Native Americas* 13, no. 1 (Spring): 4–5.

INDIA

Indigenous Peoples Displaced by Dams

With more than 300 million of its 1.2 billion people lacking electricity and those who have access to the grid using more of it for air-conditioning and other appliances, India has undertaken a crash program to increase power generation by any means possible, from coal, to solar, to hydroelectric dams. Many of India's available hydroelectric sites occupy areas that require eviction of indigenous peoples. For decades, protests of such evictions have been increasing as population increases across India.

Indigenous peoples accounted for one-third of those evicted from their homes—a total number estimated at about 65 million between 1950 and 2005, according to the Internal Displacement Monitoring Centre of Geneva, Switzerland—for development projects such as dams, highways, mines, power plants, and airports. Fewer than one-fifth of them were resettled (Mohanty 2017). "Relocation," the

Indian social activist Medha Patkar addresses a rallying crowd in New Delhi during 2006. Patkar had been on a hunger strike to protest the Sardar Sarovar dam in the central Indian state of Madhya Pradesh. (Manpreet Romana/Stringer/Getty Images)

euphemism most commonly used for such evictions, involves more than simply moving a group of people off of lands that many of them have occupied and used for centuries. It uproots cultures and languages as well as individual peoples. Eviction also destroys employment and collective identity.

During the last half century, India has built about 3,300 large dams, and "many of them have led to large-scale forced eviction of vulnerable groups. The situation of the tribal people is of special concern, as they constitute 40 to 50 percent of the displaced population. The brutality of displacement due to the building of dams was highlighted dramatically during the agitation over the Sardar Sarovar Dam in Gujarat" (Salve 2014).

The Sardar Sarovar Dam

India began aggressively constructing new hydroelectric dams in about 1980, with aid from the World Bank, the World Trade Organization, and several countries, including the United States. Indigenous peoples facing eviction from their lands are Adivasi living on the site of the Sardar Sarovar dam, which was under construction in the Gujarat state of western India, near the village of Vadgam. "Adivasi" is used to describe tribal peoples who have lived in India since before the Aryans migrated into the subcontinent from the north during the second millennium BCE. Adivasi live not only in India, but also in neighboring countries of South

Asia. A large percentage of Nepal's population is also Adivasi, who comprised 8.6 percent of India's population—104 million people, according to its 2011 census.

The Sardar Sarovar dam is the centerpiece of a development project in the Narmada Valley costing several billions of dollars and involving several dams, reservoirs, and canals that provide power for parts of three states. Construction on the project, begun in the middle 1980s, was finished in 2006. By the time it was finished, this system of dams had displaced about 320,000 people.

As Thakkar Himanshu wrote in *Cultural Survival*, "The dam alone displaced more than 41,000 families (over 200,000 people) in the three states of Gujarat, Maharashtra, and Madhya Pradesh. Over 56 percent of the people affected by the dam [were] Adivasis" (Himanshu 1999). People affected by the project organized under the banner of the Narmada Bachao Andolan (NBA) in a Save the Narmada Movement, "making their last stand supported by thousands of people taking part in a Gandhian-style 'satyagraha' or passive resistance at the [Narmada] dam site" (Raj 1999).

India moved the Adivasis after a 10-year bureaucratic process known as the Narmada Waters Dispute Tribunal Award of 1979 that was meant to resolve disputes related to the use of water in the area, but, as Himanshu wrote, this process

> had no time for consultation with the affected people. Much worse, the five large volumes of the tribunal transcripts do not contain the word Adivasi or any reference to them, although some of the stipulations of the tribunal are incidentally favorable to Adivasis. For example, the tribunal has stipulated that the affected people must be resettled as a community. (Himanshu 1999)

Relocation Sites Unsuitable for Use

Land given the indigenous peoples was too small for the 7,500 families who were forced to move and was unsuitable for cultivation. The sites set aside for the "community" had no access to clean drinking water or sanitation facilities, nor available grazing land, fodder, or firewood. People died for lack of food and water. The Adivasi were never compensated for loss of their land and economic base. The people also were not settled as a community. Even families were split up. "Even fathers and sons have been given lands in distant places. There have been scores of cases where brothers have been given lands in places a long distance away from each other," wrote Himanshu (1999).

"Some of the sites," according to one observer, "are on dry riverbeds which turn into fast-flowing drifts during the monsoons. 'When the waters recede they leave ruin. Malaria, diarrhea, sick cattle stranded in the slush'" (Raj 1999).

> Many of the resettled are people who have lived all their lives deep in the forest with virtually no contact with money and the modern world. "Suddenly they find themselves left with the option of starving to death or walking several kilometers to the nearest town, sitting in the marketplace (both men and women) offering themselves as wage labor, like goods for sale." (Raj 1999)

The Adivasis went back to court, and in 2017, Madhya Pradesh state was ordered by India's Supreme Court to compensate 681 Adivasi families whose members had been displaced by the Sardar Sarovar dam 6 million rupees each (about $90,000). Another 1,358 families won compensation of 1.5 million rupees each (about $22,500). "You have been struggling for compensation for 38 years. We are giving it to you in one shot," Chief Justice J. S. Khehar told counsel for rights group Narmada Bachao Andolan (NBA), or Save the Narmada Movement, which had filed the petition (Mohanty 2017). They were only a very small part of the human evictions provoked by the project. The families who won compensation had for the most part refused to leave their lands. Many others who did move found the new land unsatisfactory but received nothing.

The Scope of Human Devastation

Activist Medha Patkar, who spearheaded the NBA, has in the past highlighted how building dams led to dislocation of tribal societies. Official figures indicate that about 42,000 families were displaced due to the Sardar Sarovar dam but nongovernmental organizations such as NBA asserted that the figure was 85,000 families or 200,000 people. By the time it was finished, the entire Narmada Valley Development Project "is expected to have affected the lives of 25 million people," according to the Indian business magazine *IndiaSpend* (Salve 2014). Other projects have impacted indigenous peoples similarly, as described by Prachi Salve in *IndiaSpend*:

> A similar example is the Tehri project, a multipurpose irrigation and power project in the Ganges valley, 250 km north of Delhi, located in the Tehri-Garhwal district of Uttaranchal. A working group for the environment appraisal of Tehri dam put the figure of expected internal displacement at 85,600 persons. (Salve 2014)

The biggest impact has been in states with high proportions of indigenous peoples—Maharashtra, Andhra Pradesh, and Madhya Pradesh. More than three-quarters of people who were displaced came from these three states. India has laws that are supposed to guarantee compensation and resettlement for people who are displaced by development projects of all types, indigenous and otherwise. The 2007 National Rehabilitation & Resettlement Policy has been replaced by the Right to Fair Compensation and Transparency in Land Acquisition, Rehabilitation, and Resettlement Bill, 2012 (formerly known as the Land Acquisition, Rehabilitation, and Resettlement Bill, 2011). The bill was passed on August 29, 2013 in the *Lok Sabha* (House of the People, Parliament's lower house) and on September 5, 2013 in the *Rajya Sabha* (Council of States, the upper house). The laws "attempt to integrate the process of land acquisition with rehabilitation and resettlement [to] bring transparency to the entire process that will put to an end to forcible land acquisitions. The act is expected to provide fair compensation and rehabilitation and resettlement to farmers, land-owners, and livelihood losers" (Salve 2014).

Protests of Dam Construction Spread

In 2012, protests also broke out against the Lower Subansiri Dam, a 2,000-megawatt hydroelectric project for the National Hydroelectric Power Corporation, near the border of Assam and Arunachal Pradesh. A farmer's organization (Krishak Mukti) that advocates for landless Adivasi, protested that the dam would increase the frequency of floods in the area. Protesters clashed with police and brought work on the dam nearly to a standstill after police opened fire, injuring several people.

"Dissent is also growing over the proposed 1,750-megawatt Demwe Lower Hydroelectric Project, positioned barely 800 meters from Parshuram Kund, a sacred Hindu site on the Lohit River in Arunachal Pradesh" ("India: Adivasis" 2012). Reports said that the 13,000-crore rupee (US$2.6 billion) project will likely involve the felling of more than 43,000 trees and threaten endangered wildlife species including the Bengal Florican and the Ganges River Dolphin. The project will destroy forests that Mishmi tribes rely on for their traditional livelihood practices, such as jhum cultivation, and involve the eviction of people from the Riverine islands of Lohit River and also from the settlements along the Dibru Saikhowa National Park ("India: Adivasis" 2012). Protests also were breaking out against other dams. For example:

> The forum's spokesperson Vijay Taram said: "In the belts inhabited by the Adi tribe [which has a population of over 150,000], 43 massive dams are coming up. We are on the verge of being annihilated by all these developmental activities. Our language, forest, rivers, culture, tradition and identity will perish. This land belonged to our forefathers, and today we are being asked to vacate our land. The compensation offered is also meagre—just 1.5 lakh rupees [US$3,000] per hectare." ("India: Adivasis" 2012)

"A huge percentage of the displaced are tribal [people]," wrote Arundhati Roy in the *Mail & Guardian* of Johannesburg, South Africa (1999).

> Include Dalits (formerly known as untouchables), and the figure becomes obscene. According to the commissioner for scheduled castes and tribes, it's about 60 percent. If you consider that tribal people account for only 8 percent, and Dalits 15 percent, of India's population, it opens up a whole other dimension to the story. . . . India's poorest people are subsidizing the lifestyles of her richest. (Roy 1999)

Roy continued:

> What has happened to all these millions? Where are they now? Nobody really knows. They don't exist anymore. When history is written, they won't be in it. Some of them have subsequently been displaced three and four times—a dam, an artillery range, another dam, a uranium mine, a power project. Once they start rolling, there's no resting place. The great majority is eventually absorbed into slums on the periphery of [India's] great cities, where it coalesces into an immense pool of cheap construction labor (who build more projects that displace more people). (Roy 1999)

The people removed from the dam sites are among India's 300 million Dalit, many of them also members of indigenous tribes.

Further Reading

Himanshu, Thakkar. 1999. "Displacement and Development: Construction of the Sardar Dam." Cultural Survival, September. Accessed October 2, 2018. https://www.cultural survival.org/publications/cultural-survival-quarterly/displacement-and-development -construction-sardar-dam.

"India: Adivasis Fight Mega-Dams." 2012. *Minority Voices* (India), February 12. Accessed October 2, 2018. http://www.minorityvoices.org/news.php/en/1136/india -adivasis-fight-mega-dams.

Mohanty, Suchitra. 2017. "Indian Families Uprooted by Dam Win Compensation After Decades-Long Battle." *Reuters*, February 10. Accessed October 2, 2018. http://www .reuters.com/article/india-landrights-court-idUSL5N1FV1LE.

Raj, Ranjit Dev. 1999. "Last Stand of Tribals Displaced by Narmada Dam." Inter Press Service, June 8. Accessed October 2, 2018. http://www.ipsnews.net/1999/06/en vironment-india-last-stand-of-tribals-displaced-by-narmada-dam/.

Roy, Arundhati. 1999. "Lies, Dam Lies and Statistics." *Mail & Guardian*, June 18. Accessed October 2, 2018. https://mg.co.za/article/1999-06-18-lies-dam-lies-and -statistics.

Salve, Prachi. 2014. "Tribals Account for a Third of Communities Displaced by Large Projects." *IndiaSpend*, June 17. Accessed October 2, 2018. http://www.indiaspend.com/cover -story/tribals-account-for-a-third-of-communities-displaced-by-large-projects-11821.

MEXICO

The Maya, Deforestation, and a Golf Course

The uprising in Chiapas by Mayas under the Zapatista banner has notable environmental roots, including protests of deforestation provoked by road and dam construction and exploitation of oil resources that have been opening their homelands to intrusion by outsiders, including the Mexican national state and its armed forces.

Several thousands of indigenous Maya and environmental activists forced the Mexican government to stop construction of a highway in Chiapas during mid-2000 because, they asserted, it would be used to militarize the area and ease exploitation of oil resources under the aegis of improving infrastructure. Mexican and U.S. oil interests have long known of significant oil reserves in the Lacandon jungle, at the center of the Zapatista rebellion.

Zapatistas Resist Oil Exploitation

The Zapatista Army of National Liberation asserted that Mexico (with United States assistance) was increasing military presence in an area believed to be rich in oil. "With the highways have come the war tanks, the cannons, soldiers, prostitution, venereal diseases, alcoholism, rapes of indigenous women and children, death, and misery," said Subcomandante Marcos, speaking for the Zapatista movement ("Militarization and Oil" 1999).

Highway construction was halted as 2,000 Maya marched into Altamirano and another 4,000 protested in Ocosingo. According to a report in *Drillbits and Tailings*, protesters and Mexican soldiers skirmished in the remote village of San Jose La Esperanza. About 5,000 people also rallied against the road and militarization in San Cristobal de las Casas. The proposed highway would have wound through the highlands downhill into the jungle lowlands near La Realidad, the capital of the Zapatistas' symbolic homeland, 125 miles south of San Cristobal. One of the reasons for road-building has been access to oil resources in the Mayas' homelands. According to *Drillbits and Tailings*, "Various reports in the [Mexican] national press, the *Oil and Gas Journal*, the U.S. Geological Survey, and the U.S. General Accounting Office have reported important petroleum reserves in the Lacandon Jungle" ("Militarization and Oil" 1999).

The Zapatistas oppose oil development in the Lacandon area because of problems afflicting the activities of Pemex (Mexico's national oil company) in nearby Tabasco, where 20 years of oil production have caused "abnormal population growth, badly skewed income distribution, tremendous escalation of the cost of living, forced relocations, and environmental destruction and extremely hazardous living conditions for people who reside in petroleum-producing areas" ("Militarization and Oil" 1999).

Refugees Intrude on Mayas

The Mayas also are resisting intrusion of refugees from other parts of Mexico. The Lacandon rain forest of Chiapas has become a dumping ground used by Mexican authorities for refugees who have been relocated from other areas, where formerly self-sufficient indigenous *campesinos* have been assigned to prefabricated housing, "fighting off malaria and the jungle, coaxing maize from the thin soil" (Weinberg 1994, 43). Expansion of the cattle-raising industry throughout Los Altos ("the highlands") since the 1940s has forced thousands of indigenous *campesinos* from their homelands in Los Altos.

Yet another major taproot of rebellion in Chiapas has been expansion of hydroelectric power, which has created a new class of *expulsados* (refugees). Most of the power is not used in the local area, where only a third of homes are hooked into the electrical grid. The power is exported northward, to urban Mexico City and *malquiladora* factories near the U.S.-Mexico border. "The massive expansion of hydroelectricity in Chiapas during the 1980s also sent waves of Maya refugees into *la selva*," wrote Bill Weinberg in *Native Americas* magazine. "The flooding of fertile Indian farmland in highland valleys was paid for with the slash-and-burn colonization of the Lacandon rain forest" (Weinberg 1994, 43). The dams flooded 500,000 acres of the most fertile farmland in Chiapas. With projects often underwritten by the World Bank, Mexico is poised to develop even more hydropower, including the Usumacinta River complex, near the border of Guatemala, which would enable the central government to sell power to Central America.

Although it is dwarfed in size by forests in the Amazon Basin and parts of Africa and Asia, deforestation in the Selva Lacandona is proportionally among the most severe in North and Central America. Experts estimate that Mexico annually loses a forested area of about 2,300 square miles, an area roughly the size of Delaware. In 1900, forests covered an estimated 77,000 square miles of the country. Today, the figure is one-tenth that amount (Althaus 2001, 6).

The Mexican Environmental Enforcement Agency during 2001 warned of a "forestry collapse" across the country and pointed to the Selva Lacandona as one of the most threatened areas. Deforestation has been stripping the countryside of some of the most biodiverse habitations on Earth (Althaus 2001, 6). The Selva Lacandona alone is home to nearly 43,000 distinct species of plants, animals, and insects. Some scientists have estimated that a single 2.5-acre patch of the forest holds 20 species of animals, 40 of birds, 30 of trees, and 5,000 of insects (Althaus 2001, 6). Animals threatened by the Selva Lacandona's deforestation include the jaguar, ocelot, howler monkey, spider monkey, and various types of tapir.

Furthermore, wrote Dudley Althaus in the *Houston Chronicle*, "At the same time, soil erosion produced by clearing the forest may accelerate life-choking sedimentation levels in the Usumacinta River system, the world's seventh largest, into which the Selva Lacandona's rivers and streams feed. The estuaries where the Usumacinta flows into the Gulf of Mexico serve as breeding grounds for economically important fish, shrimp and other sea life" (Althaus 2001).

The Tepoztlán Golf Course "Water War"

The State of Morelos (south of Mexico City) is the legendary birthplace of the Mexica (Aztec) god Quetzalcoatl, as well as Emiliano Zapata, the Mexican revolutionary. More than seven decades after his assassination, Zapata's image became an icon for indigenous residents of Tepoztlán, a village of 28,000, as they rose in rebellion against a proposal to construct a golf course that would have used more water than everyone else in the town combined. Given the intensity of local indigenous opposition, the project eventually was suspended.

Morelos is a generally dry place, so the main environmental line of conflict involved usage of water. An environmental assessment disclosed that the golf course would have required 5 million gallons of water per day, while the entire town consumed 3 million gallons daily (Weinberg 1996, 37). The 18-hole golf course was to have been developed as a centerpiece for 800 luxury homes, a heliport, and a "data center and business park" (Weinberg 1996, 34). Plans called for golf notable Jack Nicklaus to design the course.

Morelos is regarded by many Mexicans as their nation's spiritual center because it is the ancient home of the Tlahuicas, who "were instrumental in the rise of the high culture of the valley of Mexico just north over the Ajusco mountains" (Weinberg 1996, 37). The Tlahuicas called their home Cuauhnahuac—"Land of Trees." Hispanicized, the name became that of the state capital, Cuernavaca.

The area also is known for its agricultural fertility but, like the valley of Mexico, it has become progressively drier over the years, at least in part because of deforestation caused by the expansion of urban areas and sugar plantations. Streams that once ran year-round now often cascade down from the hills only during the summer rainy season. "If we don't protect nature, we will die," said Lazaro Rodriguez, a Tepozteco. "If there are no trees, there is no rain, and the land dies" (Weinberg 1996, 41).

When Mexico City elites targeted their town for development, the Tepoztecos rose en masse against authorities, costing one man his life and prompting the imprisonment of several other people. The encroachment of farms and ranches upon the communal lands of the village's native people played a role in sparking the rebellion of Zapata in 1910. When the golf course was first proposed in 1995, local people also expressed concern that the heavy use of pesticides required to maintain the greens and fairways would pollute the local water table.

Golf courses are heavy consumers of pesticides and herbicides, averaging five to ten times the amount, per acre, used in agriculture (Weinberg 1996, 35). At one point, local people occupied a luxury home built for Guillermo Occelli, a brother-in-law of former Mexican president Carlos Salinas. They hung a banner on the house reading (in Spanish): "Tepoztlan: Communal Land. House of the People" (Weinberg 1996, 39).

Further Reading

Althaus, Dudley. 2001. "Upsetting the Balance: Nature's Way." *Houston Chronicle,* September 27: 6.

"Militarization and Oil in Chiapas, Mexico." 1999. *Drillbits and Tailings* 4, no. 16 (October 8).

Weinberg, Bill. 1994. "Flooding the Jungle." *Native Americas* 11, no. 2 (Summer): 43.

Weinberg, Bill. 1996. "The Golf War of Tepoztlan." *Native Americas* 13, no. 3 (Fall): 32–42.

PERU

Protests of Gold-Mining Pollution

Opposition to natural-resource development (principally oil extraction and gold mining) has sparked major civil unrest in Peru by peoples who have experienced previous environmental devastation, notably from Yanacocha, Latin America's largest gold mine. Major problems have included mercury poisoning of local water supplies. Elsewhere in Peru, activists have organized to demand cessation of air pollution from a lead smelter and have resisted the extraction of oil and natural gas in the Peruvian Amazon, where drug-taking tourists also have posed problems.

Protesters Block a Highway

Hundreds of protesters during 2001 blocked a major highway in northern Peru, alleging that local water supplies in the province of Cajamarca, 530 miles (850 kilometers) northeast of Lima, had been contaminated by toxic mercury from Yanacocha. Denver-based Newmont Mining Corporation, one of North America's biggest gold miners, holds majority control of Yanacocha. Along with Newmont, the Minera Yanacocha joint-venture includes Condesa, a subsidiary of Peruvian Minas Buenaventura, funded by the International Finance Corporation of the World Bank Group.

The protesters demanded that mining be halted pending investigation of water purity. "We reject the environmental contamination from Yanacocha," student leader Jorge Malca told Canal N cable television ("Water Supplies" 2001). Energy and Mines minister Jaime Quijandria told RPP radio that it was "simply and totally impossible" for the water to have been contaminated with mercury ("Water Supplies" 2001). Suspicious fish, cattle, and human deaths also have been reported among the people of Cajamarca, Peru, where Newmont is using cyanide to extract gold from ore (Chatterjee 1998).

In less than ten years, a rural agricultural and dairy-producing region in northern Peru has been overwhelmed by a multinational mining operation whose four open-pit gold mines are the most profitable in all of South America. Spread across 25,000 hectares (63,000 acres) of mountaintops, Yanacocha became the world's second-largest gold mine, and has been growing steadily. The joint-venture company owns mineral rights to an additional 125,000 hectares, including Mount Quilish, the main source of potable and agricultural water for the city of Cajamarca's 130,000 people, as well as 300,000 more people in nearby areas (Global Response 2001).

At the mine sites, huge piles of low-grade ore are soaked in a toxic cyanide solution that leaches out gold and silver. Although Yanacocha managers claim cyanide and other toxic metals cannot escape the mine site into the local watershed, mining expert Robert Moran said, "All the sites I've ever worked at experience some degree of leakage" (Global Response 2001). Mine contamination already has resulted in three major fish kills in area rivers and trout farms.

Local Residents Demand Reparations

Local residents demanded reparations from the World Bank for mercury poisoning that affected as many as 300 villagers 375 miles north of Lima on June 2, 2000. The contamination occurred after a truck carrying 330 pounds of mercury to Lima spilled it over a 27-kilometer (16-mile) portion of a road near Choropampa, 53 miles southwest of the mine. Mercury, a by-product of the gold mine, is routinely trucked to Lima for use in medical instrumentation and other applications ("Mercury Spill" 2000).

About 330 pounds of the poisonous liquid leaked from the truck, after which villagers gathered the poisonous substance. Some believed it still might contain gold or have value for other reasons. Others thought the mercury might have medicinal uses, and still others were simply curious. Many kept some of the mercury in their homes until it made many children ill ("Mercury Spill" 2000).

The spill eventually sickened more than 400 people, many of them children, contaminated 80 homes, and cost the company $12 to $14 million during a year-long cleanup. A 38-year-old woman was flown to Lima in a coma and placed in intensive care. She was examined by a critical-care specialist flown in by Newmont. According to the International Finance Corporation, Minera Yanacocha and Newmont sent medical and toxicology experts to the scene and provided mercury testing for local residents in their homes. The World Bank undertook an investigation of the incident and issued a report accusing the company of transporting the mercury without appropriate safety measures, in violation of international standards governing the handling of hazardous substances ("Mining in Peru," n.d.).

"This spill is just one more disaster brought by the mine. Many people are sick because they weren't told what the mercury was after it spilled. There were no safety precautions in place," said Segunda Castrejon, president of the Rondas Campesinas Femeninas del Norte del Peru ("Mercury Spill" 2000). On June 23, 2000, Indigenous Peoples' Day in Peru, hundreds of people marched in the city of Cajamarca protesting the mine and its impacts on the indigenous peoples who live near it. Local people demanded compensation for the families of those affected, as well as closure of the mine. "The people are demanding compensation for this. The company is getting rich off this mine while the local people suffer the impacts," said Julio Marin, speaking for Rondas Campesinas.

La Oroya's Environmental Problems

People living near the Peruvian village of La Oroya face environmental problems similar to those living near the Yanacocha gold mine: pollution of the water supply from cyanide, air pollution, and noisy truck traffic near the mine. The pollution destroys pasturelands and sickens livestock, debilitating local agriculture. Ore smelters and refineries foul the air with sulfur dioxide and destroy pasturelands needed for livestock.

The effects of a lead smelter in La Oroya owned by the United States-based company Doe Run have been endured by the community for generations. Nearby rivers contain levels of lead, iron, zinc, copper, and arsenic that exceed the limits for environmental health set by Peru's governmental agencies, according to a September 2000 study by the environmental organization CooperAcción, which also studied air quality in the area, and found concentrations of lead in the air were 800 percent above acceptable levels. The effect on people in the area is serious—local residents have high levels of lead in their blood. Jose de Echave, the deputy director of CooperAcción noted that its recent study of lead poisoning near La Oroya found that 90 percent of the people tested were far above the acceptable lead levels set by the U.S. federal government ("Mining in Peru," n.d.).

"Mining and Communities: Oral and Written Testimonies," published by CooperAcción, describes some of the environmental damage. One local farmer, who grew up only 300 meters from the smelter at La Oroya, said that the fumes routinely burned his throat and nose. Another local inhabitant, who said that the last good harvest he could remember was in 1919 or 1920, blamed the smelter for ruining agriculture in the area. "The smoke fell like a snowfall of arsenic dust on the land, rocks, and pasture. The animals got sick, it was a disaster. How could we live there? There was no harvest, and the animals died, like that, in groups. It was as though they were poisoned" ("Mining in Peru," n.d.).

Marches in Madre de Dios

Indigenous peoples from the Peruvian state of Madre de Dios marched in the city of Puerto Maldonado on July 18, 2000 to demand that the government deny further mining and logging concessions within indigenous territories. Meanwhile, Exxon-Mobil, which staked a claim in the nearby Candamo Valley in 1996, announced that it will continue exploration for oil and gas in the area. The valley is a complex ecosystem that is home to at least 20 isolated indigenous communities, including the Yoro, Ese'eje, Mascho-Piro, and Amahuaca. Scientists consider the area to be "a complete Amazon in miniature" because of its biodiversity and abundant plant and animal life ("Indigenous Peruvians" 2000).

"By all Peruvians our state is called 'the biodiversity capital of Peru' for its biological richness. It is also well known that this is a land of nearly 20 indigenous peoples, a cultural diversity that has a hand in the biological diversity. What is more, we believe that it is this cultural diversity that guarantees the biological diversity," said a manifesto written in defense of the indigenous peoples of Madre de Dios ("Indigenous Peruvians" 2000).

The indigenous peoples of Peru are facing increasing pressure from international investment in mining, oil, and natural gas, industries that impinge on their traditional lands, cultures, and ways of life. Mobil, which has since merged with Exxon, leased 3.7 million acres (1.5 million hectares) of rain forest in the Peruvian state of Madre de Dios. The area, which is known in the industry as Block 78, includes the 350,000-acre (141,600-hectare) Candamo Valley. In April 2000, ExxonMobil began a second stage of exploration in the well once owned by Occidental

---

Cholera's Climatic Connection

In January 1991, an epidemic of cholera began in Chancay, on Peru's coast, near Lima. It was the first time that cholera had appeared in the Western Hemisphere in more than 100 years and was the beginning of an epidemic that within 15 months would afflict 500,000 people, killing almost 500. The epidemic was researched by Rita Colwell, who was looking for the vector that was spreading the disease as part of her PhD dissertation research at the University of Washington. Colwell was the first to associate the spread of cholera with the onset of warming water provoked by El Niño conditions on the Pacific Coast of South America (and, by implication, the disease's presence in warmer water worldwide). She established a relationship between cholera and seaborne plankton. "In one fell swoop, Colwell had solved one of the great mysteries of this ancient disease," wrote Paul Epstein and Dan Ferber in *Changing Planet, Changing Health* (2011).

Further Reading

Epstein, Paul R., and Dan Ferber. 2011. *Changing Planet, Changing Health: How the Climate Crisis Threatens Our Health and What We Can Do About It.* Berkeley: University of California Press.

Petroleum. The study will take six months to complete, during which the company will decide whether to continue exploration ("Indigenous Peruvians" 2000).

According to the manifesto:

> One question that we must ask the people of Madre de Dios, independent of whether they are indigenous river dwellers, colonists who come here to live and coexist in this rich biological resource, or the people and enterprises that have adapted their economic activities to the conditions of biodiversity, is this: looking at sustainable development, is it possible to stop the current reality, where only the interests of the loggers decide the future of our state, people who only see the trees, but not the population that lives with a wealth of natural offerings such as ours? ("Indigenous Peruvians" 2000)

Further Reading

Chatterjee, Pratap. "Gold, Greed & Genocide in the Americas California to the Amazon." *Abya Yala News: Journal of the South and Meso American Indian Rights Center (SAIIC)* 11, no. 1 (1998): 6–9. Accessed November 28, 2018. https://saiic.nativeweb .org/ayn/document/1527.

Global Response: Environmental Education and Action Network. 2001. October.

"Indigenous Peruvians Mobilize While ExxonMobil Further Explores Rainforest." 2000. *Drillbits and Tailings* 5, no. 12 (July 20).

"Mercury Spill Poisons Villagers Near the Yanacocha Mine in Peru." 2000. *Drillbits and Tailings* 5, no. 11 (June 30).

"Mining in Peru." n.d. Oxfam America. Global Programs: South America.

"Water Supplies Alleged Contaminated by Peru Gold Mine." 2001. *Reuters*, September 28.

Wyss, Jim. 2018. "The Pope Is Visiting a Peruvian Town 'Absolutely Destroying Itself' for Gold." *Miami Herald*, January 22. Accessed October 2, 2018. https://www.miamiher-ald.com/news/nation-world/world/americas/article195206009.html#storylink=cpy

RUSSIA

Oil and Reindeer Don't Mix

The Khanty and affiliated Evenk have inhabited northwestern Siberia's forests and swamps for thousands of years, maintaining their traditional ways of life, which include hunting, fishing, and herding reindeer. Living in "chums," which are similar to North American Indian tipis, the Khanty inhabit land in the West Siberian Taiga (boreal forest), north of the Middle Ob on the river Pim. For centuries, reindeer breeders and fishers knew how to sustain themselves within this barren land, how to withstand the cold of wintertime and swarms of mosquitoes in the summer. Now their traditional way of life is in danger due to exploitation of fossil fuels ("Reindeer Herders" 1997).

Oil Drilling Versus Reindeer

Oil was found in the area first during the 1960s; by the late 1990s, according to one account, "derricks, roads, pipelines, and the workers' estates are eating their way through the land of the Khanty people" ("Reindeer Herders" 1997). The invasion of oil-drilling is antithetical to the reindeer culture. In the Khantys' homeland, oil spills blacken the wetlands, newly constructed roads trap water, causing flooding and ruining the forests, and fires caused by oil workers' carelessness send columns of smoke into the air.

Any student of the Plains Indians and the buffalo during the late 19th century in North America will recognize parallels in Vershina Khandy's description of changes in the land endured, nearly a century later, by Siberia's Evenk people, who are neighbors of the Khanty. The following sounds very much like the 19th-century changes across the U.S. Great Plains:

> The ancestral lands of the Vershina Khandy [Upper Khanda] Band of the Evenk were very strongly affected in the late 1970s and early 1980s by the construction of the Baikal-Amur Railroad. Migrations of wild ungulates from northern territories into the Khanda River Valley for wintering first reduced and then discontinued altogether, as the railroad [was] built without taking into account the ecological peculiarities of the territory. In addition, the construction project attracted numerous people (tens of times more numerous than the local population), who started hunting and fishing. As a result, the Evenk found it much more difficult to gain their daily bread, since all their incomes used to come from hunting and fishing. Then, in the 1980s, along the entire eastern boundary of the Evenk traditional territory, the KI-450 Correctional Labor Colony Administration cut the forest clean, and roads were built in areas previously inaccessible, ones that had served as reserves. As a result, there was the Baikal-Amur Railroad and actively developed areas along it on the north and the easily accessible territory cut and burned by forest fires on the east. (Khandy 2001)

Following these intrusions, only a 37-kilometer (23-mile) strip of the Khanda River Valley remained within the southern portion of the Upper Khanda Evenk Band's traditional lands, allowing fewer wild animals to pass freely to their winter feeding grounds, where the Evenk traditionally hunted them. When the RUSIA Petroleum Company built a road from the settlement of Magistral'nyy to its gas field, this migration route was cut off. Therefore, the band's traditional lands were totally isolated from the migration routes of their food sources (Khandy 2001).

Traditional Life May Be Lost

The RUSIA Petroleum Company allocated some compensation for the harm it had done to the Evenks' traditional economy, but the funds were sufficient to employ only two Evenks as professional hunters. The company's executive dealing with environmental matters, F. T. Selikov, said that the Evenk had been offered relocation at the expense of the company to other population centers of the district, but they had refused to move. They understood that by moving, their traditional way of life would be lost (along with ownership and use of the land that once supported it) as the Evenk were forced to assimilate into the general population.

The response of the indigenous people sounded much like that of some American Indians on the plains of North America a century earlier. They had, in fact, been reading indigenous American history. "As for the professional hunter salaries, the opinion is unanimous," read one statement composed by a Khanty spokesman. "It is not even an attempt to solve the problem, but glass beads of the 17th–18th-century merchants and industrialists in a modern interpretation" (Khandy 2001).

> If the condensate field development [of RUSIA Petroleum] begins reaching industrial proportions and no measures are taken, the Upper Khanda Band will cease to exist. Their southern neighbors, the Evenk band living in Vershina Tutury [Upper Tutura], Kachug District, Irkutsk Oblast, will not be left in peace either, if the field is developed and a gas pipeline is built. Being involved in public monitoring of the development of the Kovyktinskoye gas-condensate field as a member of the Baikal Environmental Wave, an Irkutsk regional nongovernmental organization, I am trying to help the Evenk, but I am afraid that this will not be enough and therefore am asking you to step in with whatever assistance is possible. (Khandy 2001)

Ecological Crisis Zones

The scope of Siberia's environmental problems was outlined by Alexei Yablokov in the *Washington Post*. The southern and central Volga regions, Bashkiria, the central and southern Urals, the Kuzbass, the oil-producing region of western Siberia, the Lake Ladoga basin, and many other regions are officially classified as ecological crisis zones. In the past several years, environmental refugees have appeared in Russia. "People have fled such heavily polluted areas as Prokopyevsk, Nizhny Tagil, Kirishi, and Angarsk, moving elsewhere in the nation in search of clear air to breathe and clean water to drink," Yablokov wrote. "Throughout the republic, according to 1989 analyses of fresh-water fish, 69 percent were extremely contaminated by mercury-based pesticides" (Yablokov 1991, C-3).

Oil and gas drilling in northern Siberia has destroyed large expanses of reindeer pasturage on which indigenous peoples built their traditional economies. Millions of barrels of oil are lost each year due to spills and other accidents related to drilling and transportation of petroleum. A team of Russian ecologists commented: "We drill more and more oil, destroy more and more wilderness areas, deprive the local peoples of all possibility of supporting themselves, then turn around and spend the dollars we make selling oil abroad to buy food for those very same people" (Yablokov 1991, C-3). The same group attributed much of Russia's environmental abuse to "the long supremacy of the totalitarian Soviet system with its innate hostility to humans and the world around them, [causing] three generations of citizens [to be] raised with a utilitarian, consumer approach to nature" (Yablokov 1991, C-3).

Oil Spills Ruin Rivers

In western Siberia, according to Survival International, oil and gas industries have polluted large areas of traditional indigenous lands. Huge flares burn off excess natural gas day and night. Oil also frequently spills into rivers, killing fish and plant life. Forests have been cut down, and reindeer pastures have been devastated by industrial vehicles. Many important fish-spawning grounds have been destroyed, as well. The Khantys, for example, watched in horror as their sacred spawning riverbeds were destroyed to mine gravel. The Evenks' and Yukagirs' lands also have been contaminated by radiation from failed nuclear tests ("Peoples of the Frozen North," n.d.).

Andrew Wiget and Olga Balalaeva, two researchers who described oil development's impacts on the Eastern Khanty, summed up the situation this way: "Siberia, like the America's Appalachian coal fields at the beginning of this century, has become a national sacrifice area" ("Putin's Oil" 2000). *Drillbits and Tailings* reported that the amount of oil spilled annually in Russia is equal to 350 accidents the size of the 1989 *Exxon Valdez* tanker disaster in Alaska, in which 40,000 tons were spilled ("Putin's Oil" 2000).

Intrusion of Pollution

The Khanty and other indigenous peoples of Siberia are very aware of the fact that their land is being despoiled to enrich other people. "If you take 100 pounds of my gold, then why can't you leave me just two?" Yaloki Nimperov wrote in a letter to Senur Markianovich Khuseinov, director of the Kamynskoye oil field that belongs to the Company Surgutneftegaz. Yaloki was so angry that he has threatened some of the oil workers with his gun. He also was angry because local buses that were being used to carry oil workers denied passage to Khanty until the indigenous people threatened to blow up a bridge. After that, the Khanty not only were allowed to ride the buses, but they were charged only half the regular price ("Reindeer Herders" 1997).

Between 1989 and 1997, 14 new production platforms and exploration bases were erected in the Khantys' indigenous lands, one of them only two kilometers from Nimperov's summer camp. Day and night, he could hear its noise. Roads to the production platforms were laid with little regard for the needs of the Khanty or the local environment on which they depend. Some roads were built through marshes and lakes. Oil workers sometimes tried to correct for this by cutting culverts to allow water exchange between parts of lakes, but the culverts were too small to do much good. Water in the culverts often froze during the winter, killing fish for lack of oxygen. When the earth thawed in the spring, the area smelled of rotten fish. Some Khanty were paid compensation while others withdrew in disgust, turning their backs on the oil platforms, moving as far as possible into what remained of the countryside, seeking shrinking pastures on which to feed their reindeer.

A Khanty elder described an earlier, more pristine, time:

> I don't want anything, only my land. Give me my land back where I can graze my reindeer, hunt game, and catch fish. Give me my land where my deer are not attacked by

stray dogs, where my hunting trails are not trampled down by poachers or fouled up by vehicles, where the rivers and lakes have no oil slicks. I want land where my home, my sanctuary and graveyard can remain inviolable. I want land where I [will] not be robbed of my clothes or boots in broad daylight. Give me my own land, not someone else's—just a tiny patch of my own land. ("Peoples of the Frozen North" 1998)

Alcoholism and Suicide

In the meantime, several once-vibrant tribes of reindeer herders across Siberia succumbed to alcoholism and suicide as their way of life was obliterated by oil exploitation. *Drillbits and Tailings* reported in its September 30, 2000, edition that Demitri, a 37-year-old reindeer herder, was one of only six out of 27 pupils in his class at school who had survived. With fewer reindeer to herd, and with hunting grounds limited and rivers polluted, two of his classmates, filled with despair, hanged themselves. According to this account, "The others died of alcohol-related incidents from drinking the vodka brought in by the oil workers" ("Putin's Oil" 2000).

"While British motorists complain about the price of petrol, the exploitation of oil is a matter of life or death to many Khanty people. If the people here [Great Britain] knew the true costs to the Khanty people of the petrol they put in their cars, they would put more energy into campaigning for a fairer deal for Siberia's tribes," wrote Stephen Corry, the Director of Survival International, a human-rights organization based in England ("Putin's Oil," 2000).

In 1994, local Russian administrators distributed parcels of land to Khanty who were living in traditional fashion. This land was not chosen because the owners' families had owned it or because it was suitable for herding reindeer. Often, the land was miles from traditional homes and ranges and entirely unsuited to traditional Khanty herding and hunting. Some of the allocated land had been destroyed by fire, denuded of moss that once made it useful for feeding animals. As a result, many Khanty moved to land they did not own (but which was still useful for herding and hunting), fearful that it could be turned into oil fields at any time ("A Reindeer Herder's Tale" 2000, 8).

Further Reading

Khandy, Vershina. 2001. "[Speaking for the] Upper Khanda Band of the Evenk." Cited in L'Auravetl'an: An Indigenous Information Center by Indigenous Peoples of Russia. United Nations Information Center, Moscow, April 27. http://www.indigenous.ru/eng lish/e_bull.htm.

Luhn, Alec. 2017. "The Reindeer Herder Struggling to Take on Oil Excavators in Siberia." *The Guardian,* March 17. Accessed October 3, 2018. https://www.theguardian.com /world/2017/mar/17/reindeer-herder-oil-excavators-siberia.

"Peoples of the Frozen North." 1998. Survival International. https://assets.survivalinterna tional.org/static/files/related_material/39_21_81_siberiabg.pdf.

"Putin's Oil Politics Threaten Siberia and Sakhalin's Indigenous Peoples." 2000. *Drillbits and Tailings* 5, no. 16 (September 30).

"A Reindeer Herder's Tale." *Survival Newsletter* 42(2000):8.

"Reindeer Herders under Siege by Oil Industry in Siberia." 1997. Institute for Ecology and Action Anthropology; Report from a Fact-finding Mission to Khanty-Mansi Autonomous Region, March 18. Accessed October 3, 2018. http://www.hartford-hwp.com /archives/56/003.html.

Yablokov, Alexei, Sviatoslav Zabelin, Mikhail Lemeshev, Svetlana Revina, Galina Flerova, and Maria Cherkasova. 1991. "Russia: Gasping for Breath, Choking in Waste, Dying Young." *Washington Post*, August 18: C-3.

UNITED STATES: NAVAJO NATION

Radioactive from the Inside Out

Uranium mined from Native American lands supplied a substantial proportion of the fuel for early nuclear power plants as well as the U.S. nuclear arsenal. The Navajos succeeded in stopping uranium mining and milling on their homeland only after several hundred people had died of its effects and many more had suffered the tortures of cancers that once were nearly unknown in their country. Following several decades of death in the mines, by about 2010 uranium mining and milling was outlawed on the Navajo Nation, as the people there heeded what tradition calls their "original instructions" to leave the soft, radioactive yellow rocks in the ground.

Navajo Cancers Accelerate

About half the recoverable uranium in the United States lies in New Mexico—and about half of that is beneath the Navajo Nation. The Navajo language has no word for "radioactivity." Initially, no one told the miners that within two or three decades, many of them would die of radiation-induced cancers. In their rush to profit from uranium mining (and the lives of the miners), very few mining companies provided ventilation in the early years. The effects of radiation were well known to scientists in the late 1940s, when the mines first opened. Some miners worked as many as 20 hours a day, entering the mines just after the blasting of sandstone had filled the mines with silica dust. Many mine owners didn't even provide toilet paper. When miners relieved themselves, they wiped with fists of radioactive "yellowcake"—uranium ore.

By 2014, 350 to 400 former Navajo underground uranium miners had died from maladies caused in large part by exposure to radiation, according to Chris Shuey, an environmental health researcher with the Southwest Research and Information Center (Knight 1999). Many more Native people also had died of a wide variety of malignant cancers, not usually from mining uranium themselves but because they lived with the "yellow dirt" in windswept waste (tailings) piles that have now blown into every crack and crevice, indoors and outdoors, for decades.

In addition to the mining and milling of nuclear fuel, New Mexico also supplied the United States with its first test site for the atomic bomb, detonated on July 16, 1945 on a 100-foot tower. The Trinity Site on today's U.S. Army's White Sands Missile Range is surrounded by two Apache tribes, several Navajo communities, and 19

Marvin Sarracino stands at the edge of a reclaimed uranium mine holding Laguna Pueblo pots made of clay from the mine. Many Native Americans who worked in or lived near uranium mines in New Mexico were kept ignorant about the dangers and died from radiation poisoning. (Bob Krist/Getty Images)

American Indian pueblos. Reports at the time by American Indians and government witnesses described a light ash that fell for days after the explosion, the effects of which were not investigated until 2014, when the National Cancer Institute started a study of radiation levels in New Mexico from that first test blast. With no prior evidence of effects, no one has been eligible for compensation under the Radiation Exposure Compensation Act, which covers other nuclear and uranium workers as well as "down-winders" who were exposed to radiation from later atomic tests (Lee 2014).

Radiation Pervades Navajo Life

Navajo uranium miners at first hauled radioactive uranium ore out of the earth as if it were coal or any other mineral. Some ate their lunches in the mines and slaked their thirst with radioactive water. Their families' homes sometimes were built of radioactive earth, and their neighbors' sheep may have watered in small ponds that formed at the mouths of abandoned uranium mines. On dry, windy days, the gritty dust from uranium waste-tailings piles covered everything in sight.

Kerr-McGee, the first corporation to mine uranium on Navajo Nation lands (beginning in 1948), found the reservation location extremely lucrative. There were no taxes at the time; no health, safety, or pollution regulations; and few other jobs for the many Navajos who had recently arrived home from service in World

War II. Labor was cheap, and uranium, in demand for stockpiles of nuclear armaments, was expensive.

The first uranium miners in the area, almost all of them Navajos, remember being sent into shallow tunnels within minutes after blasting. They loaded the radioactive ore into wheelbarrows and emerged from the mines spitting black mucus from the dust, coughing so hard that many of them experienced headaches. Such mining practices exposed the Navajos who worked for Kerr-McGee to between 100 and 1,000 times the level of radon gas later considered safe. Officials for the U.S. Public Health Service (PHS) estimated these levels of exposure after the fact; in the earliest days, no one was monitoring the Navajo miners' health.

Carrie Arnold wrote in *Environmental Health Perspectives*, published by the National Institutes of Health, that the miners and their families "were not told that the men who worked in the mines were breathing carcinogenic radon gas and showering in radioactive water, nor that the women washing their husbands' work clothes could spread radionuclides to the rest of the family's laundry" after they had worked in the 521 now-abandoned uranium mines on the reservation. The mines ranged in size from "dog holes" that could accommodate only a single man to large mines from which radioactive ore was extracted in carts on rails (Arnold 2014).

Arnold wrote that health workers were allowed to interview uranium miners only after they

> had to strike a Faustian bargain with the mining companies: They could not inform the miners of the potential health hazards of their work. Seeing it as the only way to convince government regulators to improve safety in the mines, the researchers accepted. The PHS monitored the health of more than 4,000 miners between 1954 and 1960 without telling them of the threat to their health. By 1965, the investigators reported an association between cumulative exposure to uranium and lung cancer among white miners and had definitively identified the cause as radiation exposure. (Arnold 2014)

The effects had been no secret, even early as 1950, when government workers monitored radiation levels in the mines that were as much as 750 times the limits deemed acceptable at that time.

More than 99 percent of the rock that the mines produced was waste, cast aside as tailings near mine sites after the uranium had been extracted. One of the mesa-like waste piles grew to be a mile long and 70 feet high. On windy days, dust from the tailings blew into local communities, filling the air and settling into water supplies. At the time, beginning during the 1950s, the Atomic Energy Commission (AEC) assured worried local residents that the dust was harmless.

The Uranium Mine Tailings' Legacy

When mining was initiated, no one considered environmentally appropriate ways to deal with its tailings piles. Even if the tailings were to be buried—a staggering task—radioactive pollution could leak into the surrounding water table. A 1976 Environmental Protection Agency (EPA) report found radioactive contamination of drinking water on the Navajo Reservation in the Grants, New Mexico, area near a uranium mining and milling facility (Eichstaedt 1994, 208).

Arnold wrote of the mine tailings' legacy:

> In a low, windswept rise at the southeastern edge of the Navajo Nation, Jackie Bell-Jefferson prepares to move her family from their home for a temporary stay that could last up to seven years. A mound of uranium-laden waste the size of several football fields, covered with a thin veneer of gravel, dominates the view from her front door. After many years of living next to the contamination and a litany of health problems she believes it caused, Bell-Jefferson and several other local families will have to vacate their homes for a third round of cleanup efforts by the U.S. Environmental Protection Agency. . . . Bell-Jefferson and her brother Peterson Bell played in and around the mines, splashing and swimming in pools of radioactive water that had been pumped out of the mines and then collected on their property. The contaminated water looked and tasted perfectly clean. Families used it for cooking, drinking, and cleaning. Hogans and corrals were built with mine wastes, as were roads. (Arnold 2014)

Uranium mine dust produced silicosis in the miners' lungs, in addition to lung cancer and other problems associated with exposure to radioactivity. By the 1960s, nearly 200 miners already had died of uranium-related causes. That number had doubled by 1990. Radioactivity also contaminated drinking water in parts of the Navajo reservation, producing birth defects and Down syndrome, both previously all but unknown among the Navajo.

The government knew of the risk at least by 1978, however, when the Department of Energy released a Nuclear Waste Management Task Force report disclosing that people living near the tailings piles ran at least twice the risk of lung cancer as the general population. Even then, the Coalition for Navajo Liberation was aware that a number of miners were dying of lung cancer. As Navajo miners continued to die, children who played in water that had flowed over or through abandoned mines and tailings piles came home with burning sores. Downwind of uranium processing mills, the dust from yellowcake sometimes was so thick that it stained the landscape half a mile away.

Radiation Pervades Everything

Dry winds blew dust from tailings piles through the streets of many Navajo communities. "We used to play in it," said Terry Yazzie of an enormous tailings pile behind his house. "We would dig holes and bury ourselves in it" (Eichstaedt 1994, 11). The neighbors of this particular tailings pile were not told it was dangerous until 1990, 22 years after the mill that produced the pile had closed and 12 years after Congress authorized the cleanup of uranium mill tailings in Navajo country. Abandoned mines also were used as shelter by animals that inhaled radon and drank contaminated water. Local people milked the animals and ate their contaminated meat.

Peter Eichstaedt wrote that miners watched members of their families die because of radiation poisoning that permeated their entire lives. Some miners were put to work packing thousand-pound barrels of yellowcake. Some of the miners ingested so much of the dust that it was "making the workers radioactive from the inside out" (Eichstaedt 1994, 11).

Through opposition to uranium mining among Indians and non-Indians alike runs a deep concern for the long-term poisoning of land, air, and water by low-level radiation. By the 1970s, these concerns had provoked demands from Indian and white groups for a moratorium on all uranium mining, exploration, and milling until the issues of untreated radioactive tailings and other waste-disposal problems were faced and solved. Doris Bunting of Citizens Against Nuclear Threats (a predominantly non-Native group that joined the efforts of the Coalition for Navajo Liberation (CNL) and the National Indian Youth Council in opposition to uranium mining) supplied data indicating that radium-bearing sediments had spread into the Colorado River basin, from which water is drawn for much of the Southwest. Thus, even though the mining and milling of uranium has stopped in Navajo country, the history of its deadly toll has yet to be completely experienced, nor told.

By 2017, more than 500 abandoned uranium mines required cleanup on the Navajo Nation at a cost of $1 billion or more (Mauk 2017, 59). Hundreds of other mine sites across the Colorado Plateau also needed remediation, all of them saturated with the same sort of radioactive waste that once was mixed with residential concrete in suburban basements at a time when uranium not only was mined in the area, but nuclear bombs were tested in the open air. "Radioactive ash from test sites had snowed across the [Navajo] reservation and Four Corners in the [nineteen] fifties and sixties" (Mauk 2017, 49). In Tuba City, on the reservation, one waste pile contains 2.25 million tons of mill waste, "which will exhale radon gas for a fair approximation [of] eternity" (Mauk 2017, 49). Thirty years into that eternity, radioactive waste is already leaking into the aquifer under Tuba City, which, like much of the area, is studded with abandoned mines and mill sites that are "swimming in poison." Waste in some of the abandoned mines registers as much as 700 times ambient background radiation (Mauk 2017, 56–57).

Further Reading

Arnold, Carrie. 2014. "Once upon a Mine: The Legacy of Uranium on the Navajo Nation." *Environmental Health Perspectives,* February 1. Accessed October 3, 2018. https://doi .org/10.1289/ehp.122-A44.

Eichstaedt, Peter. 1994. *If You Poison Us: Uranium and American Indians.* Santa Fe, NM: Red Crane Books.

Knight, Danielle. 1999. "Native Americans Denounce Toxic Legacy." Inter Press Service News Agency, June 14. Accessed November 28, 2018. http://www.ipsnews .net/1999/06/environment-health-native-americans-denounce-toxic-legacy/.

Lee, Tanya H. 2014. "H-Bomb Guinea Pigs! Natives Suffering Decades after New Mexico Tests." Indian Country Today Media Network, March 5. Accessed October 3, 2018. https://newsmaven.io/indiancountrytoday/archive/h-bomb-guinea-pigs-natives-suf fering-decades-after-new-mexico-tests-jpZAFe1gFEmRCGfiq42BDg/.

Mauk, Ben. 2017. "States of Decay: A Journey through America's Nuclear Heartland." *Harper's,* October: 48–59.

Chapter 6: Toxic Chemicals and Uranium

OVERVIEW

This chapter overlaps with "Indigenous Peoples," chapter 5, evoking memories of a scene following the negotiation of the Stockholm Convention, which outlaws most of the "Dirty Dozen" Persistent Organic Pollutants (POPs); Inuit activist Sheila Watt-Cloutier evoked tears from some delegates with her note of gratitude on behalf of Inuit people for enacting a ban on a number of persistent organic pollutants that contaminate the Inuit homeland in the Arctic. The treaty, she said, had "brought us an important step closer to fulfilling the basic human right of every person to live in a world free of toxic contamination. For Inuit and indigenous peoples, this means not only a healthy and secure environment, but also the survival of a people. For that I am grateful. *Nakurmiik.* Thank you" (Cone 1996, 2005, 202).

It's a long road back, however. Once they become part of the food chain, persistent organic pollutants are very difficult to dislodge. Following the negotiation of international law to outlaw these pollutants came the decades-long battle to enforce the ban and to rid the Arctic food chain of their effects. Persistent organic pollutants tend to become locked in the ecosystem, and as of this writing Inuit still consume traditional foods and breastfeed their babies with caution.

Travel nearly anywhere in North America and see that toxicity has become heritage. In Alaska, the Salt Chuck Mine, a source of copper, gold, palladium, and silver between 1916 and 1941, contaminated the Kasaan (Alaska) harvesting grounds for fish, clams, cockles, crab, and shrimp. For decades, the Native people were unaware that their harvests were saturated with effluvia from mine tailings. Even after the area was declared a Superfund site, Pure Nickel, Inc. sought to reactivate mining in 2012.

In many other places, however, contamination continues. Canada may be worse than the United States, with its emphasis on such "unconventional" oil sources as tar sands in Alberta, the mining of which affects many Native peoples. The Arctic, an environment that seems so pristine to the unknowing, naked eye is laced with so many persistent organic pollutants. Across the Arctic, a world based on ice is melting away as temperatures rise. Across North America, a struggle is underway involving Native peoples to protect a habitable environment, a vitally important issue for individual and community survival.

El Salvador Bans Mining to Protect Water

During March 2017, El Salvador became the first country to explicitly ban all forms of mining to preserve its dwindling supply of fresh water. With the country's more than 6.5 million people packed into 8,125 square miles—smaller than Massachusetts, at 10,565 square miles—the pollution of water (especially from cyanide leaching used to extract gold from rocks) has become acute enough to justify a ban on income-producing mining that is popular with urban residents and farmers, as well as the Catholic Church. El Salvador is already the second-most environmentally degraded nation in the Western Hemisphere (after Haiti), according to the United Nations.

Mining was banned by an overwhelming majority of El Salvador's legislators. "It's a wonderful moment for the first country to evaluate the costs and benefits of metallic mining and say no," said Andrés McKinley, a mining and water specialist at Central American University in San Salvador. "Mining is an industry whose primary and first victim is water," said McKinley, who added that El Salvador faced a significant scarcity. "We are talking about an issue that is a life-or-death issue for the country," he said. "Today in El Salvador, water won out over gold," said Johnny Wright Sol, a legislator from the center-right Arena Party (Palumbo and Malkin 2017).

Further Reading

Palumbo, Gene, and Elizabeth Malkin. 2017. "El Salvador, Prizing Water over Gold, Bans All Metal Mining." *New York Times*, March 29. Accessed October 11, 2018. https://www
.nytimes.com/2017/03/29/world/americas/el-salvador-prizing-water-over-gold-bans-all
-metal-mining.html.

The toll of chemicals is worldwide, from their use in war against the Kurds in Iraq to the use of the Marshall Islands' people as nuclear guinea pigs. Other examples include Vietnam's legacy of Agent Orange's Deformations Across Generations, the United States' "Cancer Alley" on the lower Mississippi River, toxic dumping in Labrador, Canada, toxic spills in Bolivia, pesticides in Mexico, and radioactivity in the Australian outback. What follows is only a small sample of what humankind is doing to our one, precious home.

Further Reading

Cone, Marla. 1996. "Human Immune Systems May Be Pollution Victims." *Los Angeles Times*, May 13: A-1.

Cone, Marla. 2005. *Silent Snow: The Slow Poisoning of the Arctic*. New York: Grove Press.

Jamieson, Alan J., Tamas Malkocs, Stuart B. Piertney, Toyonobu Fujii, and Zulin Zhang. 2017. "Bioaccumulation of Persistent Organic Pollutants in the Deepest Ocean Fauna." *Nature Ecology & Evolution* 1 (February). Accessed October 3, 2018. http://www.doi
.org/10.1038/s41559-016-0051.

Watson, Traci. 2017. "Voyage to the Bottom of the Sea: Where Tons of Toxins Are." *USA Today*, February 14: 5-A.

AUSTRALIA

Radioactivity in the Outback

When uranium mining was initiated on their lands, Australia's Aborigines were promised that it would be their ticket to the modern world. Decades later, promised jobs were nearly nonexistent, housing was substandard, and stretches of customary Aboriginal homelands, piled high with waste tailings, were unusable because of residual radioactivity. In addition to problems associated with uranium mining, Australian Aborigines also have reported health problems stemming from nuclear testing in the neighboring South Pacific during the mid-20th century, after which fallout that the native peoples called a "black mist" was carried over their homelands by prevailing winds. Elsewhere in Australia, native peoples are resisting industrial-scale gold mining that may replace sacred sites with open pits.

Resisting Uranium Mining

The Mirrar Aborigines of Australia's Northern Territory have resisted development of new uranium mining within their territory, contending that similar projects had shown that the proposed Jabiluka mine could destroy their way of life. Environmentalists argued that the nearby Ranger uranium mine provided a disastrous environmental precedent and had a severe impact on local Aborigines. Mine workers came to greatly outnumber Aborigines, who suffer from chronic alcohol abuse,

A government-operated uranium mine near Kakadu National Park, Australia. (Johncarnemol-la/Dreamstime.com)

several other health problems, and lack of adequate housing—problems which afflict Native peoples who have been deprived of their traditional economic systems the world over ("Australia" 2001). The Mirrar Aborigines fear that they will face a similar fate if the Australian government approves a proposal by Energy Resources, an Australian company, to develop Jabiluka. Development of the mine also would leave the Mirrar with millions of tons of radioactive waste.

The Mirrar agreed to allow initial uranium mining after they were led to believe that approval was their only way to secure legal rights to their land. An Australian Senate inquiry and the United Nations have criticized the tactics used by Australian authorities to obtain this agreement. Between 1979 and 1988, the Nabarlek uranium mine in West Arnhem Land, owned by Queensland Mines Ltd., extracted, stockpiled, and processed 11,000 tons of ore. This open-pit mine was constructed despite opposition by many local Aborigines, who staged a sit-in on its access road and later took Queensland Mines to court. The mine was less than one mile from an area of special significance to Aborigines, the Gabo-djang, the Dreaming Place of the Green Ants.

A Flood of Radioactive Rubbish

During March 1981, after heavy rain from a tropical cyclone, radioactive material was released from the mine's tailings dumps into a nearby creek. After the mine was closed, required cleanup work was not completed, leaving local Aborigines with a pile of radioactive rubbish. Given such experiences, many Australian Aborigines have opposed new proposals to mine uranium on or near their traditional lands. According to Vincent Forrester, writing under the aegis of Australia's Sustainable Energy and Anti-Uranium Service in 1997, the Aborigines' reasons for opposing new mining include:

- seepage from existing tailings dams
- concentration of radioactive contaminants in water systems
- soil erosion
- radon gases escaping from the tailings
- the fact that cyclones could disperse contaminated dust from strip-mining operations
- the fact that return of the tailings to the pit at the end of mining operations poses long-term effects in the Alligator Rivers area
- major geological faults in the wall-rock of the pit area
- the presence of a geological fault under the north wall of the Ranger tailings dam
- contaminated water release into Magela Creek

Aborigines and environmentalists have called upon uranium miner Energy Resources of Australia Ltd. to rescind plans for the Jabiluka mine, which would adjoin the Kakadu National Park, an area made famous by the Crocodile Dundee films. The Kakadu Park houses an extraordinary ecosystem that the Aborigines endowed with spiritual significance from ancient times. Along with some of the richest uranium deposits in the world, the area also is home to communities of Australian Aborigines who comprise one of the world's oldest surviving indigenous populations.

Energy Resources of Australia (ERA) is majority owned by global mining corporation Rio Tinto, whose chairman said during the late 1990s that development of the mine was only a remote possibility. The company refused to back away from the project altogether, however. Rio Tinto's chairman of the board, Sir Robert Wilson, said that the company was not pursuing the mine at present because current and foreseeable market conditions indicated that investment in Jabiluka was "economically unattractive" ("Hotspots: Australia" 2001). Wilson left open the possibility that market conditions could change.

The Mirrar people regard the area as their ancestral home and point to the damage done over the years by the Ranger mine, which has left 20 million tons of radioactive tailings in spoil heaps around its operations. According to Friends of the Earth, there have been 120 breaches of the mine's operational guidelines, most recently in May 2000, when 2 million gallons of radioactive liquid contaminated with manganese, uranium, and radium was released. Some of this contamination escaped into the Kakadu wetlands (Brown 2001).

Ed Matthew, from Friends of the Earth in London, said: "This [Jabiluka] mine is on land unjustly wrested from the Aboriginal people and inside a World Heritage site. If Rio Tinto proceeds with this mine, it will be telling us that there is no place on Earth the company is not prepared to plunder" (Brown 2001). Extralegal means have been used by the Mirrar to protect their country and sacred sites, including a blockade during 1998. Despite the blockade, construction work at the mine was delayed for little more than a few hours. ERA used helicopters to fly its workers into the mine compound for several weeks during April and May, when blockaders cemented cars into place, blocking a mine gate. Police later cleared the obstruction. Subsequently, the blockade was cleared with bulldozers.

Protests Using Mass Trespass

Protesters then resorted to a mass trespass on the Mineral Lease area, occasionally locking themselves to trucks, gates, and mining equipment. Similar protests also took place at the Ranger Mine. Mass trespass actions often resulted in large numbers of arrests. As many as 118 people were taken into custody on one June morning alone. The last of several protests took place during the week preceding the Australian federal election on October 3, 1998, producing more than 90 arrests as protesters walked onto the lease area wearing masks depicting Australian Prime Minister John Howard. A few days later, after Howard won reelection, the blockade camp was dismantled as the monsoon season set in.

Yvonne Margarula, a leader of the Mirrar people, has been active in the antimining movement, along with Jacqui Katona, another Mirrar leader. Both said that the Mirrar people oppose the mine for two reasons: firstly, it will devastate a broad area that includes many sacred places; secondly, the Mirrar have a justifiable fear of the mine's potential for radioactive poisoning of their land, which is likely to result from release of radon gas into the atmosphere.

By the year 2000, most Australian Aborigines were united against further uranium mining on their traditional lands. This opposition was reflected in a statement by senior traditional owners of the Jabiluka mineral lease, the Gundjehmi people, and the Alliance Against Uranium, a global coalition of numerous Aboriginal and environment groups from all over Australia. Katona, who works for the Mirrar people in Kakadu National Park, said:

> [The] Mirrar Gundjehmi, Mirrar Erre, Bunitj, and Manilakarr clan leaders have many concerns about mining in their homelands. A new mine will make our future worthless and destroy more of our country. We oppose any further mining development in our country. We have no desire to see any more country ripped up and further negative intrusions on our lives. ("Traditional Owners" 1997)

For Children and Grandchildren

Katona believes that "stopping the Jabiluka mine is the first step in changing the future for our community. We have a responsibility to our children and grandchildren and their children to strengthen their heritage by acting now. This is our future. Without industrial domination. Without aggression. With meaningful positive change. For us, by us" ("Traditional Owners" 1997).

According to Vincent Forrester, chairperson of the Northern Territory National Aboriginal Conference, "There is simply no proper information given to Aboriginal people living in the area about the effects of uranium mining on the land. The monitoring scientists have made no attempt to interpret their findings to the effected Aboriginal people" (Forrester 1997). Forrester also said that no substantial study had been done of radiation levels in Aborigines' diets in the uranium-mining regions. Because they lack crucial information, he said, "We can only guess what amount of radiation they have in their bodies or in the food chain" (Forrester 1997). Closed uranium mines also pose problems for Aborigines. One such mine, Rum Jungle, was abandoned in 1971. Its tailings dam had been breached by monsoon rains that have polluted the Finniss River with radioactive materials. Aborigines who live in the area can no longer safely use the affected land.

Some Australian Aborigines also may have sustained health damage following dumping of radioactive materials from other mines. Fifteen thousand gallons (60,000 liters) of radioactive liquid was sprayed onto the ground at the Beverley uranium mine in South Australia, about 300 miles (520 kilometers) north of the city of Adelaide, early in 2002. The spill was one of 24 spills of radioactive liquid at the mine during the previous two years. Australian environmental groups are calling for the closure of the mine run by U.S.-based Heathgate Resources.

Radioactive leaks have become a constant problem on Mirrar lands. Early in 2002, a uranium leak from the Jabiluka and Ranger uranium mines contaminated Swift Creek in Kakadu National Park. Resident Mirrar people said that Energy Resources of Australia Ltd. (ERA), the company that owns and operates the two mines, waited six weeks to notify them of the leak ("Hotspots: Australia" March 2002).

During spring 2002, yet another large leak was revealed at the Ranger mine, raising renewed protests from the Mirrar people. Australia's Office of the Supervising Scientist released a report which said the internal management at the scene in charge of the mine had failed when a uranium leak occurred earlier this year. Andy Ralph from the Gundjehmi Aboriginal Corporation, which represents Mirrar traditional owners, said that the latest leak was seven times larger than the one discovered earlier in the same year. "There is a concern that a lot have bypassed the Magela Creek and bypassed their wetland filter and did not actually get filtered," he said ("Traditional Owners" 2002).

Further Reading

"Australia: Aborigines Fight Mine." 2001. Survival International Update, July.

Brown, Paul. 2001. "Gift of Life." The Guardian, February 14. Accessed November 29, 2018, https://www.theguardian.com/society/2001/feb/14/guardiansocietysupplement3.

Davison, Helen. 2017. "Jabiru: The Kakadu Mining Town Facing Closure Seeks a Fresh Start." The Guardian, July 23. Accessed October 3, 2018. https://www.theguardian.com/aus tralia-news/2017/jul/24/jabiru-kakadu-mining-town-facing-closure-seeks-fresh-start.

Forrester, Vincent. 1997. "Uranium Mining and Aboriginal People." The Sustainable Energy and Anti-Uranium Service, Inc. (Australia).

"Hotspots: Australia." 2001. Drillbits and Tailings 6, no. 7 (October 31): 3.

"Hotspots: Australia." 2002. Drillbits and Tailings 7, no. 1 (January 31).

"Hotspots: Australia." 2002. Catherine Baldi, ed. Drillbits and Tailings 7, no. 3 (March 29).

"Traditional Owners Concerned at More Leaks from Ranger Mine." 2002. Australian Broadcasting Corporation Indigenous News, April 24.

"Traditional Owners Statements: Statement from the Gundjehmi Aboriginal Corporation." 1997. Sustainable Energy and Anti-Uranium Service (Australia), April.

BOLIVIA

Logging Concessions, Oil Exploration, and Toxic Spills

A large percentage of Bolivia's people share indigenous heritage, often mixed with Spanish roots, and a declining proportion still live off the land. Those who do are facing increasing pressure due to commercial deforestation and oil exploration, as well as toxic spills of waste associated with mining of antimony, gold, silver, and zinc. Roughly 1,000 tons of mining waste was being dumped into Bolivian rivers every day by the middle 1990s, the effects of which continued well into the 21st century.

Protesting Logging Concessions

By the year 2000, the Bolivian government had allocated more than a million hectares of primary rain forest as part of new logging concessions, none of which were negotiated with local indigenous peoples' consent. These new logging concessions included 27 on lands recognized under the Bolivian constitution as indigenous territories. At the same time, growing areas of indigenous lands in Bolivia have been facing increasing pressure due to oil exploitation and the mining of several metals and minerals, both of which provoke environmental damage from cyanide and arsenic associated with breaches of mining-waste reservoirs.

On July 31, 1997, Bolivia's forest superintendent granted 85 new forest concessions, "effectively eliminating large stretches of primary forest, which constitute zones of traditional and cultural usage that are indispensable to the survival of the Indigenous People" ("Bolivian Rainforests" 1999). Appeals of these decisions were denied in a number of administrative channels, then appealed to the Bolivian Supreme Court of Justice, which refrained from blocking them.

According to local observers, "The concessions effectively eliminate 500,000 hectares of Guarayo territory, more than 140,000 hectares of Chiquitano de Monte Verde territory, more than 15,000 hectares of Yaminahua Machineri territory, more than 17,000 hectares of indigenous multiethnic territory, and more than 28,000 hectares of indigenous territory [in the] Isiboro Secure National Park" ("Bolivian Rainforests" 1999). In all, a total of more than 700,000 hectares of legally recognized indigenous territory was assigned for exploitation by transnational lumber businesses. Following the allocation of these concessions, conflicts increased between displaced indigenous peoples and logging companies, whose interests often were supported by local police forces, Bolivian army troops, and agents of the national superintendent of forests.

Bolivia's Declining Forests

The logging concessions came at a time when Bolivia's remaining forests were steadily declining in the area. During the late 1990s, Bolivia's national territory included roughly 440,000 square kilometers of rain forests, comprising 57 percent of the country's lowlands. At the same time, the rate of deforestation had reached 168,000 hectares per year as log exports increased steadily. Deforestation continued in decades hence, especially in parts of Bolivia that lie within the Amazon River watershed. Deforestation in Bolivia also was being aggravated by oil exploration. Inhabitants of San Ignacio de Moxos reported that Repsol, an oil company owned mainly by Spanish nationals, had advanced 90 kilometers into the Multiethnic Indigenous Territory in the Amazon forest, using a road previously opened by loggers. The Multiethnic Indigenous Territory is inhabited by several indigenous peoples, including the Trinitary, Mojeño, and Chimán. Repsol was reported to have drilled two exploratory wells without legally required environmental permits ("Bolivia" 2000). The wells are said to have affected an area inhabited by the Quichua and Aymara indigenous peoples.

Industrial-Scale Pollution

Areas of Bolivia's remaining forests have been increasingly threatened by industrial-scale pollution. For example, a rupture in a dike at a Compañía Minera del

Sur (Comsur) Bolivian mine, which is owned by Gonzalo Sanchez de Lozada, the president of Bolivia, caused 235,000 tons of pollutants (including arsenic and cyanide) to be spilled into the Yana Machi River and other tributaries of the Pilcomayo River. More than 8,000 indigenous Mataco and Chiriguano peoples live along the Bolivian portion of the Pilcomayo (which also traverses parts of Paraguay and Argentina), sustaining themselves principally on the sabalo, a fish that is rapidly becoming extinct due to the arsenic and other toxic residues ("Major Toxic Spill" 1996).

Comsur is one of several mines in Bolivia that feed waste from mining operations into polluted lakes behind dams (another is the Inti Raymi mine). Approximately three dozen other mines that extract antimony, gold, silver, and zinc dump their wastes directly into rivers that feed the Pilcomayo. Studies estimate that an average of 1,000 tons of mining waste was being dumped into Bolivian rivers every day ("Major Toxic Spill" 1996).

In the southern Bolivian town of Tarija, near the Argentine border, indigenous peoples assembled a protest march "in defense of life and the environment" and to demand that Comsur compensate them for damage caused to crops irrigated by water from the Pilcomayo. "We want to avoid more deaths, pollution of crops, and the displacement of people living along the Pilcomayo due to the contamination of its waters," said Julio Rodriguez, president of the Tarija civic committee ("Major Toxic Spill" 1996).

A study by the state university of Tarija reported that arsenic poisoning of the Pilcomayo by an accident at the Comsur mine probably had caused the deaths of three miners who drank water and ate fish from the river. A government commission that visited the area of the spill downplayed the significance of the accident, however, claiming that thanks to quick action by Comsur, no plants or animals were harmed ("Major Toxic Spill" 1996).

Climate Change, Pollution, and Declining Lake Levels

Climate change and toxic contamination are fundamentally altering Bolivia's aquatic topography, changing its largest highland lakes. Lake Poopó, the country's second largest at 1,000 square miles, had dwindled to a few polluted marshes by 2016, while an ecological battle has been joined to preserve what remains of Lake Titicaca, the largest. Before widespread mining, Lake Poopó was almost 1,000 square miles in size. By the year 2010, only a few marshes remained in the salty desert of the high plateau, which is littered with dead animals and abandoned boats.

In both cases, the lakes' deterioration has been caused by several human-induced problems, including mining wastes, excessive sedimentation, diversion of rivers (usually for irrigation), and global warming, which has increased evaporation and often decreased precipitation and melting mountain snow that used to feed the lakes. All of these factors also have taken a toll on animal species. Poopó, in particular, has been plagued by massive fish mortality since millions of them died there during 2014. As the lake's size shrinks, remaining waters become more toxic. What was once shoreline is now marked by abandoned, rotting boats that once were part of a vibrant fishing economy. "We're no longer the men of the lake. If the lake goes, we will too. For generations, we've called ourselves hunters, fishermen, and gatherers. That's why we call ourselves 'Men of the Lake,' but we're losing this identity now," said Simiano Valero, who used to fish in Lake Poopó (López 2016). By December of 2015, Lake Poopó had nearly disappeared.

The site of the ailing Lake Poopó connects to Lake Titicaca through the Desaguadero River. Despite its harsh environment, this region long has been an important source of biodiversity, with approximately 200 animal and plant species. "You would think there is no life in the altiplano, but it is home to a great quantity of wild fauna and flora," said ornithologist Carlos Capriles (López 2016). Biodiversity decreases as the lakes shrink and the remaining water becomes more toxic due to mining wastes and other pollution. Fish die, and birds migrate, if they are able. The lakes are shallow and have no outflow, so remaining water becomes less hospitable to many species of plants and animals as their size shrinks and pollution rises.

Mining, the area's main economic activity since the Spanish invasion, has expanded to include more than 300 sites that dump untreated wastewater into the lakes' tributaries, including, according to one account, "heavy metals like cadmium, zinc, arsenic, and lead. The Poopó Lake Basin Program (Programa de la Cuenca del Lago Poopó), created in 2010 to address its pending disappearance, calculates that 2,000 tons of solid minerals enter the lake on a daily basis" (López 2016).

Toxic Mining Waste

As an illustration of the sheer volume of mining waste accumulated over many years, Bolivia is treating 4.5 million tons of toxic mining waste at a Chinese concentration plant in the Andean region of Potosi.

> The biggest problem in the area is at San Miguel, where waste has been accumulating for decades, causing the contamination that sparked indignant protests by locals demanding that the hazard be cleaned out of their area. The 4.5 million tons of concentrated waste in San Miguel contains sulfurous minerals as well as oxidized deposits containing various amounts of quartz, pyrite, chalcopyrite, copper, arsenic, and other minerals that generate acidic waters, the Bolivian mining authority said. ("Bolivia to Treat" 2013)

Carlos Valdez of the Associated Press reported from Lake Titicaca that "gulls swept down to feast on hundreds of dead and dying giant frogs floating in the rancid waters along a southeastern shore of Lake Titicaca, where the algae-choked shallows reek of rotten eggs. The die-off was the most striking sign yet of the deteriorating state of South America's largest body of fresh water" (Valdez 2015). Local fishermen have been put out of work as farmers find their crops stunted by toxic irrigation water.

"Most pollution on the Bolivian side, including such toxic heavy metals as lead and arsenic, originates in El Alto, a fast-growing city of 1 million people near La Paz that sits 600 feet (200 meters) above the lake and just 25 miles (40 kilometers) away," wrote Valdez (2015). In addition to the chemical pollution, waste from the more-than 50 percent of the people along the lake's shores who lack proper plumbing flows into the lake, compounding its accelerating toxicity.

Further Reading

"Bolivia: Indigenous Peoples' Forests Menaced by Oil Exploration." 2000. *World Rainforest Movement Bulletin* 35 (June). Accessed November 29, 2018. https://wrm.org .uy/articles-from-the-wrm-bulletin/section1/bolivia-indigenous-peoples-forests -menaced-by-oil-exploration/.

"Bolivian Rainforests Allocated without Indigenous Consent." 1999. Global Response. Worldwide Forest/Biodiversity Campaign News, August 1.

"Bolivia to Treat 4.5 Million Tons of Mining Waste at Chinese Plant." 2013. Fox News Latino, October 12. Accessed October 3, 2018. http://latino.foxnews.com/latino/news /2013/10/12/bolivia-to-treat-45-million-tons-mining-waste-at-chinese-plant/print.

Center for Legal Studies and Social Research (CEJIS). Accessed November 29, 2018. https:// www.iwgia.org/en/iwgia-partners/24-center-for-legal-studies-and-social-research-cejis.

López, Aldo Orellana. 2016. "Bolivia's Second Largest Lake Disappears, Due to Desertification and Contamination." Mongabay, February 15. Accessed October 3, 2018. https://news.mongabay.com/2016/02/bolivias-second-largest-lake-disappears-due -to-desertification-and-contamination/.

"Major Toxic Spill at Mine Owned by Bolivian President." 1996. *Drillbits and Tailings*, November 7.

Valdez, Carlos. 2015. "Frog Die-Off in Lake Titicaca Puts Spotlight on Unchecked Pollution That Threatens Livelihoods." Associated Press in *U.S. News & World Report*, June 25. Accessed October 3, 2018. https://www.usnews.com/news/world/articles/2015/06/25 /unchecked-pollution-befouling-majestic-lake-titicaca.

CANADA

Labrador's Toxic Legacy

Inuit people have discovered that parts of their homelands are laced with toxic "hot spots" left behind by abandoned military installations and mines, all imports from the industrial south. Several of these hot spots are located at or near the 63 military sites in Canada, Greenland, and Alaska that comprise the Distant Early Warning (DEW) system of radar sites. At these sites, according to the Arctic Monitoring and Assessment Program's report, *Arctic Pollution Issues: A State of the Arctic Environment*, an estimated 30 tons of PCBs were used, and "an unknown amount has ended up in their landfills" ("Communities Respond," n.d.).

Who Pays for Cleanup?

Under an agreement reached in 1998, Canadian taxpayers, not the U.S. government, are paying most of the $720 million cleanup bill for 51 decommissioned U.S. military sites across Canada. Cleanup of cancer-causing PCBs, mercury, lead, radioactive materials, and various petroleum by-products is expected to take nearly 30 years. Under the arrangement, the United States was absolved of legal responsibility for environmental damage in Canada in exchange for $150 million in U.S. weapons and other military equipment. The cleanup of all 51 American military sites has revealed pollution that newspaper reports in Canada characterized as "staggering" (Pugliese 2001, B-1). For example, at Argentia, Newfoundland, 70 miles southeast of St. John's, an abandoned U.S. Navy base left behind PCBs, heavy metals, and asbestos, as well as landfills laced with other hazardous wastes. Waste fuels also have contaminated the water table in the area.

The 10,000-acre U.S. naval base at Argentia, opened in 1941, was used as a major naval staging area for European operations during World War II. It hosted a 1941 meeting between British prime minister Winston Churchill and U.S. president Franklin Roosevelt as they negotiated the Atlantic Charter that created the North American Treaty Organization (NATO). In the midst of World War II, the base employed 20,000 people.

The naval base was closed in 1994, as plans were made to convert the area into a private port, pending cleanup of PCBs and other contamination. The Canadian government asked the U.S. Department of Defense to cleanse the area, but funding was stalled because treaty arrangements contained no clause for environmental responsibility. The U.S. Defense Department said it would only clean up areas that posed "an imminent health and safety hazard" due to PCB leaks at the base's ship-building facility and a local landfill that was leaching contaminants into surrounding soil and water. John Maher, mayor of Placentia (a nearby town) and a member of the Argentia Redevelopment Commission, said that redevelopment of the site was important for revival of the area's depressed economy. According to a United Press International report, Maher said use of the site, with its intact infrastructure and ice-free harbor, could aid a province that has been suffering economically following the collapse of Atlantic fisheries. "We haven't anything else to fall back on," said Maher. "It's a gold mine just sitting here that could be a major contributor to the economy of Newfoundland" ("U.S. Closes" 1994).

Abandoned DEW sites in the Arctic were contaminated with discarded batteries, antifreeze agents, solvents, paint thinners, PCBs, and lead. According to news accounts, "[Canadian] Defence Department scientists have established that PCBs have leaked from the DEW line sites into surrounding areas as far away as 20 kilometers and, in some cases, the chemicals have been absorbed by plant and animal life" (Pugliese 2001, B-1). Many of the DEW line locations were established in areas where native people hunt and fish. Alaska Community Action on Toxics works with indigenous communities that face toxic contamination from Cold War sites, including the Yu'pik community on Saint Lawrence Island. Alaska Community Action on Toxics also has provided the first comprehensive map of more than 2,000 hazardous waste sites in Alaska.

Pollution from Europe and Russia

The Inuit also endure pollution from the European and Asian side of the Arctic Ocean. Pesticide residues and other pollutants spill into the Arctic Ocean from north-flowing Russian and Siberian rivers. Decaying Russian nuclear submarine installations on the White Sea have polluted the ocean with nuclear waste, including entire reactor cores from scrapped ships. While many of the former Soviet Union's worst polluters have gone out of business, some prosper despite the fact that their effluent is adding to the Canadian Arctic's toxic overload. The worst offender is the Norilsk nickel smelter, located in northern Siberia. Traces of heavy metals from Norilsk's industries have been detected in the breast milk of Inuit mothers.

Geoffrey York, a reporter for the Toronto *Globe and Mail*, described the industrial city of Norilsk, population 230,000, "the world's most polluted Arctic metropolis":

> Looming at the end of the road is a horizon of massive smokestacks, leaking pipes, rusting metal, gigantic slag heaps, drifting smog, and thousands of denuded trees as lifeless as blackened matchsticks. Inside malodorous smelters, Russian workers wear respirators as they trudge through the hot suffocating air, heavy with clouds of dust and gases. . . . Soviet [era] research in 1988 found that Norilsk Nickel had created a 200-kilometer corridor of dead forests to the southeast of the city. (York 2001, A-11)

It took scientists several years to find another PCB dump at Saglek Bay on the remote northern coast of Labrador during the late 1990s, when they conducted tests on fish and birds there. The cause was another abandoned United States military installation—in this case, a Cold War–era Air Force radar station. The toxic chemicals had been used in electrical transformers, then dumped, spreading toxicity into to the soil along about 35 miles of shoreline. Wildlife showed evidence of contamination many years after the base had closed and the use of PCBs had been outlawed.

Nature Slowly Begins to Heal

After that, however, nature slowly began to heal itself. "On a recent return visit," according to a report in Toronto's *Globe and Mail*,

> researchers were astonished at how rapidly animals in the area have been able to recuperate from their toxic exposures. The levels of PCBs in some of the fish and birds had fallen by up to 95 percent. The discovery is heartening to the scientists, who say it provides some of the strongest evidence that fragile Arctic ecosystems are able to cleanse themselves after heavy contamination by one of the most dangerous . . . chemicals ever made. The north is dotted with dozens of other toxic hot spots at old military installations, and the finding suggests many of them should be able to revert to a more natural state, if proper cleanups are undertaken. (Mittelstaedt 2009)

"We are quite delighted, actually, to see this change. . . . We're moving dramatically quickly in the right direction," said Ken Reimer, director of the environmental sciences group at Royal Military College of Canada in Kingston and the lead researcher on the project (Mittelstaedt, 2009).

The relatively quick natural recovery may be notable worldwide because PCBs and other "persistent organic pollutants" (POPs) are generally very difficult to remove from natural systems. They accumulate quickly in animals' fatty tissues and become more toxic in an exponential manner as one animal eats another, up the food chain. The process is called "biomagnification," and is a major reason why Inuit mothers have been warned for many years not to breastfeed their babies.

The U.S. Air Force base had opened in 1974 but was phased out 20 years later as its radar equipment became obsolete. Once it was closed, scientists located the source of the contamination, which was not difficult to contain. Roughly 22,500 cubic meters of soil and debris that had been laced with PCBs was removed, transported to Quebec, and incinerated, destroying the PCBs.

Scientists meanwhile measured contamination at and near the site, as well as offshore. According to the *Globe and Mail* account, "They also sampled shorthorn sculpin, a bottom-dwelling fish, and black guillemots, a pigeon-sized marine bird. Levels of the chemical in the animals were so high that scientists worried about their ability to reproduce and have immune systems capable of fighting off infections" (Mittelstaedt 2009). Contamination was still present in sea creatures, enough that Labrador Inuit warned against eating sculpin and guillemots. By 2007, however, PCB levels in local wildlife and sediments under Saglek Bay had fallen sharply. "The levels right now are getting really close to levels that no longer pose risks to wildlife in the area," said Tom Sheldon, director of environment for the Nunat-siavut government, which represents Inuit in the area (Mittelstaedt 2009).

Further Reading

"Communities Respond to PCB Contamination." n.d. PCB Working Group, IPEN.

Mittelstaedt, Martin. 2009. "Fragile Labrador Ecosystem Overcomes a Toxic Past." *Globe and Mail*, December 21. Accessed October 3, 2018. http://www.theglobeandmail.com/news/national/fragile-labrador-ecosystem-overcomes-a-toxic-past/article4296700/.

Pugliese, David. 2001. "An Expensive Farewell to Arms: The U.S. Has Abandoned 51 Military Sites in Canada. Many Are Polluted, and Taxpayers Are Paying Most of the $720 Million Cleanup Cost." *Montreal Gazette*, April 28: B-1.

"U.S. Closes Last Military Base in Canada." 1994. United Press International, September 29. Accessed October 3, 2018. http://www.upi.com/Archives/1994/09/29/US-closes-last-military-base-in-Canada/7951780811200/.

York, Geoffrey. 2001. "Russian City Ravaging Arctic Land." *Globe and Mail*, July 25: A-1, A-11.

IRAQ

The Kurds' Genocide via Poison Gas

More than 200 attacks using poison gases have been attributed to the Iraqi armed forces retaliating for the indigenous Kurds' support of Iran during a war with Iraq in the 1980s. In addition to killing between 50,000 and 200,000 indigenous people in northern Iraq, the attacks have caused deformations in many survivors at the DNA level, ensuring that genetic defects will be passed from generation to generation. In addition, formerly fertile farm fields and water sources have been poisoned (Goldberg 2002, 61).

A Long History of Chemical Warfare

Iraq's regime under Saddam Hussein had a long history of using chemical agents against the Kurds, who number 30 million in Iraq, Iran, and Turkey. Iraq's armed forces used the blister agent mustard gas as early as 1983 and tabun (a nerve gas that can kill within minutes) beginning in 1985, as it faced attacks from "human waves" of Iranian troops and their loyalists. The United Nations found in 1986 that Iraq had contravened the Geneva Convention by using chemical weapons against Iran. The toxic offensive against the Kurds in northern Iraq appears to have begun in 1988, after Kurdish guerrillas joined the Iranians.

In August 1988, shortly after the cease-fire that ended the Iran-Iraq war, the Iraqi government launched a major military offensive against the Kurds in northern Iraq, using mustard gas and other nerve-destroying agents that long have been illegal under several international protocols. A Physicians for Human Rights team concluded that bombs containing mustard gas and at least one unidentified nerve agent had been dropped on Kurdish villages in northern Iraq ("Nerve Gas" 1993). "These chemical weapons attacks were part of a genocidal campaign carried out against Kurdish civilians," said Kenneth Anderson, director of the Arms Project of Human Rights Watch and a member of a forensic team that visited Iraqi Kurdistan in June 1992 ("Nerve Gas" 1993).

The Kurdish city Halabja (population 45,000), 15 miles from Iran's border, suffered widespread (and by the year 2002, well-documented) attacks that killed

An Iraqi Kurd sits on a makeshift hospital bed after having fled a chemical gas attack in Oshnavieh, Iran, on August 5, 1988. (Kaveh Kazemi/Getty Images)

many thousands of people, left others disabled for life, and ruined what was once a fertile agricultural area. Estimates of the number of civilians killed in and near Halabja ranged from 3,200 to 5,000 ("Saddam's Iraq," n.d.). Hamish de Bretton-Gordon reported for Al Jazeera, "On the morning of March 16, 1988, Iraqi war planes and artillery pounded the Kurdish town of Halabja in northern Iraq with mustard gas and the deadly nerve agent sarin. Some 5,000 people—mainly women and children—died on the day, and up to 12,000 have lost their lives since" (Bretton-Gordon 2016).

Effects of Poison Gas at Ground Level

"First came the blast of the bombs dropped from the jets, followed by a smell of apples. Then the birds the family kept caged in the yard dropped dead. Minutes later, human beings began to die, too," reported Gabriele Barbati (2013). "We took refuge in the basement," Minira Abdul Qader recalled from her house in Halabja, "but when jets came again at night, we decided to flee the city, like everybody else. I remember people lying still on the fields, and me asking my brother why they were sleeping there." Qader survived the attack but has become blind as a result of the attack. "I have been in pain for 25 years because of my eyes and some breathing problems. I wish I had died the day of the attack" (Barbati 2013). Many small children born after the attacks suffered microcephaly, a neurodevelopmental disorder in which children are afflicted by abnormally small brains.

The same chemical agents were used 25 years later in the Ghouta gas attack in Syria. A first wave of bombings broke windows, forcing people inside and underground. A second wave involved heavier-than-air chemical weapons—sarin and mustard agents—which seeped into underground hiding places and killed people en masse. A third wave of aerial bombings blasted the remains beyond recognition to destroy evidence of the poisoning.

A BBC report said that "Chemical weapons were also used during Iraq's 'Anfal' offensive, a seven-month scorched-earth campaign in which an estimated 100,000 to 180,000 Kurdish villagers were killed or disappeared and hundreds of villages were razed" ("Saddam's Iraq," n.d.). Many of the victims were buried alive.

On the late morning of March 16, 1988, reported Jeffrey Goldberg in the *New Yorker*, an Iraqi air force helicopter flew over Halabja, a city of 80,000 people, taking pictures with still and video cameras. A half-hour later, Iraqi ground troops fired artillery shells filled with poison gases into the city. People in the city recalled smelling garlic, sweet apples, and garbage. Soon, sheep, goats, cows, and birds fell onto their sides, dying (Goldberg 2002, 53).

Residents of Halabja hid in shelters, prepared for a bombardment, but soon they became very ill. Some felt painful sensations in their eyes, "like stabbing needles" (Goldberg 2002, 53). Children began to vomit. The people realized that their basements and other shelters were becoming gas chambers. The gas was heavier than the ambient air, so it followed people who hid below ground level. Others emerged from their shelters, running, weakened by nausea, trying, to no avail, to avoid the spreading clouds of gas. Leaves began to fall from the trees, even though it was early spring.

People fleeing from Halabja began to die as they ran—children and old people first. "They were running, then they would stop breathing and die," said one eyewitness (Goldberg 2002). Some people stripped off their clothes and died laughing madly (Goldberg 2002, 53). Others became disoriented and died where they stood. Many children went blind as they ran. "The children were crying: 'We can't see. My eyes are bleeding!'" (Goldberg 2002, 54). Several people tried to avoid the gas by jumping into ponds; they died, and their decomposing remains later poisoned the water table, making the land useless for farming.

More Widespread Gas Attacks

The nerve gases were dropped on small villages as well as larger cities. Eyewitnesses have said that Iraqi warplanes dropped three clusters of four bombs each on the village of Birjinni on August 25, 1988. In Birjinni, a small village of about 30 houses, a mosque, and a school, observers recalled "seeing a plume of black, then yellowish smoke, followed by a not-unpleasant odor similar to fertilizer and also a smell like rotten garlic. Shortly breathing, their eyes watered, their skin blistered, and many vomited—some of whom died. All of these symptoms are consistent with a poison-gas attack" ("Nerve Gas" 1993). At least four people were killed during the attack on Birjinni, two in an orchard and two brothers in a cave where they sought refuge. The remaining villagers fled.

The gas attacks in Birjinni and Halabja were similar to many others in Kurdistan during the ensuing several months. While the Iraqi government later denied the attacks, several international eyewitnesses confirmed them with photographic evidence of the dead and deformations among the living. Physicians for Human Rights also detected trace elements of the gases in the area years after the attacks ("Nerve Gas" 1993). Alastair Hay, a consultant to PHR and senior lecturer in Chemical Pathology at the University of Leeds, said, "This discovery . . . confirms eyewitness accounts and medical examinations of Kurdish people that nerve gas as well as mustard gas were used against them" ("Nerve Gas" 1993).

Deaths, Deformations, and Ecological Damage

In addition to deaths, deformations, and ecological damage, the genocidal effects of the nerve gases used against the Kurds took another form: infertility. By the year 2000, miscarriages were outnumbering live births in the Kurdish territories, and many children born alive were deformed, with harelips, cleft palates, and other afflictions (Goldberg 2002, 62). Residents of cities attacked by the nerve agents also were afflicted by cancer rates (notably colon cancer) five to ten times those of unaffected regions. Local hospitals lacked radiation or chemotherapy capabilities, so cancers often were cut out by surgeons.

After the attacks, surviving men often were taken away by the Iraqi army and probably killed, leaving behind legions of widows doing their best to look after sick, deformed children. Goldberg reported that "Most of the Kurds who were murdered . . . were not killed by poison gas; rather, the genocide was carried out, in large part, in the traditional manner, with roundups at night, mass executions, and anonymous burials" (Goldberg 2002, 62).

ISIS Gas Attacks

Twenty-five years after the poisonings in Halabja and other Kurdish towns, some of the same chemical killers were used by the Islamic State (ISIS or ISIL) in its barbaric dominance of the same area. "Some 28 years after Halabja, the Iraqi Kurds are once more under chemical attack from a tyranny, in this case under the banner of ISIL. On many occasions in the past two weeks [as of March 16, 2016], the Peshmerga, the fighting force of the Kurdistan Regional Government of Iraq, has been under chemical bombardment from ISIL" (Bretton-Gordon 2016). Homemade "dusty" mustard—though not as toxic as the liquid produced by Saddam and the Syrian regime—has still killed a number and injured hundreds. In 2015, "ISIL . . . used the mustard agent a number of times against the Peshmerga in the Mosul Dam area and against civilians in the northern Syrian town of Marea" (Bretton-Gordon 2016). As of this writing, the genocide against the Kurds continues.

Further Reading

Barbati, Gabriele. 2013. "25 Years after Worst Chemical-Weapon Massacre in History, Saddam Hussein's Attack on Halabja in Iraq, the City Is Reborn." *International Business Times*, March 15. Accessed October 3, 2018. http://www.ibtimes.com/25-years-after -worst-chemical-weapon-massacre-history-saddam-husseins-attack-halabja-iraq-city.

Bretton-Gordon, Hamish de. 2016. "Remembering Halabja Chemical Attack." Al Jazeera, March 16. Accessed October 3, 2018. http://www.aljazeera.com/indepth/opinion /2016/03/remembering-halabja-chemical-attack-160316061221074.html.

Goldberg, Jeffrey. 2002. "The Great Terror." *New Yorker*, March 25: 52–73. Accessed November 29, 2018. https://www.newyorker.com/magazine/2002/03/25/the-great-terror.

"Nerve Gas Used in Northern Iraq on Kurds: Medical Group Proves Use of Chemical Weapons through Forensic Analysis." 1993. Physicians for Human Rights, April 29. http:// www.phrusa.org/research/chemical_weapons/chemiraqgas2.html.

"Saddam's Iraq: Key Events. Chemical Warfare: 1983–1986." n.d. BBC News. Accessed October 3, 2018. http://news.bbc.co.uk/2/shared/spl/hi/middle_east/02/iraq_events /html/chemical_warfare.stm.

MARSHALL ISLANDS

Nuclear Guinea Pigs

The Marshall Islands, an array of coral atolls located in the Pacific Ocean, bore the brunt of open-air nuclear testing by the United States until such tests were banned in 1963. At least 66 nuclear devices were set off over the islands. Some blasts provoked the relocation of people from entire atolls that were obliterated, reduced to radioactive cinders by the explosions.

At a meeting of the Atomic Energy Commission held during 1956, Marshall Islanders were assigned a role as nuclear guinea pigs. Merrill Eisenbud, an Atomic Energy Commission official, was quoted as saying:

> Utirik [one of the islands] is . . . by far the most contaminated place in the world, and it will be very interesting to go back and get good environmental data. Data of this type has never been available. While it is true these people do not live the way Westerners do, civilized people, it is nevertheless also true that these people are more like us than mice. (Ewen 1994, 140)

"Fallout Fell Like Snow"

Before their island was bombed during the 1940s, the people of Bikini Atoll were removed to Rongerik Island, where, after the atomic-bomb tests, the coconut trees stopped bearing fruit, no small thing in an area where the local traditional economy was based on them. In December 1947, Navy doctors who visited the refugees reported finding "a starving people" (Wypijewski 2001, 44). A few months after that, the Bikini residents were moved again, this time to Kwajalein, where they were hired out as menial labor at a U.S. base. After 1952, downwind of Bikini, stillbirths and miscarriages rose by a factor of 11 times (Wypijewski 2001, 44).

When hydrogen bombs were tested during the 1950s, the wind blew from the testing grounds to nearby islands, where "Native children whirled with delight as fallout fell like snow" (Wypijewski 2001, 42). After the hydrogen bomb Bravo was detonated March 1, 1954, three inches of radioactive ash accumulated on Rongelap, 120 miles downwind of Bikini. During October and November 1962, nine atmospheric nuclear tests were conducted at Johnston Atoll, including four tests at high altitude. Four decades later, the atoll still was polluted by plutonium from those tests. Johnston Atoll also was used to store hundreds of barrels of Agent Orange after the Vietnam War. According to a report from the Pacific Concerns Resource Center, which is based in New Zealand, "Many of these drums have leaked, polluting the environment with dioxin" ("Johnson Atoll" 2000).

Kwajalein, about 100 miles from Bikini, is the world's largest atoll, with a land area of 5.6 square miles and a lagoon area of 1,100 square miles. Most of the island is occupied by a U.S. base that resembles a Midwestern small town (except for its missile-launching facilities), "the picture of 1950s suburbia recreated on the unsubmerged rim of a sunken volcano" (Wypijewski 2001, 43). The base is run by U.S. governmental contractors, including Boeing, TRW, Raytheon, and Lockheed.

Diseases, Alcoholism, and Suicide

To make way for this base, the native people who once lived on Kwajalein were removed to a small adjacent island, Ebeye, with a land area of about one-sixth of a square mile. By the late 1990s, the population of this sliver of land was estimated at more than 13,000—a density resembling that of Manhattan Island, except that everyone lived on one level (Wypijewski 2001, 44–45). The island, which receives more than 100 inches of rain a year, was short of clean water and subject to occasional cholera outbreaks, tuberculosis, and venereal disease. Alcoholism was endemic. More than half the population was under the age of 17, and 10 percent of the deaths on Ebeye were suicides, most of them young boys (Wypijewski 2001, 43–45).

Even after more than 70 years, Tony deBrum experienced goose bumps when he recalled the hydrogen bomb blasts during the middle 1950s; he watched them while fishing with his grandfather in shallow water off a tiny island where they lived. He remembered seeing a four-mile-wide fireball rise from Bikini Atoll and "the sky turns blood red. Wind and thunder follow" (Zak 2015). He was describing "the largest American nuclear-weapons test—the biblical, 15-megaton detonation on Bikini Atoll, 280 miles northwest of his island. Its flash was also seen from Okinawa, 2,600 miles away. Its radioactive fallout was later detected in cattle in Tennessee" (Zak 2015).

While tourists seek the perfect wave along miles of beaches 70 years after the United States developed its nuclear arsenal on and near Bikini Atoll, survivors, their children, and their grandchildren suffer very high rates of diabetes and thyroid abnormalities. In a place where the United States has spent billions of dollars, children play in landfills. Conversations center on rising seas provoked by global warming. Rising late-winter king tides surge onto the islands, forcing hundreds to evacuate as the rising tides pull houses into the ocean.

Another nearby atoll, Ebeye, was used to resettle people who had lived too close to the blasts. From there, wrote Dan Zak:

> More than 10,000 people are crammed into a tenth of a square mile of livable space on Ebeye. The island is crawling with children. A third of its residents are unemployed, and over half are under 20 years old. Government buildings stand on crumbling stilts with exposed rebar, the concrete spalled away by a constant salty wind off the ocean. Raw sewage pools in the streets. There are occasional outbreaks of cholera and dengue fever. The hospital has an on-again, off-again insect problem. (Zak 2015)

"We . . . Blew It to Hell"

"We located the one spot on Earth that hadn't been touched by the war and blew it to hell," the legendary comedian Bob Hope once reportedly said about the Bikini Atoll (Zak 2015). Blasts a thousand times as strong as those that devastated Hiroshima and Nagasaki "hurled into the sky a massive plume of pulverized coral that drifted eastward and fell like ashy snowflakes on the people of Rongelap and Utirik atolls. Several days later . . . children had eaten the ash" (Zak 2015).

"We have basically destroyed a culture," said Glenn Alcalay, an anthropology professor at New Jersey's Montclair State University. Documentation of cancers has been scanty, as the United States federal government has shown little willingness to shoulder responsibility, although an extensive account in the *Washington Post* said, "Everyone seems to have a relative whose cancer falls on the Energy Department's list of ailments traceable to radiation" (Zak 2015). People express fear of islanders who lived close to the bomb-test zones and often will not have children with them, fearing radioactivity-induced mutations. Alcoholism and suicide are rampant. Residents of the Marshall Islands are allowed to enter the United States without visas, however, and many have departed to find work there. Students from the Marshall Islands attend U.S. colleges and universities on Pell grants, even as the United States still uses the local ocean as a landing ground for unarmed intercontinental ballistic missiles (ICBMs) launched from Vandenberg Air Force Base in California, 4,300 miles away.

More than 60 years after the blasts, exiled Bikini Atoll islanders still refused to return, saying they are too fearful to ever go back because of nuclear contamination. *The Guardian* reported:

> Bikini islanders and their descendants have lived in exile since they were moved for the first weapons tests in 1946. When U.S. government scientists declared Bikini safe for resettlement, some residents were allowed to return in the early 1970s. But they were removed again in 1978 after ingesting high levels of radiation from eating foods grown on the former nuclear test site. ("Bikini Atoll" 2014)

"I won't move there," said Evelyn Ralpho-Jeadrik of her home atoll, Rongelap, which was engulfed in fallout from Bravo and evacuated two days after the test. "I do not believe it's safe, and I don't want to put my children at risk" ("Bikini Atoll" 2014).

Unfinished Business

The Marshall Islands Nuclear Claims Tribunal awarded more than $2 billion to residents of the islands for personal injury and land-damage claims, but the United States has refused to pay, instead giving the islanders small individual payments averaging about $140 a month. Marshall Islands' president Christopher Loeak called on the United States to resolve the "unfinished business" of its nuclear testing legacy, saying compensation provided by Washington "does not provide a fair and just settlement" for the damage caused ("Bikini Atoll" 2014).

Further Reading

"Bikini Atoll Nuclear Test: 60 Years Later and Islands Still Unlivable." 2014. *The Guardian*, March 1. Accessed October 3, 2018. https://www.theguardian.com/world/2014/mar/02/bikini-atoll-nuclear-test-60-years.

Ewen, Alexander. 1994. *Voices of Indigenous Peoples: Native People Address the United Nations.* Santa Fe, NM: Clear Light.

"Johnston Atoll to Be Used as Site for Contaminated U.S. Military Waste." 2000. Pacific Concerns Resource Center (New Zealand), May 5. Accessed October 3, 2018. http://www.converge.org.nz/pma/jmil.htm.

Wypijewski, JoAnn. 2001. "This Is Only a Test: Missile Defense Makes Its Mark on the Marshall Islands." *Harper's*, December: 41–51.

Zak, Dan. 2015. "A Ground Zero Forgotten: The Marshall Islands, Once a U.S. Nuclear Test Site, Face Oblivion Again." *Washington Post*, November 27. Accessed October 3, 2018. http://www.washingtonpost.com/sf/national/2015/11/27/a-ground-zero-forgotten/.

MEXICO

Farmworkers Living with Pesticides Around the Clock

The Yaqui are an indigenous, farming people who live and work in and near the Yaqui Valley in Sonora, Mexico, spanning the border between the United States and Mexico. Beginning after World War II, due to lack of available water and financing, many of the Yaquis became unable to support their own farms, though they had been doing so for centuries. The Yaquis were then forced to lease their lands to outsiders, mainly corporate farmers, who were heavy users of pesticides, herbicides, and fungicides. The use of these chemicals, usually applied by aerial spraying, by tractor, and by hand, inflicted widespread contamination of land, water, and people.

Concurrently, valley farm operations became mechanized, and irrigation and transport systems were established. The result was a "Green Revolution," with farming becoming big business. Yaqui families from the nearby mountain foothills moved into the valley for employment, while some valley residents moved into the foothills to maintain family-scale farms (Guillette et al. 1998).

Pesticides Pervade Valley Life

Farmers in the valley reported that two crops a year usually were planted, with pesticides applied as many as 45 times per crop. Compounds included multiple organophosphate and organochlorine mixtures, as well as pyrethroids. Thirty-three different compounds were used for the control of cotton pests alone between 1959 and 1990. This list included DDT, dieldrin, endosulfan, endrin, heptachlor, and parathion-methyl, to name a few examples. As recently as 1986, 163 different pesticide formulations were sold in the southern region of the state of Sonora. Substances banned in the United States, such as lindane and endrin, are readily available to farmers living in the Mexican parts of the valley.

In the valley, pesticide use has long been widespread and continues throughout the year, with little governmental control. Contamination of the resident human population has been documented, with women's breast milk concentrations of lindane, heptachlor, benzene hexachloride, aldrin, and endrin all above limits of the Food and Agricultural Organization of the United Nations after one month of lactation (Guillette et al. 1998). During 1990, high levels of several pesticides were found in the cord blood of newborns and in breast milk of valley residents. Children are usually breastfed, then weaned onto household foods.

Foothill Families Avoid Pesticides

Household insect sprays were usually applied each day throughout the year in the lowland homes. In contrast, the foothill residents maintained traditional intercropping for pest control in gardens. They usually controlled insects in their homes by swatting them. Most of these people were exposed to pesticides only when the government sprayed DDT each spring to control malaria (Guillette et al. 1998).

Angel Valencia, a spiritual leader of the Yaqui tribe in Sonora, Mexico, in the village of Potam, described the effects of these chemicals among valley residents. Valencia spoke as a representative of the Arizona-based Yoemem Tekia Foundation, an affiliate of the International Indian Treaty Council.

> I have seen with my own eyes the effects of daily contact with these pesticides—it burns their skin, they lose their fingernails, develop rashes and in some cases they have died as a result of exposure to these poisons. . . . The tragedy of this situation makes me both sad and angry—to think of what has been done to the innocent children who are the future of the Yaqui people. They will not be able to grow and develop, as they deserve to. (Valencia 2000)

Pesticide Exposure of Yaqui Children

During the 1990s, Elizabeth Guillette, an anthropologist and research scientist at the University of Arizona, studied the impacts of pesticide exposure on Yaqui children. Guillette's studies confirmed the observations of Valencia, who said that exposure to pesticides had "a serious impact on the health and physical and mental development of the children of our villages" (Valencia 2000). Prior to Guillette's research, researchers at the Technological Institute of Sonora in Obregón, Mexico, had shown that children in Sonora's Yaqui Valley often were born with detectable concentrations of many pesticides in their blood and were exposed again through consumption of their mothers' breast milk.

"I know of no other study that has looked at neurobehavioral impacts—cognition, memory, motor ability—in children exposed to pesticides," said neurotoxicologist David O. Carpenter of the State University of New York at Albany. "The implications here are quite horrendous," he said, because the magnitude of observed changes "is incredible—and may prove irreversible" (Raloff 1998). "Although the children exhibited no obvious symptoms of pesticide poisoning, they're nevertheless being exposed at levels sufficient to cause functional defects," observed pediatrician Philip J. Landrigan of Mount Sinai Medical Center in New York.

In Guillette's study, children of the agrarian region were compared to children living in the foothills, where pesticide use is minimal. The study selected two groups of four- and five-year-old Yaqui children who resided in the Yaqui Valley of northwestern Mexico. These children shared similar genetic backgrounds, diets, water-mineral contents, cultural patterns, and social behaviors. The major difference was the level of their exposure to pesticides. Guillette adapted a series of motor and cognitive tests into simple games the children could play, including hopping, ball catching, and picture drawing.

The study was constructed in this manner to minimize variables that can affect the outcome of a pesticide study on child growth and development. The population had to meet the requirements of similar genetic origin, living conditions, and related cultural and social values and behaviors, all of which are necessary for comparable study and reference groups.

Guillette had assumed that any differences between the two groups would be subtle. Instead, she recalled, "I was shocked. I couldn't believe what was happening" (Luoma 1999). According to an account by Jon R. Luoma in *Mother Jones*:

> The lowland children had much greater difficulty catching a ball or dropping a raisin into a bottle cap—both tests of hand-eye coordination. They showed less physical stamina, too. But the most striking difference came when they were asked to draw pictures of a person. . . . Most of the pictures from the foothill children looked like recognizable versions of a person. The pictures from most of the lowland children, on the other hand, were merely random lines, the kind of unintelligible scribbles a toddler might compose. . . . It appeared likely they had suffered some kind of brain damage. (Luoma 1999)

During a follow-up trip in 1998, two years after her initial visit, Guillette found that both groups of children (who at that point were in primary school) had improved in drawing ability. While the lowland children's drawings looked more like people than before, the foothill children were drawing far more detailed images. The lowland youngsters were still evidencing some motor problems, particularly with balance. "Some of these changes might seem minute, but at the very least we're seeing reduced potential," Guillette said. "And I can't help wondering how much these kinds of chemicals are affecting us all" (Luoma 1999).

Pesticide Exposure's Effects on Child Behavior

No differences were found in physical growth patterns of the two groups of children. Functionally, however, "the exposed children demonstrated decreases in stamina, gross and fine eye-hand coordination, 30-minute memory, and the ability to draw a person" (Guillette et al. 1998). Guillette gave children red balloons for successful completion of tasks. "Well over half of the lesser-exposed children could remember the color in the object, and all remembered they were getting a balloon. Close to 18 percent of the exposed children could not remember anything," and only half could remember they were getting a balloon. "It was quite a contrast," she said (Mann 2000, C-9).

Guillette said she noticed that exposed Yaqui children would walk by somebody and strike them without apparent provocation. Otherwise, they tended to sit in groups and do nothing. Foothill children, by contrast, were always busy, engaged in group play. "I'd throw the ball to a group of kids. In the valley, one child would get the ball and just play with it himself," she said. The foothills children played with the ball as a group (Mann 2000, C-9). Yaqui mothers from the valley also reported more problems getting pregnant and higher rates of miscarriages, stillbirths, neonatal deaths, and premature births.

While it has not been possible to determine the Yaquis' exposure to specific pesticides, Guillette said, "We know for sure there has been DDT exposure." The Mexican government

> does not know what's being used. The farmer does not give out the information. Pesticides are tied to bank loans, and the banks won't reveal what is being used with certain crops. I just assume everything. The other problem is they get a little of this and a little of that and mix it up. It is very important to remember that the situation is no different agriculturally than what you find in California, the Midwest, or the East Coast in the U.S. (Mann 2000, C-9)

Guillette noted:

> Many of these contaminants have similar reactions in the body. Many disrupt the endocrine system, which regulates body functions, and that's the main reason I looked at subtle changes. The shift may seem slight, but when they occur within a total society, they can have major implications. To me, the approach should not be treatment of the disease or trying to teach compensation for the deficit but to look at the basic problem of contamination. (Mann 2000, C-9)

"Valley children appeared less creative in their play. They roamed the area aimlessly or swam in irrigation canals with minimal group interaction. Some valley children were observed hitting their siblings when they passed by, and they became easily upset or angry with a minor corrective comment by a parent" (Guillette et al. 1998). These aggressive behaviors were not noted in the foothills. "Some valley mothers stressed their own frustration in trying to teach their child how to draw," said Guillette and colleagues (Guillette et al. 1998).

Eating, Sleeping, Breathing Pesticides

Exposure to toxic pesticides is one of the greatest risks faced by indigenous migrant workers in Mexico, where tobacco growers and other agricultural companies use them on an industrial scale. Many of the workers are denied safety equipment and access to showers and facilities to wash their clothes after having come into contact with pesticides. In addition, many of the workers live in the same pesticide-laced fields that they harvest, exposing them to contamination around the clock.

During 1993, for example, an estimated 170,000 field workers arrived in the valleys of Sinaloa during each planting season. Among these workers, roughly 5,000 have been found to suffer from toxic contamination as a result of the handling of, or prolonged exposure to, pesticides that are used in cultivation (Diaz-Romo and Salinas-Alvarez, n.d.).

Of the 35,000 agricultural laborers who worked in the San Quintin Valley of Baja California during 1996, 70 percent were indigenous people. The majority of the indigenous migrant workers who work in the agro-industrial fields of northern Mexico are Mixtecos, Triquis, and Zapotecs from Oaxaca, Nahuas, Mixtecos, and Tlapanecos from Guerrero and Purépechas from Michoacan. According to Estela Guzmán Ayala, women (34 percent) and children under 12 years of age (32 percent) constitute two-thirds of the indigenous labor force in the agricultural regions in northern Mexico (Diaz-Romo and Salinas-Alvarez, n.d.).

A report on the travails of the Huicholes said:

> To arrive at the tobacco fields the Huicholes make a journey from the sierras under subhuman conditions, arriving hungry and thirsty. The "valuable and appreciated" human merchandise includes pregnant women, babies incapable of crying, mute from pain, who have recently been born to malnourished mothers or mothers with tuberculosis. Vulnerable elders and even the "strong" men arrive at these centers in weak condition. (Diaz-Romo and Salinas-Alvarez, n.d.)

Favored workers are given purified water, while the remainder are forced to drink water from irrigation channels that draw from the pesticide-laced Santiago River or local wells that also are contaminated with the chemical cocktail used in the tobacco fields. As they toil in the heat, the workers become drenched with sweat, allowing their bodies to absorb pesticide residues more easily. Nicotine in the tobacco also causes skin irritation and hives, a condition called Green Tobacco Sickness. Child laborers are particularly susceptible to the effects of pesticides and the Green Tobacco Sickness (Diaz-Romo and Salinas-Alvarez, n.d.).

The harvesting families often spend the entire day and night in the fields, living and sleeping in boxes or under blankets or sheets of plastic, beneath the strings of drying tobacco leaves, further exposing themselves to toxic chemicals and tobacco residues. Most have no potable water, drainage, or latrines. Occasionally, "The Huicholes use the empty pesticide containers to carry their drinking water without paying notice to the grave dangers that this represents, since the majority cannot read the instructions on the labels which may be written in English" (Diaz-Romo and Salinas-Alvarez, n.d.).

Further Reading

Diaz-Romo, Patricia, and Samuel Salinas-Alvarez. n.d. "A Poisoned Culture: The Case of the Indigenous Huicholes Farm Workers." *Abya Yala News: The Journal of the South and Meso-American Rights Center*. Accessed October 4, 2018. https://saiic.nativeweb.org /ayn/files/original/f2dda1bd26659f3c2dc59a3dd3f7403f.pdf.

Gamlin, Jennie. 2016. "Huichol Migrant Laborers and Pesticides: Structural Violence and Cultural Confounders." *Medical Anthropology Quarterly*, September: 303–320.

Guillette, Elizabeth A., Maria Mercedes Meza, Maria Guadalupe Aquilar, Alma Delia Soto, and Idalia Enedina Garcia. 1998. "An Anthropological Approach to the Evaluation of Preschool Children Exposed to Pesticides in Mexico." *Environmental Health Perspectives* 106, no. 6 (June). Accessed November 29, 2018. https://www.ncbi.nlm.nih.gov /pmc/articles/PMC1533004/ .

Luoma, Jon R. 1999. "System Failure." *Mother Jones*, July/August. Accessed November 29, 2018. https://www.motherjones.com/politics/1999/07/system-failure/.

Mann, Judy. 2000. "A Cautionary Tale about Pesticides." *Washington Post*, June 2: C-9.

Raloff, J. 1998. "Picturing Pesticides' Impacts on Kids." *Science News* 153, no. 23 (June 6): 358.

Valencia, Angel. 2000. Statement of Angel Valencia, Yoemem Tekia Foundation, at the Tucson, Arizona, POPs Negotiations, March 23. Native News: *The Mail* Archive, March 27.

UNITED STATES

Dying Young in "Cancer Alley"

The nickname "Cancer Alley" has been applied to a stretch of the Mississippi River between Baton Rouge and New Orleans, Louisiana, that includes several mostly low-income, mainly African-American communities in the shadows of petrochemical and plastic industries. Fourteen of 15 U.S. plants that produce vinyl chloride monomer (VCM) and ethylene dichloride (EDC), the basic building blocks used to make polyvinyl chloride (PVC), are in Louisiana and Texas. A large number of the incinerators that burn discarded PVC products also have been located in low-income minority communities. Dioxin and dioxin-like pollutants, including PCBs, are produced and released into the air and water as hazardous waste during the manufacture of VCM, the incineration of vinyl products, and the burning of PVC in accidental fires.

The manufacture of polyvinyl chloride produces copious amounts of dioxins, which "have been linked to immune-system suppression, reproductive disorders, a variety of cancers, and endometriosis," according to one observer, who continued, "Dioxins are an unavoidable consequence of making PVC. Dioxins created by PVC production are released by on-site incinerators, flares, boilers, wastewater treatment systems and even in trace quantities in vinyl resins" (Costner 1995). Products manufactured with PVCs "create dioxins when burned, leach toxic additives during use . . . and are the least recyclable of all major plastics" (Cray and Harden 1998). "The production of the carcinogenic monomer is what results in the highest levels of dioxin release," said Charlie Cray, a Greenpeace toxics campaigner ("Shintech: The Battle" 1999).

Suing to Prevent Toxicity

During 1987, 106 residents of Reveilletown, Louisiana, a small African American community about 10 miles south of Baton Rouge, filed a lawsuit against Georgia Pacific and Georgia Gulf arguing that they had suffered health problems and property damage. After settling out of court for an undisclosed amount, Georgia Gulf relocated the remaining families and then tore down every structure in town. Management at Dow Chemical's neighboring factory in Plaquemine followed suit soon afterward, buying out all the residents of the small town of Morrisonville (Bowermaster 1993).

The PVC plants incinerate dioxin-laced wastes, and during the process, according to Greenpeace, a portion of the originally discharged dioxins are emitted undestroyed and new dioxins are created as by-products of incineration. The presence of copper and other metals in PVC-industry wastes can act as a catalyst to further increase dioxin formation (Duchin 1997). Under usual operating conditions, several chemicals are burned in a constantly changing mixture, creating a variety of synergistic chemical-thermal reactions and emissions.

During September 1998, Greenpeace activists joined the people of Covenant, Louisiana, three-quarters of whom are black, in celebration of the news that the Japanese chemical giant Shintech would not, following their protests, build an enormous 3,000-acre polyvinyl chloride (PVC) factory in their town. Shintech had been trying since 1996 to locate three factories and an incinerator near homes and schools in Covenant. Local people argued against location of the plant in their town on civil-rights grounds; they argued that Louisiana authorities would violate federal civil-rights laws if they licensed the Shintech plant in a predominantly African-American community where pollution was already making people sick (Cray and Harden 1998). This struggle was undertaken with the idea of providing other poor, minority communities with legal precedents that would be useful in fending off expansion of environmentally intrusive industries.

Resisting a PVC Plant

The proposed PVC plant would have been one of the largest of its type in the world. According to 42 U.S. Code § 7422, local residents worried most about anticipated emissions of polyvinyl chloride (PVC), ethylene dichloride (EDC), and vinyl chloride monomer (VCM) that "may reasonably be anticipated to result in an increase in mortality or an increase in serious irreversible or incapacitating reversible illness. Vinyl chloride is a known human carcinogen which causes a rare cancer of the liver." According to the company, the proposed Shintech manufacturing plant would have released about 600,000 pounds of toxic chemicals into the air per year and would have poured nearly 8 million gallons of toxic wastewater each day into the Mississippi River, which provides drinking water for the city of New Orleans (Public Notice, n.d.).

Greenpeace's campaign against dioxins began during the 1980s. In 1988, the Greenpeace ship *MC Beluga* toured Cancer Alley; at the same time, Greenpeace also released a report, *We all Live Downstream*, documenting serious chemical pollution in the Mississippi River. During the middle and late 1990s, Greenpeace activists, often swimming at night in polluted water, sampled effluent from several U.S. vinyl producers and found dioxins at each of the 27 sites tested.

One sample, obtained at the Vulcan Chemicals facility in Geismar, Louisiana, contained six parts per million TEQ dioxin, a level as high as the historic levels in wastes left from Agent Orange production (Dioxin Deception 2001). "These data are important, since they add to the growing body of evidence pointing to the life cycle of polyvinyl chloride (PVC) plastic as one of the largest single sources of the nation's total dioxin burden" (Duchin 1997).

Residents in Covenant turned their backs on arguments that the plant would provide them jobs and economic prosperity. They argued that the town already hosted a bevy of high-technology industries that emit various toxins, yet 40 percent of the townspeople lived below the poverty line. Most of the plants were highly automated with little need for workers possessing marginal skills. Instead, the plants tended to hire a few people (usually from the outside) with computer skills and a working knowledge of physics and chemistry.

After the EPA rejected its air permit, the Louisiana Department of Environmental Quality was required, by law, to convene public hearings on Shintech's application. Nearly 300 people attended the hearings in Convenent on December 9, 1996. Testimony continued for more than eight hours, according to one account, at least 95 percent against Shintech. "Enough is enough," testified Patricia Melancon, president of the local group St. James Citizens. "This is a low-income area, and less than half of the adults in this community have a high-school diploma. We will get all of the pollution but none of the jobs" (Cray 1996).

Another Town, Same Issues

After withdrawing from Covenant, Shintech switched its building plans to Plaquemine, 30 miles north, where its executives hoped that local opinion would be more accommodating. Shintech said it planned to buy vinyl chloride monomer (feedstock finished PVC) from a Dow Chemical plant in Plaquemine. People in Plaquemine then organized People Reaching Out To Eliminate Shintech's Toxins (PROTEST). In their opposition to new PVC manufacturing there, the members of PROTEST cited already-high cancer rates in the town, falling property values, and a lack of emergency-evacuation routes should an accident occur. Liz Avants, speaking for PROTEST, said, "Every day, we hear about more cases of cancer and other health effects" ("Shintech: The Battle" 1999).

On March 5, 1999, 60 Greenpeace activists from 22 different nations converged on the Louisiana State House to call attention to global health and environmental threats caused by Louisiana's numerous polluting PVC production facilities. Dressed in T-shirts with the slogan "Love Louisiana But Not PVC" in 16 languages, the Greenpeace Toxic Patrol called on Governor Mike Foster to clean up the state's Cancer Alley. The group then marched with local environmental activists from the capitol to the governor's mansion to deliver a "lunch" of contaminated fish and water from some of the state's most polluted waterways ("Louisiana's Cancer" 1999).

Greenpeace's "Toxics Patrol"

On June 22, 1999, Greenpeace launched a "Toxics Patrol" bus trip to more than a dozen of Louisiana's chemical facilities whose emissions have made Louisiana a "global toxic hot spot." The bus tour began with stops at three controversial facilities: Rhodia, the nation's first napalm incinerator; Formosa, which makes components of vinyl; and Dow Chemical, another vinyl producer, which planned to join with Shintech on a newly proposed vinyl plant in West Baton Rouge. With representatives of several state and local environmental groups, Greenpeace displayed posted signs at chemical facilities warning that toxic pollution "does not stop at the fence" ("With Public's" 2010).

"Louisiana ranks number one in the nation in per-capita toxic releases to the environment, and her citizens are bearing a terrible health burden for it," said Greenpeace toxics campaigner Damu Smith. "Our Toxics Patrol is out to expose some of the state's worst toxic offenders. . . . Louisiana is at the center of the nation's growing problem of environmental racism and injustice" ("With Public's" 2010).

Acclaimed author Alice Walker, actress Alfre Woodard, actor Mike Farrell, Reverend Al Sharpton, and members of Congress Maxine Waters and John Conyers toured "Cancer Alley" on June 9, 2001 ("Celebrities to Tour," n.d.). After the walking tour, the delegation convened in New Orleans for the first-ever National Town Meeting on Environmental Justice with residents, government officials, and representatives of the chemical industry. During the town meeting, the delegation heard personal testimonies of industrial-contamination victims who lived in "Cancer Alley" communities along the Mississippi River between New Orleans and Baton Rouge who were experiencing a variety of illnesses and social problems associated with toxic pollution. The town meeting was coordinated and sponsored by Greenpeace ("Celebrities to Tour," n.d.).

Bill Moyers's two-hour special report, *Trade Secrets*, aired on the Public Broadcasting System on March 26, 2001. It described the lives and deaths of some "Cancer Alley" workers, such as Ray Reynolds, 43 years of age, who was shown in the living room of his house a few miles from the chemical plant where he'd worked for 16 years. Reynolds was shown dying of toxic neuropathy that had spread from his nerve cells to his brain. Another worker, Dan Ross, made his living for 23 years producing the raw vinyl chloride that is basic to the manufacture of PVC plastic. In 1989, Ross was told he had a rare form of brain cancer.

Corporate documents displayed by Moyers indicated that manufacturers of PVCs knew as early as 1959 that "500 parts per million is going to produce rather appreciable injury when inhaled seven hours a day, five days a week for an extended period" (Moyers 2001). As the years went by, the level at which exposure to PVC and other organochlorines were believed to cause injury to the human body crept downward. In 1959, workers were regularly exposed to at least 500 ppm during their work shifts. Some workers described standing in clouds of the chemical at levels much above 500 ppm. Some X-rays showed workers' bones dissolving, obvious evidence that exposure was cursing them with early deaths in "Cancer Alley."

Further Reading

Bowermaster, J. 1993. "A Town Called Morrisonville." *Audubon*, July/August 1993: 42–51.
"Celebrities to Tour 'Cancer Alley,' Louisiana; Alice Walker, Alfre Woodard, and Mike Farrell among Speakers at National Town Meeting on Environmental Justice." n.d.

Costner, Pat. 1995. "PVC: A Primary Contributor to the U.S. Dioxin Burden." Comments submitted to the U.S. EPA Dioxin Reassessment. Washington, D.C. Greenpeace U.S.A., February.

Cray, Charlie. 1996. "Hundreds Oppose Shintech Proposal in Louisiana. Citizens and Other Interest Groups Cite Health Concerns." Greenpeace. December 9.

Cray, Charlie, and Monique Harden. 1998. "PVC and Dioxin: Enough Is Enough." *Rachel's Environment and Health Weekly* 616 (September 18). Environmental Research Foundation, Annapolis, Maryland.

"Dioxin Deception: How the Vinyl Industry Concealed Evidence of Its Dioxin Pollution." 2001. Greenpeace. March 27.

Duchin, Melanie. 1997. "Greenpeace's Secret Sampling at U.S. Vinyl Plants: Dioxin Factories Exposed." April. Accessed November 29, 2018. http://pvcinformation.org/assets/pdf/DioxinFactoriesExposed.pdf.

"Louisiana's Cancer Alley an International Threat." 1999. Environment News Service, March 5.

Moyers, Bill. 2001. *Trade Secrets: A Moyers Report*. Program transcript. Public Broadcasting Service, March. Accessed October 4, 2018. http://www.pbs.org/tradesecrets/transcript.html.

Public Notice, Air Permit Application, Shintech Corporation, 1997 (December), and Shintech Application to Discharge Process Wastewater, n.d.

"Shintech: The Battle Continues." 1999. *Earth Action Network, Inc.*, March–April. https://www.thefreelibrary.com/SHINTECH%3A+THE+BATTLE+CONTINUES.-a054233048.

Wade, Elliott. 2018. "Oil over Lives: Louisiana's Cancer Alley Calls for Change," Allons, August 28. Accessed October 4, 2018. https://www.thevermilion.com/allons/oil-over-lives-louisiana-s-cancer-alley-calls-for-change/article_7f2f23c6-aa73-11e8-92c5-779fddb12b94.html.

"With Public's 'Right-To-Know' in Jeopardy, Greenpeace Kicks Off Bus Tour of Louisiana's Worst Chemical 'Hot Spots.'" 2010. Greenpeace, July 6. Accessed November 29, 2018. https://www.greenpeace.org/usa/news/greenpeace-posts-signs-in-calc/.

VIETNAM

Agent Orange's Generational Deformations

At the time of the first Earth Day, in the spring of 1970, the United States was pouring dioxin (as an active ingredient of Agent Orange) on the jungles of Vietnam, Laos, and Cambodia, in an attempt to defoliate the jungles and deny Viet Cong insurgents places to hide from aerial bombing. The guerillas were said to be "fish" in a "sea" of rural peoples that would be stripped bare of vegetative cover by defoliants. Between 1962 and 1971, at least 12 percent of southern Vietnam's land area was doused liberally with nearly 18 million gallons of 2,3,7,8-tetrachlorodibenzo-p-dioxin (TCDD), the most potent of dioxin's many varieties (Schecter et al. 2001, 435).

United States armed forces dropped more bombs (measured by weight) on Vietnam than it dropped in the entire Pacific Theater during World War II. By 1971, more than 600 pounds of bombs *per person* had been rained on Vietnam.

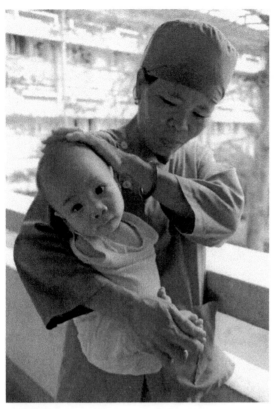

A Vietnamese nurse holds an armless one-year-old baby allegedly affected by Agent Orange, a pesticide used during the Vietnam War that caused congenital malformations. (AFP/Getty Images)

Between 12 percent (U.S. figure) and 43 percent (a National Liberation Front figure) of South Vietnam's land area was sprayed at least once with defoliants, usually Agent Orange (Johansen 1972, 4).

Eighteen million gallons of Agent Orange were sprayed over vast tracts of Southeast Asian forests between 1962 and 1971 in concentrations up to 1,000 times as potent as dioxin-based herbicides sold over the counter in the United States (McGinn 2000, 26). Large areas of the countryside became unfit for human habitation during the war and for several years thereafter. As noncombatants fled poisoned, defoliated, and bombed-out rural areas, the population of Saigon, now Ho Chi Minh City, increased tenfold between 1954 (when the war began with French intervention) and 1970, from 300,000 to 3 million people (Johansen 1972, 4).

"Monster Babies"

Most of the herbicides, including Agent Orange, were applied to areas without prior hazard testing. Agent Orange was applied in large amounts, often haphazardly, by troops taking part in "Operation Ranch Hand," whose participants proclaimed, "Only we can prevent forests," an allusion to Smokey the Bear's slogan, "Only you can prevent forest fires" (Johansen 1972, 4). Samples collected between 1970 and 1973 documented elevated levels of TCDD in milk samples, as well as in fish and shrimp. Nursing mothers who had been heavy consumers of fish were found to have the highest levels in their blood (Schecter et al. 2001, 435).

Soon after spraying of Agent Orange and other herbicides began during the late 1960s, reports increased of deformed births in unusually large numbers. Areas sprayed with Agent Orange later reported very high incidences of certain birth defects: anencephaly (absence of all or parts of the brain), spina bifida (a malformed vertebral column), and hydrocephaly (swelling of the skull).

The Saigon newspaper *Tin Sang* published descriptions of "monster babies" born to mothers in areas that had been sprayed. The newspaper reported that one woman "reported that her newly pregnant daughter was caught in a chemical strike and fainted, with blood coming out her mouth and nostrils and later from her vulva. She was taken to a hospital where she was later delivered of a deformed fetus" (Johansen 1972, 4). The same day (October 26, 1969), *Dong Nai,* another Saigon newspaper, published a photograph of a stillborn fetus with a duck-like face and an abnormally twisted stomach. A day later, the newspaper reported that a woman in the Tan An district who had been soaked with Agent Orange had given birth to a baby with two heads, three arms, and 20 fingers (Johansen 1972, 4). Many other similar accounts were published before the South Vietnamese government shut down the newspapers for "interfering with the war effort" (Johansen 1972, 4). The South Vietnam health ministry also began to classify accounts of deformed births as state secrets.

When these accounts were presented to U.S. Department of Defense officials, they were, at first, dismissed as unconfirmed enemy propaganda. When the accounts persisted and became more specific and numerous, the United States armed forces finally stopped using Agent Orange. The U.S. Air Force later found a "significant and potentially meaningful" relationship between diabetes and bloodstream levels of dioxins in its ongoing study of people who worked with the defoliant Agent Orange during the Vietnam War (Brown 2000, A-14; Institute of Medicine 1994). Members of the U.S. armed services who were exposed to high levels of dioxins were found to be more prone to development of diabetes compared to those with lower levels of exposure. People with the highest exposure levels developed diabetes most rapidly. While it once dismissed reports of cancers caused by Agent Orange as groundless, three decades later the U.S. Army was giving a special medallion—the Order of the Silver Rose—to soldiers who have been afflicted.

Agent Orange and Risk of Diseases

By mid-2001, the U.S. Department of Veterans Affairs was soliciting applications or compensation from Vietnam veterans with any of a large number of "presumptive disabilities": chloracne, Hodgkin's disease, multiple myeloma, non-Hodgkin's lymphoma, soft-tissue sarcoma, acute and subacute peripheral neuropathy, spina bifida, and prostate cancer ("Bulletin Board" 2001). The same request for claims asserted that diabetes mellitus soon would be included in its list of dioxin-induced pathologies.

At roughly the same time, a panel advising the U.S. Environmental Protection Agency, well-stocked with industry representatives, was still arguing whether dioxin should be classified as carcinogenic for human beings. After ten years of work, during the summer of 2000, the EPA released a 3,000-plus-page *Draft Dioxin Reassessment* (Environmental Protection Agency 2000), which concluded that "TCDD (and possibly other closely related structural analogs, such as the chlorinated didenzofurans) are carcinogenic in humans and can cause immune-system alterations; reproductive, developmental, and nervous system effects; endocrine disruption; altered lipid metabolism; liver damage; and skin lesions" (Schecter et al. 2001, 436). The EPA study confirmed many other studies that had linked TCDD and other forms of dioxin to "cancer and cancer mortality at relatively high levels in chemical workers and in toxicity studies" (Fingerhut et al. 1991; Flesch-Janys et al. 1995; Flesch-Janys et al. 1998; Becher et al. 1998; *Report on Carcinogenesis* 1998).

Effects on Succeeding Generations

Men who sprayed Agent Orange in Vietnam between 1962 and 1971 were followed to determine whether exposure to dioxin affected their children. Nervous-system disorders were found to be widespread. Spontaneous abortions, birth defects, and developmental delays also were noted; paradoxically, men who received low doses of dioxin tended to give birth to more children with these problems than those who had been exposed at higher levels (Wolfe et al. 1995). Newspaper reports more than 40 years later indicated that medical problems had been passed down to U.S. veterans' children and grandchildren (Liewer 2016, 1-A, 5-A).

Dioxin levels remained very high for several decades in some areas of Vietnam that had been sprayed with Agent Orange more than three decades earlier (Schecter et al. 1992). Tests of people in the city of Biên Hòa (population 390,000), about 20 miles north of Ho Chi Minh City (formerly Saigon), in particular evidenced dioxin readings 135 times higher than levels in Hanoi, which was not sprayed. Levels in Hanoi were measured at two to three parts per billion TCDD (roughly background level in today's world), while blood levels of 271 ppb were found in the blood of people living in Biên Hòa (Schecter et al. 2001, 435).

The research of Arnold Schecter (an environmental scientist at the University of Texas School of Public Health) and colleagues clearly indicated how residents of Biên Hòa—some of whom had not been born during the war—continued to acquire contamination. The dioxin first dumped in the area by U.S. armed forces was bioaccumulating up the food chain, from phytoplankton to zooplankton and then to fish consumed by people.

Effects Pervasive Decades Later

Thanh Xuan, a "peace village" near Hanoi, housed a hundred children who are mentally or physically disabled, some with stunted limbs or twisted spines. Most arrived at the "peace village" unable to walk, speak, or read. Across Vietnam, rates of birth defects, miscarriages, and other complications were still uncommonly high almost three decades after the spraying of Agent Orange ended during 1971. Many of the deformed children in the "peace village" were born to parents who were sprayed during the war.

"If I wasn't here, I don't know what I would do," lamented Nguyen Kim Thoa, 15, sitting in her bedroom beneath a Britney Spears poster. A reporter described Thoa's delicate features, "wrapped in a shroud of spongy skin tumors and charcoal splotches sprouting bristles" (Verrengia 2000). "I wasn't able to go to school at home," Thoa said. "The children always made fun of me. In their eyes, I was a freak. Here, I have friends and teachers who love me" (Verrengia 2000). Thoa's father served in the Vietnamese Army between 1978 and 1980 along the Cambodian border, an area that was heavily sprayed during the war.

Hoang Dinh Cau, chairman of a national Vietnamese panel that investigates the war's ongoing health consequences, estimates that about 1 million Vietnamese people have been afflicted with dioxin poisoning, including 150,000 children. Thirty years after the war, some rice paddies that were abandoned after spraying have not been reclaimed, as "soaring forests with 1,000 different tree species shriveled, replaced by weedy meadows that livestock won't graze. Farmers call the new growth 'American grass'" (Verrengia 2000).

In a sparsely decorated bedroom of a two-story concrete building, Bui Dinh Bi recalled his days with communist forces in Quang Tri, in what was then called South Vietnam. During the early 1970s, after he was exposed to Agent Orange, Bui's skin lesions changed from mosquito-bite-like bumps to tumors of a type that covered his body 30 years later. Bui and his wife had eight children. The first was stillborn, and then the next five died in infancy. Their two surviving children are mentally disabled (O'Neill 2000). Bui lived with 29 other veterans and 70 children in Friendship Village, near Hanoi, one of about a dozen similar communities that the Vietnamese government established for veterans and children afflicted with dioxin toxicity.

The *Washington Post* reported in April 2000 that "Canadian researchers have found high levels of dioxins in children [who] were born long after the spraying ceased in 1971. The lingering contamination is so severe in some areas that if they were in the United States, they would be declared Superfund sites, requiring an immediate cleanup effort" (O'Neill 2000).

Further Reading

Becher, H., K. Steindorf, and D. Flesch-Janys. 1998. "Quantitative Cancer Risk Assessment for Dioxin Using an Occupational Cohort." *Environmental Health Perspectives* 106 (1998): 663–670.

Brown, David. 2000. "Defoliant Connected to Diabetes." *Washington Post*, March 29: A-14. Accessed October 4, 2018. https://www.washingtonpost.com/archive/politics/2000/03/29/defoliant-connected-to-diabetes/281227f4-094f-44fd-b35e-212f825f6f8f/.

"Bulletin Board: Vietnam Veterans Benefit from Agent Orange Rules." 2001. *Indian Country Today*, May 16: D-4.

Fingerhut, M. A., W. E. Halperin, D. A. Marlow, L. A. Piacitelli, P. A. Honchar, M. H. Sweeney, . . . A. J. Suruda. 1991. "Cancer Mortality in Workers Exposed to 2,3,7,8-tetrach lorodibenzo-p-dioxin." *New England Journal of Medicine* 324, no. 4 (January): 212–218.

Flesch-Janys, D., J. Berger, P. Gurn, A. Manz, S. Nagel, H. Waltsgott, and J. H. Dwyer. 1995. "Exposure to Polychlorinated Dioxins and Furans (PCDD/F) and Mortality in a Cohort of Workers from a Herbicide-Producing Plant in Hamburg, Federal Republic of Germany." *American Journal of Epidemiology* 142, no. 1 (December): 1165–1175.

Flesch-Janys, D., K. Steindorf, P. Gurn, H. Becher. 1998. "Estimation of the Cumulated Exposure to Polychlorinated Dibenzo-p-dioxins/furans and Standardized Mortality Ratio Analysis of Cancer Mortality by Dose in an Occupationally Exposed Cohort." *Environmental Health Perspectives* 106 (April): 655–662.

Institute of Medicine. 1994. *Committee to Review Health Effects in Vietnam Veterans of Exposure to Herbicides: Veterans and Agent Orange.* Washington, DC: National Academy Press.

Johansen, Bruce. 1972. "Ecomania at Home; Ecocide Abroad." University of Washington *Daily*, May 24: 4.

Liewer, Steve. 2016. "A Toxic Legacy." *Omaha World-Herald,* June 5: 1-A, 5-A.

O'Neill, Annie. 2000. "Damaged Lives: Vietnamese Veterans and Children: While World Leaders Debate the Effects of Agent Orange, a Multinational Project Reaches Out to People at the Center of the Storm." *Pittsburgh Post-Gazette*, November 5. Accessed November 29, 2018. http://old.post-gazette.com/magazine/20001105agentorange1.asp.

Report on Carcinogenesis: TCDD. 1998. Bethesda, MD: National Institute of Environmental Health, National Toxicology Program.

Santillo, David. 2000. "World Chemical Supplies Contaminated with Toxic Chemicals." Greenpeace Listserv, March 19.

Schecter, A., O. Päpke, M. Ball, D. C. Hoang, C. D. Le, Q. M. Nguyen, . . . J. Spencer. 1992. "Dioxin and Dibenzofuran Levels in Blood and Adipose Tissue of Vietnamese from Various Locations in Vietnam in Proximity to Agent Orange Spraying." *Chemosphere* 25, no. 7–10 (October–November): 1123–1128.

Schecter, Arnold, Le Cao Dai, Olaf Päpke, Joelle Prange, John D. Constable, Muneaki Matsuda, . . . Amanda L. Piskac. 2001. "Recent Dioxin Contamination from Agent Orange in Residents of a Southern Vietnamese City." 2001. *Journal of Occupational and Environmental Medicine* 43, no. 5 (May): 435–443.

U.S. EPA Draft Dioxin Reassessment. 2000. Environmental Protection Agency.

Verrengia, Joseph B., and Tini Tran. 2000. "Vietnam's Children Feeling Effects of Agent Orange." Associated Press in *Amarillo Globe-News*, November 20.

Wolfe, W. H., J. E. Michalek, J. C. Miner, A. J. Rahe, C. A. Moore, L. L. Needham, and D. G. Patterson Jr. 1995. "Paternal Serum Dioxin and Reproductive Outcomes Among Veterans of Operation Ranch Hand." *Epidemiology* 6, no. 1 (January): 17–22.

Chapter 7: Pollution of Air and Water

OVERVIEW

Pollution has spread with the advent of fossil-fueled industry around the Earth. Where at first London, Los Angeles, and Pittsburgh were famous for their smogs (the word was born in Great Britain as an amalgam of "smoke" and "fog"), today the worst air in the world envelops New Delhi and Beijing. The world's industrial heartland has been "outsourced" to Asia. London and Mexico City still have toxic air when atmospheric conditions are stagnant, but the "brown cloud" today belongs to southern and eastern Asia.

Mining of valuable minerals provides many other sources of pollution, and here the toxic frontier has been outsourced around the world as well. Siberia, which is sparsely populated and part of the largest land mass on Earth, is home to many Native peoples who share one major problem: the Russian state's appetite for resources—oil, natural gas, uranium, and others—which lie on or under their lands. From the Yamal peninsula on the Arctic Circle, northeast of the Ural Mountains, to Sakhalin Island, north of Japan, indigenous peoples of Siberia have been confronted with large-scale development of oil and gas, uranium, and other resources that has been carried out largely heedless of environmental contamination.

Twenty-six distinct indigenous peoples live in Siberia, ranging in number from fewer than 200 (the Oroks) to as many as 34,000 (the Nenets); their total population is more than 160,000. Two other, larger indigenous peoples, the Sakha (formerly called the Yakuts) and the Komi, have their own republics within the Russian federation. Most of the indigenous peoples who live in the tundra rely on reindeer herding, while those who live in the forest and near the sea subsist on hunting and fishing. The reindeer herders are nomadic, following the reindeer in a cyclical pattern of migration, living in movable houses made from reindeer skins (*urangas*). The hunters live in permanent settlements (which traditionally have been made of bark) or in semi-subterranean huts covered with earth and moss. Today, roughly 10 percent of the indigenous people continue to migrate (30 years ago, 70 percent migrated); most of the others live in settlements, with almost half the population involved in herding, fishing, and hunting ("Peoples of the Frozen North," n.d.).

The sparsely populated lands of the Amazon basin also have become frontiers of pollution from mining waste as the tendrils of industrialization reach the most remote locations. Some small Pacific Ocean islands now have nearly been eaten away in pursuit of copper. People living near the Peruvian village of La Oroya face

Air pollution in New Delhi, India, is the dirtiest on Earth, as effluent from coal power generation (above) mixes with vehicle exhaust and agricultural fires. (Ajay Bhaskar/Dreamstime.com)

environmental problems similar to those living near the Yanacocha gold mine: pollution of the water supply from cyanide, air pollution, and noisy truck traffic near the mine. The pollution destroys pasturelands and sickens livestock, debilitating local agriculture.

People who live with pollution are protesting its degradation of their lives. China, with its 1.4 billion people, has some of the worst air and water pollution on the planet. More often in recent years, Chinese people have been organizing to pressure local, regional, and national governments to clean up pollution. Often, they organize in defiance of a ruling Communist Party that asserts control over all media and political activity. More often, with an assist from social media, environmental organizers have been outwitting the control system, as government has been forced to deal with pollution. Air pollution is a lively subject in Delhi. Given recent publicity, many people believe that the world's worst urban air pollution is in Beijing. Actually, China's capital runs second in this noxious sweepstakes as the world capital of smog to the Ganges River valley around Delhi, according to the World Health Organization. Delhi's highest court has called Delhi's atmosphere "like a gas chamber" (Sudarshan 2015, 11). That's an apt and brutally honest assessment of living conditions in many of the world's industrial workshops today.

This chapter will examine sewage pollution in Brazil, China's fight for clean air, smog in India, Italy and its struggle to combat air pollution, Japan's leadership in cleaning its air and water, Mexico's tainted water and air pollution, the effects of industrialization on the air and water in Russia, and the return of smog in the United Kingdom

Further Reading

"Peoples of the Frozen North." 1998. Survival International. https://assets.survivalinterna tional.org/static/files/related_material/39_21_81_siberiabg.pdf.

Sudarshan, V. 2015. "In Delhi, Just Waiting to Exhale." *The Hindu*, December 17: 11.

BRAZIL

Rio de Janeiro's World-Famous Sewage

Due to Rio de Janeiro's role as host of the 2016 Summer Olympics, the city also became known worldwide as home to some of the world's most significant sewage. Rio de Janeiro (often known simply as "Rio") is the second-most populous metropolitan area in Brazil, with 16.5 million people (after São Paulo, with 21.2 million). Rio also has the sixth largest metropolitan population in the Western Hemisphere and is one of the most densely settled urban areas anywhere. A large number of people have migrated into Rio during recent years, and many of them have taken up residence in shantytowns without municipal services, including sewage pipes or treatment. Sewage from these shantytowns and many other urban neighborhoods is routinely piped (or simply flows downhill) into Guanabara Bay, where the Olympics staged some of its outdoor events (swimming and diving were held in an indoor pool).

"From the top of Rio de Janeiro's Sugarloaf or the Christ the Redeemer statue that towers above the city, Guanabara Bay looks picturesque. But the ailing estuary, site of the Olympics' sailing and rowing, is anything but pristine," wrote Lindsay Konkel for the *National Geographic* (2016). Athletes practicing for the 2016 Games complained that the water in the bay was littered with trash and that its chemicals irritated their skin and caused stomach problems. Those who competed on boats were careful to avoid falling into the water or even being splashed with it.

A Toxic Brew

The toxic brew in Rio's water can cause water mains to burst after heavy rainfalls. Alfonso Stefanini, an environmental consultant there, wrote in the *Rio Times*:

> Not even "Alerta-Rio," the City Halls environmental warning system's loudest conch blow, will keep the "dudu" and "kaka"—also two popular nicknames in Brazil—from hitting the ceiling fans in the low-lying sea margin zones throughout the metropolitan area where the majority of the state's population lives. . . . Only after people are critically hurt, property heavily damaged, and entire city blocks turned upside-down, ex post facto mitigation measures are put in place, but always after companies and government entities come to a dickering halt as to who was to blame; in other words, "Whoever smelled it, dealt it." (Stefanini 2016)

Rio's water pollution problems are actually quite routine for large, growing urban areas in many developing countries. "According to the United Nations," wrote Konkel, "up to 90 percent of wastewater in developing countries flows untreated into waterways used for bathing, drinking, or fishing" (Konkel 2016).

In addition to its iconic tourism and festivals, its Sugarloaf Mountain and the Christ the Redeemer statue, Rio de Janeiro hosts a large number of major industrial plants that pour sewage, oil residues, and heavy metals into the bay. It also has lost many mangrove swamps that acted to filter out the lower pollution levels of earlier years. The bay also acts as the final deposit for pollution flowing downstream from several rivers.

Neighboring Sepetiba Bay also is subject to many of the same environmental problems, which are generated by 1.3 million people living on its shores. Rio and surrounding towns and cities, like a number of other major urban areas, are also subject to air pollution from cars and trucks, as well as industrial activity that dumps heavy metals (mainly zinc and cadmium) from factories in Santa Cruz, Itaguaí, and Nova Iguaçu. Urbanization, including construction of large apartment blocks that lack sewage systems, also have despoiled the Marapendi and Rodrigo de Freitas lagoons. As authorities have failed to enforce existing environmental laws prohibiting clandestine discharge of sewage into these waters, these and other effluents have contributed to algae proliferation that has drastically reduced oxygen levels and caused massive fish kills.

A Partnership, but Few Changes

A partnership of public and private interests was convened in 2008 to decontaminate the waters of these lagoons to a level that would allow safe use by people and a revival of animal life. Plans have been announced to remove contaminated sludge from the lagoons and create cleansing conduits between these bodies of water and the ocean. Little actual progress was made in time for the 2016 Summer Olympics, however, when rowing competitions were held in the lagoons.

"With thousands of liters of raw human sewage pouring into the ocean every second from Rio de Janeiro, August's Olympic Games have trained a spotlight on Rio's spectacular failure to clean up its waterways and world-famous beaches," commented Jenny Barchfield on NBC News. "Just across the Guanabara Bay, its sister city of Niteroi is showing that a real cleanup is possible" (Barchfield 2016).

> In Niteroi, 95 percent of sewage is treated, and authorities say they are on track for 100 percent within a year, even though Rio's failure to do its part means that sludge still flows in from across the bay. Rio has not only broken promises made to fix its sewage problem in time for the upcoming Summer Games, but the state has been downplaying expectations, even suggesting it might be 2035 before a full cleanup happens. (Barchfield 2016)

So what is in Guanabara Bay? The *National Geographic* tested the waters and found that a lot of the pollution comes from raw sewage. Roughly half of the houses in the Guanabara Bay drainage basin—Rio de Janeiro and the surrounding cities—remain unconnected to sewage treatment plants. That means waste from millions of people flows untreated into the bay. Experts say the percentages of households without a sewage system may be much higher in slums and shantytowns where some houses have no access at all to toilets or tap water. Sewage overflows worsen during the rainy season from October to April, when rainfall can overwhelm the sewers that carry waste to sewage-treatment plants. (Konkel 2016)

A "Petri Dish of Pathogens"

The land area around Guanabara Bay contains about 17,000 industries—oil and gas terminals, fuel refineries, pharmaceutical factories, and others—that emit a daily average of 150 metric tons of waste. In addition, human waste includes a "petri dish of pathogens" in feces that may contain viruses and bacteria. Adenovirus can cause several problems, "from diarrhea to respiratory, eye, and skin infections," said Kristina Mena, an associate professor at the University of Texas Health Science Center at Houston School of Public Health (Konkel 2016). Some of the viruses in Guanabara Bay have become antibiotic-resistant, including at least one strain of "superbug" that scientists classify as carbapenemase-producing bacteria (CPB) that were found at five Rio-area beaches, including the famous Copacabana. These bacteria are usually detected in hospital waste.

Not everyone who comes into contact with potentially dangerous pathogens becomes ill. The human immune system combats many of them. Some athletes at the Olympics adopted a common-sense approach. "We just have to keep our mouths closed when the water sprays up," Afrodite Zegers, 24, a member of the Dutch sailing team, which practiced in Guanabara Bay, told the *New York Times* (Irby, 2016). "There were roughly 12 [wind]surfers last year that fell ill and had very similar symptoms to me. Nausea, dizziness, vomiting, diarrhea," Ace Buchan, one of the world's top surfers, told Public Radio International. "That's nearly a quarter of all the competitors, male and female" (Irby 2016).

In the Guanabara Bay area, Mena says the viral counts are higher than they have ever been, in her experience there. "This is such an extreme situation, both in the range of microorganisms present and the extent of exposure. There are going to be health risks," she said (Konkel 2016). Residents of the Rio area do suffer from many waterborne illnesses, "especially Rio's poorest people who live near the most polluted parts of the bay and have the least access to sanitation," said Ricardo Igreja, an infectious disease doctor at the Federal University of Rio de Janeiro (Konkel 2016). Local waters also routinely test high for E. coli from human waste.

Many of the threats may be long-term (not obvious after immediate contact), including contamination by PCBs, industrial chemicals, heavy metals, and hydrocarbons (oil by-products). Eating seafood from local waters is a common source, said Abílio Soares Gomes, a marine biologist at the Federal Fluminense University in Niterói, a Rio suburb (Konkel 2016). Historical accounts indicate that local waters were rich in dolphins, whales, and many more species of fish than are present today. Remaining sea life contains high levels of mercury, other industrial chemicals, and heavy metals that compromise their hormonal and immune systems. Oil spills have reduced populations of maritime creatures, as have algal blooms that reduce waterborne oxygen levels.

After the Olympics, not much changed in Rio. Will Carless of Public Radio International (PRI) noted that "black, fetid water laps against the edge of Rio de Janeiro's Guanabara Bay. Greasy sand is so blanketed in trash and filth that it's hard to tell where the beach ends and the water begins" (Carless 2017). About all that Rio's authorities did was dispatch "eco-boats" to collect and contain garbage. They did little but impede fishermen.

Even after the city of Rio de Janeiro received several grants totaling millions of dollars over 20 years from the Inter-American Development Bank, the World Bank, and the International Olympic Committee for wastewater treatment, by 2016 only a third of the flow was being processed. "Not that the treatment plants don't work," wrote Stefanini (2016). "The problem is that the majority of the waste is not collected from its source. Meanwhile in places like California, one of the top ten biggest economies in the world like Brazil, companies are fighting to own every last droplet of sewage to turn it into usable water and valuable material residues" (Stefanini 2016).

Further Reading

Barchfield, Jenny. 2016. "Rio's Sewage Problem May Not Be Fixed until 2035." NBC News, June 7. Accessed October 4, 2018. http://www.nbcwashington.com/news/national-in ternational/Rio-Sewage-May-Not-Be-Treated-Until-2035--382094531.html.

Carless, Will. 2017. "Rio's Water Cleanup Barely Works and It's Crimping Impoverished Fishermen." Public Radio International, February 9. Accessed October 4, 2018. https://www.pri.org/stories/2017-02-09/rio-s-water-cleanup-barely-works-and-it-s -crimping-impoverished-fishermen.

Irby, Kate. 2016. "Olympians Will 'Literally Be Swimming in Human Crap,' Scientists Say." July 28. Accessed October 4, 2018. http://www.miamiherald.com/news/nation-world /world/article92337342.html#storylink=cpy.

Konkel, Lindsay. 2016. "What's in Rio's Bay and Beaches?" *National Geographic,* August 15. Accessed October 4, 2018. http://news.nationalgeographic.com/2016/08/what-s -in-rio-s-bay-/.

Stefanini, Alfonso. 2016. "Rio de Janeiro and the Lords of Sewage." *Rio Times,* November 1. Accessed October 4, 2018. http://riotimesonline.com/brazil-news/tag/sewage/.

CHINA

Resistance to Air, Soil, and Water Pollution

China, with its 1.4 billion people, has some of the worst air and water pollution on the planet. Recently, Chinese citizens have organized locally, regionally, and nationally in an attempt to pressure the government to try and put an end to pollution.

China's "Pollution Census"

In December 2017, China took account of its pollution sources in a national "census" for a second time in a decade to establish a baseline for assessing and measuring the impact of environmental taxes that begin in 2018—one sign that the central government has been working on addressing at least some of the country's monumental environmental problems. Polluting industries will be required to acquire discharge permits beginning in 2020.

Scientists reported in *Nature* that the 2017 pollution census

targeted nearly 6 million industrial, agricultural, and residential sources and centralized pollution-control facilities. However, large inconsistencies between the census data and those collected from other sources created big problems for the nationwide control system for total emissions. For instance, environmental statistics put the total discharge of organic pollutants in water at 15 million tonnes, whereas the census figure was double that. (Hu et al. 2017)

Further Reading

Hu, Qing, Xuetao Zhao, and Xiao Jin Yang. 2017. "China's Decadal Pollution Census." *Nature* 543 (March 23): 491. Accessed October 11, 2018. http://www.nature.com/nature/journal/v543/n7646/full/543491d.html

This presents a challenge, given the ruling Communist Party's control over media and political activity. Social media have helped environmentalists around this issue, and the government has been forced to clean up the pollution.

Environmental Protests Swell

Modern environmental protest in China originated in 2007 in Xiamen, an affluent city on China's southeast coast, with street demonstrations against a factory producing paraxylene (PX), which is added to fabrics. Demonstrators blocked PX-producing plants and then expanded their campaign to a nuclear reprocessing plant that they said was dangerous to their health. At one protest, thousands of people in Xiamen forced the government to suspend construction of a proposed plant that manufactures PX. *Forbes Asia* reported, "Faced with unprecedented numbers of protesters from among the urban and middle class rather than disenfranchised farmers, the provincial government quietly changed its plans and built the plant in poorer, smaller Zhangzhou, the site of [a] recent [chemical] accident" (Hoffman and Sullivan 2017). In April 2015, a factory in Zhangzhou, Fujian Province, had exploded, requiring, according to one account, "the attention of the Chinese army's anti-chemical warfare unit and the evacuation of 30,000 people" (Hoffman and Sullivan 2015).

China is the world's number-one producer and consumer of PX, a petrochemical. *Forbes Asia* said in 2015 that "the last eight years have seen large-scale and violent anti-PX protests in the southern provinces of Guangdong and Yunnan, the eastern provinces of Zhejiang and Fujian, the northern province of Liaoning, and the western province of Sichuan" (Hoffman and Sullivan 2015).

At another protest of PX production, help on April 4, 2015 in Guangdong Province, police arrested 18 people. These incidents and many others were described via social media on Weibo, China's often-censored analogue to Twitter. More than a thousand people marched in Shanghai in opposition to planned construction of a chemical plant in mid-June 2015, as well.

The government had told the public that PX projects were safe, but events told people otherwise. A month before an explosion at a plant using PX in Zhangzhou during 2013, the state-controlled *People's Daily* had said that PX was "no more harmful than a cup of coffee" (Hoffman and Sullivan 2017). The demonstrators were anxious over the explosions but also asserted, without evidence, that long-term exposure to PX is carcinogenic.

Demonstrations in Xiamen and elsewhere in China were aided by use of social media that operated outside of state censorship. A documentary titled *Under the Dome*, made by Chai Jing, a former anchorwoman on state television, described the impact of Beijing's smog on her child, who was born with a lung tumor. In March 2015, it received 300 million views in one week before censors shut it down. The film had originally been approved by censors, but at 300 million views, its support got astoundingly out of hand.

Social Media Aids Protests

Another example of unauthorized protest took place against a waste incinerator in Luoding, in southern Guangdong Province (northeast of Hong Kong), which was scrapped in 2015 following large protests against anticipated air pollution from cement manufacturing. The protests were so large that police could not suppress them. On April 8, city officials announced revocation of the facility's operating permit "in response to public demands" (Dominguez 2015). Similarly, at about the same time, in northern China's Inner Mongolia region, officials agreed to shutter several chemical plants after protests against their pollution turned violent. Also in 2015, a very large explosion and fire fed by burning oil at a Fujian Province chemical plant southwest of Shanghai provoked evacuation of about 30,000 people from a densely populated residential area; this was followed by protests from many who asked why such a plant should be located among such a large number of homes.

Some environmental bloggers on Weibo have millions of followers. Deng Fei, for example, "asked his followers to post photographs of polluted rivers and lakes in their hometowns. The outpouring of images prompted a series of press articles and shamed several local governments into cleaning up their waterways" (Hook 2013).

During mid-April 2015, residents in China's southern Guangdong Province protested smog from a coal-fired power plant in the city of Heyuan. Reuters reported that "Xinhua [China's official news agency] said thousands had taken to the streets, but a man at the Heyuan government's publicity office said only about 200 people had joined the protest" (Martina and Shuping 2015).

"For the first time, delivering clean air and safe water became equally important as securing economic growth," Shuo Li, a Greenpeace policy officer and Alexander von Humboldt fellow, said (Dominguez 2015). China's Institute of Environmental Planning has recognized that failure to address pollution of air, water, and soil in China contributes to social dissatisfaction and sometimes to violent conflict with authorities. "There is a huge gap between how fast the environment is being improved and how fast the public is demanding it to be improved, and environmental problems could easily become a tipping point that leads to social risks," said the officially sanctioned *China Environmental News* (Dominguez 2015).

"War on Air Pollution"

In March 2015, Chinese premier Li Keqiang declared "war on air pollution" at the opening of the National People's Congress in Beijing. He vowed that the government and the ruling Communist Party will fight environmental pollution "with all our might" (Dominguez 2015). During the National People's Congress (NPC) in March 2015, President and Communist Party chairman Jinping Xi pledged that the government would "punish with an iron hand any violators who destroy China's ecology or environment" (Hoffman and Sullivan 2015).

Initially, political rhetoric was not matched by progress. Air quality improved little in China's major urban areas, especially in winter, when air is often stagnant and much of the electric power is provided by low-quality coal. The winter of 2016–17 brought periods of nearly asphyxiating smog in and around Beijing, making its air the second dirtiest in the world, after New Delhi, India. In addition, even state media admitted that almost 60 percent of China's underground water is polluted.

Leslie Hook, writing in The United Kingdom's *Financial Times Magazine,* described one protest in Kunming, "a leafy town in southern China famous for its mild climate." Mayor Li Wenrong found himself confronted in 2013.

> Hundreds of angry protesters were marching outside the city government building, the second such demonstration in as many weeks. . . . They wanted the mayor to halt the construction of a giant new petrochemical plant that was under way in a neighboring county. In particular, the protesters were concerned about the pollution the plant would release as a by-product of producing paraxylene, a chemical used in plastics. (Hook 2013)

More than a thousand people took to the streets of Shanghai in mid-June 2015 to protest construction of a new chemical plant, shortly after a similar number had rallied in Tianjin, a northern industrial city, against a steel plant because of effluvia that activists said was carcinogenic. In all of these places, environmental issues appeared to be uniting blue-collar workers and the white-collar middle class, as well as young and old.

Entrenched Public Distrust

Forbes Asia reported:

> [Communist] Party leaders fret about political stability and potential challenges to the regime; pollution is one of their greatest concerns. But the Chinese government is failing to address the underlying cause of this discontent—an entrenched public distrust of officialdom—and, in the long term, is risking the possible "joining up" of environmental protests into a widespread movement. . . . These environmental protests are striking at the heart of the Chinese governance model of "adaptive authoritarianism" and exposing its limitations. The Party's strategy in dealing with major environmental disputes that bring together local communities across all ages and classes has often been one of short-term appeasement. But when governments are known to make ad-hoc concessions to quell disorder, it encourages further episodes of contention. (Hoffman and Sullivan 2015)

Policies have changed in China at the national level to recognize degrading the environment as a crime. For example, Sara Hsu wrote in *Forbes*, "An amended environmental protection law passed at the beginning of 2014 allows authorities to detain company bosses for 15 days if they continue to pollute despite warnings or fail to carry out environmental impact assessments. The law also has made it easier for unofficial groups to take legal action on behalf of public interest" (Hsu 2016). Some local jurisdictions also have recognized environmentalism on a legal basis.

According to *Forbes*, "Cities like Tianjin, home to a deadly chemical warehouse explosion last year [2015], have stepped up pollution fees and reduced coal pollution from coal-burning plants and boilers" (Hsu 2016). Some local governments have not accommodated protestors. "For example, in January 2013, authorities in Dalian tried six activist leaders from the 2010 anti-PX protests in that city for 'libel' and 'deliberately concocting false information to terrorize the public'" (Hoffman and Sullivan 2015).

> Environmental activists continue to be taken into custody for disturbing the peace, sometimes on trumped-up charges. In December of last year, two environmental activists, Mr. Xu and Ms. Tian, were arrested for alleged prostitution, after inquiring into nickel mining and its connection to wetland degradation in Ningde. In June of this year, some of the tens of thousands of protesters against a planned chemical plant in Hubei province were beaten, and eight individuals were arrested for starting the protest. (Hsu 2016)

Cleaning Up a Monumental Task

At the same time, little has improved in many "cancer villages," so-called because many people who live in them have developed several types of cancer after sustained exposure to heavy metals in soil and water. China's lung-cancer rate has been rising. Outside of India's Ganges Valley, which is the most densely populated area on Earth, most of the world's dirtiest urban air continues in Eastern China. Beijing has sustained several "red alert" days for air pollution, during which motor traffic and industrial activity have been curtailed. "In fact, 20 of the 30 most polluted cities in the world are in China," reported the Climate Institute (Xu 2008). Cleaning it up is a monumental task.

Even after Chinese president Xi Jinping pledged to reduce environmental devastation in China as part of a "war on pollution" and authorities passed laws to permit class action suits by its victims, justice for poisoning has been a difficult struggle. An example was provided by parents of 300 children in Dapu, a village of 62,000 in Central China's Hunan Province, who sued the Meilun chemical company after their children fell ill to acute lead poisoning. One boy, Wang Yifei, five, was described as no longer able to "recognize the sound of his mother's voice. Bony and pale, vanishing beneath a winter coat, he spoke mostly in grunts and screams, the language of his malady. He stumbled as he walked, never certain of the ground beneath him" (Hernández 2017).

The parents sued in 2012 and spent considerable time and money assembling a case, only to find local Communist Party officials colluding with company executives and judges appointed by party members to whittle down the defendants' case. Of 53 parents who sued, all but 13 accepted small amounts of money (usually about $1,500) to withdraw. The judge adopted very strict rules of evidence that were met by only two of the remaining defendants, who, after four years of legal proceedings, were awarded 13,000 renminbi each (about $1,900), not enough to begin paying legal bills for their poisoned children. "There's no way to win," said Wang Jiaoyi, Wang Yifei's father. "There's no such thing as justice" (Hernández 2017).

Blue Skies in Beijing

By late 2017, air pollution in parts of China was subsiding, as a Greenpeace analysis of Chinese government data in Beijing and 27 other cities in northeastern China indicated a drop in pollution levels by 33 percent. In Beijing itself, pollution fell 53 percent during the last three months of 2017, compared to the previous year at the same time. Greenpeace said that 160,000 fewer people had died prematurely of air pollution in China in 2017 vis-à-vis 2016. An intense campaign against air pollution that was in its fourth year was beginning to show observable results in a turquoise-blue sky. The drop in pollution also was aided by favorable air circulation, with fewer days of stagnant air.

However, according to the *New York Times*:

Pollution levels fell less precipitously or rose elsewhere, suggesting that a concerted effort . . . to shift heating to natural gas from coal may have simply shifted the harmful effects to regions far from the capital. In the northern province of Heilongjiang, on the border with Russia, pollution levels rose 10 percent. In a statement with its analysis, Greenpeace argued that the results demonstrated the need for more government action, noting that nationwide the drop in pollutants was only 4 percent. (Myers 2018)

Further Reading

Dominguez, Gabriel. 2015. "Environmental Activism Gaining a Foothold in China." DW (Deutsche Welle), April 15. Accessed October 5, 2018. http://www.dw.com/en /environmental-activism-gaining-a-foothold-in-china/a-18384605.

Hernández, Javier C. 2017. "'No Such Thing as Justice' in Fight over Chemical Pollution in China." *New York Times*, June 12. Accessed October 5, 2018. https://www.nytimes .com/2017/06/12/world/asia/china-environmental-pollution-chemicals-lead-poison ing.html.

Hoffman, Samantha, and Jonathan Sullivan. 2015. "Environmental Protests Expose Weakness in China's Leadership." *Forbes Asia*, June 22. Accessed October 5, 2018. http://www.forbes.com/sites/forbesasia/2015/06/22/environmental-protests-expose -weakness-in-chinas-leadership/#1145d1db2f09.

Hook, Leslie. 2013. "China's Environmental Activists." *Financial Times Magazine* (UK), September 20. Accessed October 5, 2018. https://www.ft.com/content/00be1b66-1f43 -11e3-b80b-00144feab7de.

Hsu, Sara. 2016. "China Wages War on Pollution While Censoring Activists." *Forbes* August 4. Accessed October 5, 2018. http://www.forbes.com/sites/sarahsu/2016/08/04 /china-wages-war-on-pollution-while-censoring-activists/#61380a8a6244.

Martina, Michael, and Niu Shuping. 2015. "Hundreds Protest against Pollution from South China Coal Plant." Reuters, April 14. Accessed October 5, 2018. http://www.reuters .com/article/us-china-environment-protest-idUSKBN0N30BK20150414.

Myers, Steven Lee. 2018. "A Blue Sky in Beijing? It's Not a Fluke, Says Greenpeace." *New York Times*, January 11. Accessed October 5, 2018. https://www.nytimes.com/2018/01/11 /world/asia/pollution-beijing-declines.html.

Xu, Christine. 2008. "China: Locals Turn to Environmental Activism." Climate Institute, April. Accessed October 5, 2018. http://climate.org/archive/topics/international-action /chinese-environmental-action.html.

INDIA

The World's Dirtiest Air

Given recent publicity, many people believe that the world's worst urban air pollution is in Beijing. Actually, China's capital runs second in this noxious sweepstakes as the world capital of smog to the Ganges River Valley around Delhi, according to the World Health Organization. In fact, the highest court in Delhi has claimed the atmosphere is "like a gas chamber" (Sudarshan 2015, 11). The Delhi airport sometimes closes for smog so thick that jets cannot safely land. Many people have persistent coughs, and some wear gas masks. Many days, the sun shines through a gaseous haze. Approaching Delhi by air from the south at 35,000 feet, one can watch as the setting sun turns the Ganges Valley brown cloud a dull yellow, then a dull orange, then dark red before the sky goes black.

A United Nations report issued in 2017 asserted that air pollution in India, especially in the Ganges Valley, was triggering neuro-inflammation, which impedes cognitive development of children's brains. At the same time, the World Bank said that air pollution was costing India at least $55 billion a year, and *the Lancet,* a British medical journal, said that the same pollution had caused 2.5 million Indian people to die prematurely in 2015 (Gettleman 2017). In 2016, another report said that air pollution had played a role in 1.1 million premature deaths per year. This joint report, by the Health Effects Institute of Boston and the Institute of Health Metrics and Evaluation of Seattle, said that India's premature deaths from dirty air had doubled in 25 years, from 1990 to 2015, as the country's population grew to 1.2 billion and more people acquired cars and coal-powered electricity (Anand 2017).

One reason for the Ganges Valley's filthy air is combustion of petroleum coke, a residual waste product of Canadian tar-sands refining that is imported from the United States. It burns hotter than coal but "also contains more planet-warming carbon and far more heart- and lung-damaging sulfur, a key reason few American companies use it. Refineries instead are sending it around the world, especially to energy-hungry India, which last year received almost a fourth of all the fuel-grade 'petcoke' the U.S. shipped out" (Webber and Daigle 2017). After widespread press coverage of this practice late in 2017, India's government said such imports would be phased out. "We should not become the dust bin of the rest of the world," said Sunita Narain, a pollution authority member who heads the Center for Science and the Environment. "We're choking to death already" (Webber and Daigle 2017, 4-D).

Delhi "Gasping for Breath"

On December 16, 2015, India's Supreme Court banned registration of luxury diesel cars and SUVs with an engine capacity of more than 2,000 cubic centimeters in the National Capital Region (which includes Delhi) until March 31, 2016. It also imposed a one-time pollution tax on small diesel cars (Mahapatra 2015, 1).

On the same front page that it reported the final negotiation of the Paris world climate accord, the *Sunday Times of India* said that pollution in Delhi and the rest of the Ganges Valley had reached the worst levels of the fall and early winter, as cooler-than-average weather drained downhill off freshly fallen snow in the foothills of the Himalayas, pooling in the Ganges Valley (Mohan 2015, A-1). Delhi was described by the newspaper as "gasping for breath" ("Winter's Officially Here" 2015, A-1). In some areas, smoky bonfires were being lit to provide warmth for the poor and homeless.

A lively debate was taking place in the public press over what do about Delhi's pollution. Coal-fired power and the dung-fueled stoves used by 600 million (for the most part rural) Indians seemed mainly off-limits. People needed to eat, and at least 300 million in India had no electricity. That left only "four-wheels"—private cars and trucks—open to regulation.

One proposal reflected Beijing's use of restrictions on when one can drive as pollution reaches asphyxiation levels, based on odd-and-even last numbers on license plates. This one was implemented in the Delhi area January 1, 2016, amid howls of protest from the rich, at least one of whom had three drivers and three cars—all with license plates ending in odd numbers (Najar 2016, A-4).

India's *Economic Times Magazine* devoted a cover story to solutions to exploding traffic. It pointed out that while the rate of vehicle ownership in Delhi has doubled between 2000 and 2015, the proportion lags many rich countries: India's car ownership in 2015 was 13 per 1,000 people, a rate that is 617 in Japan and 439 in the United States (Seetharaman 2015, 8).

Smog Costs India $80 Billion a Year

A year after Delhi's air was likened to that of a gas chamber, it got even worse. The United States Embassy in New Delhi measures air quality on its roof (the same is done at the Chinese embassy in Beijing); on Monday, November 7, 2016, the reading on a scale of zero to 999 (twice the hazardous level of 500 and 16 times the safe limit of 60) was off the chart, completely beyond the 0–999 range. According to Cable News Network's James Griffiths, "In Khan Market, shops that specialized in antipollution masks were doing brisk business as people queued up to buy some modicum of protection from the toxic smog [and] hundreds of people gathered to protest outside Parliament in the city center as others voiced their displeasure online, posting to Twitter with the slogan #MyRightToBreathe" (Griffiths 2016). More than 5,000 schools closed in the Delhi area because of the smog, and construction work stopped for five days.

> Some worry that the worsening air pollution in New Delhi and the rest of India will soon take a toll on the country's economy. Bloomberg reported that environmental degradation was costing India $80 billion per year, which is nearly 6 percent of the country's gross domestic product (GDP). Workers' absences may increase if air quality does not soon improve, while school closures and temporary restrictions on construction, coal-fired power plants, and driving could cause losses in all manner of industries. (Feltman 2016)

The *New York Times* reported that New Delhi's particulate level that day was as dangerous to the city's 20 million residents as smoking more than two packs of cigarettes every day (Barry 2016). "Open a window or a door, and the haze enters the room within seconds. Outside, the sky is white, the sun a white circle so pale that you can barely make it out. The smog is acrid, eye-stinging and throat-burning, and so thick that it is being blamed for a 70-vehicle pileup north of the city," reported Ellen Barry (2016).

The Pervasiveness of India's Dirty Air

A Greenpeace report indicated that 90 percent of India's cities had dirty air. Damini Nath reported in *the Hindu* that "in an analysis of 2015 data for 168 cities by

Greenpeace India, 154 were found to have an average particulate matter level higher than the national standard. None of the cities studied had air quality matching the standard prescribed by the World Health Organization" (Nath 2017). Cities were ranked based on an annual average of particulate matter (PM) counts, specifically their average PM10 (particles less than 10 microns in diameter). These include the very harmful fine particles, PM2.5 or less, which are most harmful to human health because they easily lodge in the lungs. According to Greenpeace, as reported in *the Hindu*, Delhi was the most-polluted city at an average PM10 at 268 micrograms per cubic meter, more than four times the 60 micrograms/cubic meter limit issued by India's National Ambient Air Quality Standards of the Central Pollution Control Board. Allahabad, Ghaziabad, and Bareilly, all also in the Ganges Valley of Uttar Pradesh State, and Faridabad, Haryana State, also were severely polluted on an annual basis.

"Due to the Himalayas and the cooler weather as well as big industrial clusters, the levels of pollution are higher in the North. Southern India has the benefit of the mixing of sea breeze. However, pollution is a national-level problem and has to be treated as such," said Sunil Dahiya, one of the authors of the report and a campaigner with Greenpeace India (Nath 2017). "Whether it is in the transport sector or industries, the uncontrolled burning of fossil fuels is the main cause of air pollution," said Dahiya.

Southern India is booming, and so is power production and pollution. Google recently announced plans to build its largest campus outside the United States in Hyderabad, a city in south India with a population the size of Los Angeles. Hyderabad's traffic jams, like those of Los Angeles, have become legendary. Nearby Bangalore has become India's Silicon Valley. Nearby, the state of Andhra Pardesh's government late in 2015 announced plans for two new coal-fired 800-megawatt super critical thermal power plants to supply rapidly rising needs for energy. Traffic on the road between Hyderabad and Vijayawada, about 150 miles to its southeast, is so thick that the trip usually takes four hours.

Pollution in the Ganges Valley early in the winter is compounded by stagnant air, as well as the celebration of Diwali, a religious festival at which millions of fireworks are set off, adding smoke and haze to automobile exhaust, smoke from coal-fired power plants, and industrial effluent, as well as open fires used for warmth by homeless people. At about the same time, many farmers scorch fields to dispose of crop remnants. The London *Guardian* reported in early November that hospitals in the Delhi area "have reported increased admissions of people suffering respiratory diseases—of which India has the highest rate in the world, with 159 deaths per 100,000 people in 2012, according to the World Health Organization" ("Indian Government" 2016). About half of Delhi's 4.4 million schoolchildren experience stunted lung development because of air pollution.

During November 2017, smog, thicker than in previous years, choked New Delhi. The *New York Times* reported, "The toxic haze blanketing New Delhi was so severe on Tuesday [November 7] that politicians announced plans to close schools, flights were delayed, and the chief minister of Delhi state, Arvind Kejriwal, said the city had 'become a gas chamber'" (Kumar and Schultz 2017), recalling the words of India's High Court two years previously. The smog had become a subject of intense debate, but little had changed in the atmosphere.

Arvind Kumar, a chest surgeon for more than three decades, said the toll on people's health was cumulative. "I don't see pink lungs even among healthy non-smoking young people," he said. "The air quality has become so bad that even if you are a nonsmoker, you are still suffering." The options for Delhi residents are three, Kumar said. "One is to stop breathing. That is not possible. Second is to quit Delhi. That is also not possible. Third is to make the right to breathe fresh air a people's movement" (Kumar and Schultz 2017).

Pollution peaked at more than 700 micrograms per cubic meter in the New Delhi metropolitan area, with its 45 million people, which is equal to smoking two and a half packs of cigarettes per day. The burning of crops has been deemed illegal, as has the sale of fireworks during Diwali, to little avail. Agricultural fires' smoke, sealed into the Ganges Valley by stagnant air, again combined with auto and truck exhaust, emissions from smokestacks, and burning of garbage to intensify the smog.

An October 2017 article in the British medical journal *The Lancet* estimated that up to 2.5 million premature deaths were caused by air pollution across India during 2015—the highest national total in the world.

Further Reading

Anand, Geeta. 2017. "India's Air Pollution Rivals China's as World's Deadliest." *New York Times*, February 14. Accessed October 5, 2018. https://www.nytimes.com/2017/02/14/world/asia/indias-air-pollution-rivals-china-as-worlds-deadliest.html.

Barry, Ellen. 2016. "Smog Chokes Delhi, Leaving Residents 'Cowering by Our Air Purifiers.'" *New York Times*, November 7. Accessed October 5, 2018. https://www.nytimes.com/2016/11/08/world/asia/india-delhi-smog.html.

Feltman, Rachel. 2016. "Air Pollution in New Delhi Is Literally Off the Charts." *Popular Science*, November 8. Accessed October 5, 2018. http://www.popsci.com/air-pollution-new-delhi.

Gettleman, Jeffrey, Kai Schultz, and Hari Kumar. 2017. "Environmentalists Ask: Is India's Government Making Bad Air Worse?" *New York Times*, December 8. Accessed October 5, 2018. https://www.nytimes.com/2017/12/08/world/asia/india-pollution-modi.html.

Griffiths, James. 2016. "New Delhi Is the Most Polluted City on Earth Right Now." Cable News Network (CNN), November 8. Accessed October 5, 2018. http://www.cnn.com/2016/11/07/asia/india-new-delhi-smog-pollution/.

Harris, Gardiner. 2014. "Coal Rush in India Could Tip Balance on Climate Change." *New York Times*, November 17. Accessed October 5, 2018. http://www.nytimes.com/2014/11/18/world/coal-rush-in-india-could-tip-balance-on-climate-change.html.

"Indian Government Declares Delhi Air Pollution an Emergency." 2015. *The Guardian* (United Kingdom), November 6. Accessed October 5, 2018. https://www.theguardian.com/world/2016/nov/06/delhi-air-pollution-closes-schools-for-three-days.

Kumar, Hari, and Kai Schultz. 2017. "Delhi, Blanketed in Toxic Haze, 'Has Become a Gas Chamber.'" *New York Times*, November 7. Accessed October 5, 2018. https://www.nytimes.com/2017/11/07/world/asia/delhi-pollution-gas-chamber.html.

Mahapatra, Dhananjay. 2015. "Small Diesel Car Buyers to Pay One-Time Pollution Tax: SC [Supreme Court]." *Times of India*, December 17: 1.

Mohan, Vizhwa. 2015. "'Historic' Climate Deal Done, Meets India's Key Concerns." *Sunday Times of India*, December 13: A-1.

Najar, Nida. 2016. "Streets, If Not the Air, Clear Out as Delhi Tests Car Restrictions." *New York Times*, January 5: A-4.

Nath, Damini. 2017. "Air Pollution a National Problem." *The Hindu*, January 12. Accessed October 5, 2018. http://www.thehindu.com/news/cities/Delhi/Air-pollution-a-national-problem/article17026779.ece.

Seetharaman, G. 2015. "Getting Past the Gridlock." *Economic Times Magazine*, December 13–19: 8–9.

Sudarshan, V. 2015. "In Delhi, Just Waiting to Exhale." *The Hindu*, December 17: 11.

Webber, Tammy, and Katy Daigle. 2017. "U.S. Exporting Dirty Fuel to Pollution-Choked India." Associated Press in *Omaha World-Herald,* December 7: 4-D.

"Winter's Officially Here, Hits Flights, Worsens Pollution." 2015. *Sunday Times of India,* December 13: A-1.

ITALY

The Smoggiest Country in Europe

Italy experiences more deaths attributed to human-caused air pollution than any other country in Europe. Every so often, the crud emerging from cars' exhaust pipes, as well as wood- and coal-burning stoves, coal-burning power plants, and industries gets a natural boost from Mt. Etna, Europe's most active volcano, which, depending on wind direction, has been ejecting ash across the area at least since Roman days. On February 28, 2017, Mt. Etna, on the eastern coast of Sicily, spewed ash into the air, adding to Europe's human-induced pollution woes (Spamer 2017).

Italy's pollution accumulates on days with stagnant, dry air, especially during winter. According to one estimate, air pollution causes more than 467,000 deaths in Europe each year, with annual health costs between €400 billion and €900 billion ("Smog Levels" 2017). According to the European Environment Agency (EEA), in 2012 more than 84,000 people died prematurely in Italy due to dirty air that is associated with respiratory illnesses. Respiratory and heart problems contributed to almost one-tenth of deaths in Italy of people less than 30 years of age (Sylvester 2015).

Pollution, while most prevalent in cities, also reaches the foothills of the Alps, which most people imagine to be pristine. The use of stoves and furnaces that burn coal and wood is a major contributor to smog in those areas during periods when the air is stagnant. Much of Italy's smog involves the most dangerous very small particles of less than 2.5 microns in diameter that lodge in the lungs and cause

A massive banner hung by Greenpeace activists at the Lazio Region building in Rome, 2018. The activists descended along the side of the building as they unfurled a banner showing the face of Nicola Zingaretti, president of the Lazio Region of Italy, covered by a mask and accompanied by an inscription that may be translated as "Clean air now!" (Corbis/Getty Images)

respiratory illnesses. Carbon monoxide and lead also contribute to degraded air quality.

Cities Restrict Use of Cars

Two of Italy's largest cities, Rome and Milan, have been restricting use of cars, motorcycles, and trucks on the smoggiest days, as the corrosive air defaces ancient structures. Rome has used a ban on cars with even-numbered license plates for up to 12 hours on some days, and odd-numbered plates on others. Cars that meet environmental standards, such as those with hybrid engines, have been exempted from the bans. In Rome, "high concentrations of particulate matter and nitrogen dioxide continue, aggravated by the weather situation of high pressure and absence of wind," the city government said (Pianigiani 2015). Northern Italy, including Milan, reported some of the worst air pollution in Europe. Turin and Brescia also reported smog, with heavy industry a major contributor. Florence, a major tourist destination, sometimes reported lower air quality than Rome or Milan.

In Milan, all-day transit passes for €1.50 (about $1.65) were issued to encourage people to leave their cars at home. Drivers in both cities were fined €150 (about $165) for violating the bans. "In these days of major emergency, we cannot remain indifferent," said Milan's mayor, Giuliano Pisapia ("Italy Smog" 2015). "For

Pope Francis and Climate-Change Diplomacy

During the summer of 2015, Pope Francis, who has become well known for directly tackling many controversial issues, made climate change a Vatican priority by issuing an encyclical (essentially a policy statement) detailing how the burdens of a changing climate worldwide fall disproportionally on the poor. The encyclical was part of a broader campaign by the pope to advocate protection of the Earth and all of creation.

Francis sought to help jump-start international diplomacy requiring nations to curb production of greenhouse gases. The encyclical was preceded by a year of meetings among prominent Vatican officials on the issue. The encyclical was issued after a 12-week campaign in the United States, during which Catholic bishops raised climate-change related issues under the rubric of environmental stewardship in their sermons and homilies, as well as news-media interviews and letters to editors. Timothy E. Wirth, vice chairman of the United Nations Foundation, said: "We've never seen a pope do anything like this. No single individual has as much global sway as he does. What he is doing will resonate in the government of any country that has a leading Catholic constituency" (Davenport and Goldstein 2015).

Further Reading

Davenport, Coral, and Laurie Goldstein. 2015. "Pope Francis Steps Up Campaign on Climate Change, to Conservatives' Alarm." *New York Times*, April 27. Accessed October 11, 2018. http://www.nytimes.com/2015/04/28/world/europe/pope-francis-steps-up-campaign-on-climate-change-to-conservatives-alarm.html.

these provisions against air pollution to be more effective, they have to regard a vast area, and not only the individual towns," Pisapia said.

Near Naples, Italy, the mayor of San Vitaliano made world news when smoggy air prompted a three-month ban on wood-fired pizza stoves until they were fitted with pollution filters, bringing the cooking of the town's iconic cuisine to a halt. Tourists were advised to take precautions. "Italy's a beautiful country. You can be moved to tears by the landscape. Then there's the food, the wine, and the art. But there are Italian cities which take your breath away in other ways," said a travel advisory issued by World Nomads, a tourism advisory service (Sylvester 2015).

"Pollution alerts indicate [that] high-pollution days often take place during summer in Rome, Florence, Milan, and Naples. Anyone with a respiratory problem will want to avoid going outside in major cities when this occurs. . . . Headaches and breathing problems are common after a day of sightseeing if you're prone to pollution-related sickness," warned World Nomads (Sylvester 2015). The tourist advisory also said that an unusually large proportion of Italians (almost one quarter of adults) smoke tobacco, another contributor to air pollution.

Tourists also were warned to watch where they put their feet: "Another city pollutant you may encounter is dog feces, as many Italians don't clean up after their pets. And thanks to all the cars and motorbikes, noise pollution is also an urban issue" (Sylvester 2015). And beaches could be a problem as well: "The beaches, while beautiful, are not always the most environmentally friendly locales, either. Industrial waste and oil and sewage spills can deter you from a romantic stroll in the sand and a dip in the Mediterranean. You'll find cleaner beaches [than elsewhere in Italy] in Elba, Sardinia, and Sicily" (Sylvester 2015).

"Italian Cities Need to Breathe Again"

By 2017, the number of cities imposing restrictions on travel had increased to nine. "Italian cities need to breathe again," the country's environmental organization, Legambiente, warned during January 2017. Legambiente said that air quality was reaching a level in some areas that required restrictions as often as 50 percent of the time. Environmental officials in Pavia, a university town in the province of Lombardy (which includes Milan, in northern Italy), advised everyone to remain indoors as much as possible and avoid using cars.

With safe limits for fine particulates having been exceeded for several straight days in Rome and Turin, driving bans were imposed again in 2017. "In the capital," according to one report, "The most polluted areas included Via Tiburtina and Via Praenestina—both narrow streets lined by tall buildings—and the busy Corso Francia." The pollution was notable in Rome's city parks as well. Legambiente's regional president, Roberto Scacchi, was quoted as having said that "the fact that smog had reached even open spaces showed 'the situation is serious, and the smog is everywhere'" ("Smog Levels" 2017). Authorities estimated that 70 percent of Rome's smog was caused by cars and trucks, with the remainder from heating or heavy industry. Pollution was being aggravated by lack of rainfall, especially in Northern Italy, where many areas had no rain for two months. Scacchi said more permanent pollution-prevention measures were required. "You can't just do a rain dance, waiting for it to come and wash away pollution," he warned ("Smog Levels" 2017).

Cremona, in Lombardy, reported the most intense pollution, with unsafe levels of fine particles during 20 days of January 2017. Turin followed with 19 days, Frosinone in Lazio at 18. Padua, Vicenza, and Reggio Emilia also experienced unsafe levels of pollutants during half of January. "The data doesn't bode well," said a country-wide Legambiente Mal'Aia (Bad Air) report ("Smog Levels" 2017). "The air quality in Italian towns needs to become a government priority, on a local, regional, and national level," warned Legambiente's national president, Rossella Muroni. "Otherwise, we will continue to condemn Italian citizens to breathe in polluted air." Environmental activists marched on government offices carrying banners reading "Our lungs are breaking! No to smog" ("Smog Levels" 2017).

Air pollution was putting half the children in Turin at increased risk of illness that was altering their DNA, according to one study.

Researchers analyzed DNA samples from children aged between six and eight in Turin, Brescia, Lecce, Perugia, and Pisa. In Turin, 53 percent of these samples from students had at least one micronucleus (an indicator of a mutation in DNA). While this does not present an immediate health risk and the mutation is not passed on to future generations, over time it can lead to tumors and other chronic diseases. "We do not want to create anxiety, but we have done this work to understand the effects of smog on children's biological systems," explained Giorgio Gilli, one of the professors who was involved in the study. ("Turin Smog" 2016)

Lawsuits Filed for Clean Air

On February 22, 2017, following one of Italy's worst episodes of air pollution, ClientEarth filed suit against the government of Lombardy, Italy's most-polluted region. The organization said that its suit could be the first of several in Italy. At about the same time, the European Commission discussed legal action against Italy and four other European Union nations (France, Germany, Italy, and Spain) "for failing to protect their citizens' health against dangerous levels of air pollution" (Matthews 2017).

The basis of the EU's legal action was nitrogen dioxide (NO_2) above legal limits. Nitrogen dioxide comes mainly from combustion of gasoline and diesel fuel and can cause or aggravate bronchitis, heart disease, coughing, and wheezing, with the worst effects usually afflicting children and elderly people. ClientEarth has been using its lawsuits to establish a human right to breathe clean air. To do that, the suit zeroes in on Lombardy, which it pinpointed as one of the most polluted regions in Europe.

The lawsuit makes an example of Lombardy and Italy as a proxy for the 22 other European countries that it asserts are violating air-quality laws. "Air pollution in Lombardy is alarmingly high. It has similar levels of pollution to the most heavily industrialized regions on the continent despite the fact it is one of the richest. It has the means to resolve this crisis but continues to dither," said James Thornton, ClientEarth's chief executive officer. "The regional government of Lombardy has a legal and moral obligation to tackle the region's toxic air crisis" (Matthews 2017). ClientEarth already had successfully sued the United Kingdom over the government's failure to reduce air pollution.

"Blocking traffic for one or two days is merely a palliative; so is stopping cars based on license plate numbers," said Nicola Pirrone, director of the Institute of Atmospheric Pollution Research at Italy's National Research Council. "Italy needs serious infrastructural investments to enhance greener transportation and greenhouse emissions" (Pianigiani 2015).

Further Reading

"Italy Smog: Milan and Rome Ban Cars as Pollution Rises." 2015. BBC News, December 28. Accessed October 5, 2018. http://www.bbc.com/news/world-europe-35188685.

Matthews, Janie. 2017. "Italy Slapped with Air Pollution Lawsuit." EURACTV.com, February 22. Accessed October 5, 2018. https://www.euractiv.com/section/climate-environment/news/italy-slapped-with-air-pollution-lawsuit/.

Pianigiani, Gaia. 2015. "Italy, Dirty Air at Record Levels, Is Putting Limits on Traffic." *New York Times*, December 24. Accessed October 5, 2018. https://www.nytimes.com/2015/12/25/world/europe/italy-air-pollution.html.

"Smog Levels Way Above Safe Limits in Northern Italy." 2017. The Local It (Italy), January 31. Accessed October 5, 2018. https://www.thelocal.it/20170131/our-lungs-are-breaking-smog-levels-way-above-safe-limits-in-northern-italy.

Spamer, Courtney. 2017. "Mount Etna's Eruption to Impact Air Quality, Visibility Across Southern Europe." AccuWeather, March 2. Accessed October 5, 2018. http://www.accuweather.com/en/weather-news/mount-etnas-eruption-to-impact-air-quality-visibility-across-southern-europe/70000985.

Sylvester, Phil. 2015. "Italy's Breathtaking Pollution and Other Health Hazards." World Nomads, June 2. Accessed October 5, 2018. https://www.worldnomads.com/travel-safety/europe/italy/pollution-other-health-hazards-in-italy.

"Turin Smog Puts Half of Children at Health Risk: Study." 2016. The Local It (Italy). December 22. Accessed October 5, 2018. https://www.thelocal.it/20161222/turin-smog-means-half-of-children-suffer-dna-mutations-health-risk-.

JAPAN

Working to Clean Air and Water

With a metropolitan area of 30 million people, Tokyo has its share of days with dirty air, especially when China's "brown cloud" blows in from the southwest. At the same time, Japan has dramatically improved environmental conditions and has also become a model for other newly industrialized countries, the largest of which is China.

Japan Campaigns Against Pollution

China has realized that Japan has something to teach about how to deal with air pollution caused by rapid industrial development. Cai Hong, writing in the *China Daily*, described how far Japan has come since the 1960s, when "Mount Fuji was obscured behind a perpetual fog of exhaust and particulate matter. Traffic police were equipped with oxygen tanks on particularly hazardous days, and many students were treated in schoolyards for inhaling photochemical smog. This is what Tokyo was in the 1960s" (Hong 2017). Beginning after World War II, accelerating from the 1950s into the 1970s, Japan developed into one of the world's most productive industrial economies and did so with very little environmental control, as, according to Hong's account, "Japan gained an unenviable reputation as the world's most toxic country. American biologist Paul Ehrlich described Japan as the developed world's 'canary in the coal mine,' a country so profoundly polluted that it became a test case for how high human tolerance levels could be" (Hong 2017).

Along the way, environmental afflictions acquired Japanese names, such as "Yokkaichi asthma" and "Minamata disease" (a form of mercury poisoning), both named after the heavily industrialized urban areas where they first appeared. During the ensuing decades, citizen movements demanded cleanups, and by 2017, Japan had become one of the cleanest countries on Earth. At the same time, a swiftly growing industrial base produced intense pollution in Chinese urban areas, especially in and near Beijing and Shanghai.

By 1970, Japan's central government had enacted new laws that turned pollution into a crime. Polluters were forced to pay fines to their victims, procedures were created to adjudicate environmental crime, and regulations on air, water, and sound pollution, as well as toxic waste disposal, were strengthened. During the first half of the 1970s, government spending on environmental enforcement doubled by the federal government and tripled at the local level; businesses' spending on pollution control rose as much as 40 percent a year (Hong 2017). Business investment in antipollution measures also increased as much as 40 percent a year during the 1970s. At the same time, output of goods and services increased, making Japan the world's second-largest economy after the United States, until China surpassed it in 2010.

"If China can lift hundreds of millions of people out of poverty, it surely can find a way to keep its cities' air clean," Hong commented (2017). "China can turn to its neighbor for experiences and technologies to deal with air and water pollution, not least because for the second straight year, Tokyo topped the list of the most livable cities in 2016 in the annual Quality of Life Survey conducted by the British lifestyle magazine *Monocle*" (Hong 2017).

At the same time that Japan has cleaned its own air and water, China's pollution has been drifting in from the southwest on prevailing winds. On occasion, flights have been canceled and people in some areas have been advised to wear facemasks and avoid outdoor activity.

> A dirty, yellow-brown smog rolled across Tokyo Bay like an inexorable wave on Sunday, March 10, [2013], engulfing Yokohama's Bay Bridge, the docks district, and the Minato Mirai district of shops, hotels, and the largest Ferris wheel in Japan in a haze of dust. What had been a bright spring morning in this coastal city just south of Tokyo had been turned into an afternoon of swirling dirt, with people caught outside choking as they held handkerchiefs to their mouths. (Ryall 2013)

The Japan Meteorological Agency indicated that the dirty air was a result of both Chinese industrial pollution and sandstorms blowing in from northern China and Mongolia. The southern Japanese island of Kyushu was the worst-affected area. Air pollution in China also triggered high mercury levels on Japan's Mount Fuji. Osamu Nagafuchi, an environmental science professor at the University of Shiga Prefecture and lead scientist on the study, said Chinese factories burning coal, which sends mercury and other toxic elements to the atmosphere, was the primary culprit. "Whenever readings were high, winds were blowing from the continent (China)," Nagafuchi said (Jordan 2013).

Japan as a Leader in Pollution-Detection Technology

In 2017, Japan's Environment Ministry strengthened its systems that monitor fine particulate matter that threatens health at 10 locations nationwide 24 hours a day, "aiming to discover the mechanism for its formation, identify the sources, and implement countermeasures" (Nozaki 2017). At the same time, efforts continue to raise the accuracy levels of dispersion forecasts for pollution that drifts over from China or is emitted within Japan.

The main target of these systems are the tiny airborne particles (PM2.5) that are created by burning, often creating sulfur oxide. Sources include factories, vehicles, and natural sources such as volcanoes. The particles measure less than one-thirtieth the width of a human hair, a size that can lodge in the lungs and provoke respiratory problems, causing asthma and bronchitis, as well as increasing lung-cancer risk. Breathing these small particles also contributes to hardening of the arteries, raising the risk of heart attacks and strokes, known to medical professionals as "cerebral infarction."

Previously, particles caught by such devices needed to be extracted and transferred to another device in a separate location for analysis. But the continuous measurement equipment has an internal device for analyzing the particles, making it possible to instantly see changes in numerical data on computers in a distant location.

Yusuke Mizuno, 39, a Horiba official who played a role in developing the new devices, said that "when trees or weeds are burned, the density of potassium becomes higher. When coal is burned, the density of lead rises. These are some of the specific trends observed. . . . By combining the data with that on wind direction, it's possible to figure out the sources of PM2.5" (Nozaki 2017). The new machines allow continuous monitoring at all times, whereas

> currently, the ministry and local governments measure PM2.5 weights in about 1,000 locations across the nation. However, because the work requires time and effort, detailed analyses of the microparticles' components are only conducted four times a year—in spring, summer, autumn, and winter—for a total of eight weeks. (Nozaki 2017)

Toshihiko Takemura of Kyushu University developed a system that calculates global air pollution caused by airborne particulate matter and its effects on climate change. Every day, Takemura provides forecast data in four levels on his website. The levels include "very high," which means people should be on alert; "high," which is roughly equal to the national environmental standard; "moderate," meaning that views are a little hazy; and "low," meaning the air is clean. These forecasts are also used by news media (Nozaki 2017).

Japan's Wind Power Platforms

Japan is working to clear the air in Tokyo with wind power platforms at sea. Japan's wind-energy industry got a boost after 2011 with "a giant floating wind turbine [that] signals the start of Japan's most ambitious bet yet on clean energy . . . from the severely damaged and leaking nuclear reactors at Fukushima" (Tabuchi 2013). A single turbine, 2,350 feet tall and situated 12 miles out to sea, rose first, supplying enough power for 1,700 homes.

"Unremarkable, perhaps," reporter Hiroko Tabuchi wrote, "but consider the goal of this offshore project: to generate over one gigawatt of electricity from 140 wind turbines by 2020. That is equivalent to the power generated by a nuclear reactor" (Tabuchi 2013). As Japan turned away from nuclear power after the accident at Fukushima in 2011, wind turbines became a symbol of a new way of supplying energy to a nation that has no fossil fuels of its own. With coastlines of several islands offering thousands of turbine sites, potential seems limitless.

"It's Japan's biggest hope," said Hideo Imamura, a spokesman for Shimizu, during a recent trip to the turbine ahead of its test run. "It's an all-Japan effort, almost 100 percent Japan-made" (Tabuchi 2013). The Japanese have designed platforms for substation and electrical transformer equipment that can expand the potential locations of wind farms offshore. By one estimate, Japan can use such platforms, according to Tabuchi, "to generate as much as 1,570 gigawatts of electricity, roughly eight times the current capacity of all of Japan's power companies combined, according to computer simulations based on historical weather data by researchers at Tokyo University, one of the project's main participants" (Tabuchi 2013).

One estimate has it that the Japanese technology, applied worldwide, could supply as much as 7.5 terawatts of electricity. According to a study by researchers at the University of Delaware and Stanford University in 2012, that would meet half the world's power demand in 2030 (Tabuchi 2013). "We're opening a new page in the history of offshore wind power," said Takeshi Ishihara, a civil engineer at Tokyo University who leads the project.

However, Paul J. Scalise, a research fellow at the University of Tokyo Institute of Social Science, said these estimates were too grandiose. "We shouldn't forget the obvious reality check. The farther from the coast they place these floating wind farms, the more expensive it becomes to build them and transmit the power back to Japan," he said. "It becomes a cost-plus benefit analysis in which you weigh the benefits of the electricity versus the cost to build and maintain the infrastructure" (Tabuchi 2013).

Others have challenged the wind farm, especially local fishermen whose range has been restricted by the platforms. Cost is also a possible obstacle. All told, the platforms will require about eight times as much money per kilowatt hour as land-based turbines. Sponsors of the project hope that design improvements will narrow that gap.

Further Reading

Hong, Cai. 2017. "China Can Learn from Japan How to Fight Pollution." *China Daily*, January 9. Accessed October 5, 2018. http://www.chinadaily.com.cn/opinion/2017-01/09/content_27895726.htm.

Jordan, Alec. 2013. "Study Points to China as Source of Mt. Fuji Air Pollution." *Tokyo Weekender*, October 7. Accessed October 5, 2018. http://www.tokyoweekender.com/2013/10/study-points-to-china-as-source-of-mt-fuji-air-pollution/.

Nozaki, Tatsuya. 2017. "Japan to Strengthen Air Pollution Monitoring." *Japan News* in *Albuquerque Journal*, February 9. Accessed October 5, 2018. https://www.abqjournal.com/946213/japan-to-strengthen-air-pollution-monitoring.html.

Ryall, Julian. 2013. "Parts of Japan Smothered in Chinese Air Pollution." *DW* (*Deutsche Welle*), December 3. Accessed October 5, 2018. http://www.dw.com/en/parts-of-japan-smothered-in-chinese-air-pollution/a-16665471.

Tabuchi, Hiroko. 2013. "To Expand Offshore Power, Japan Builds Floating Windmills." *New York Times*, October 24. Accessed October 5, 2018. http://www.nytimes.com/2013/10/25/business/international/to-expand-offshore-power-japan-builds-floating-windmills.html.

MEXICO

A Parched, Polluted Capital Sinks

When the Spanish conquistador Hernán Cortés first laid eyes on Tenochtitlan, the capital of the Aztecs in 1519, he said it was the grandest city he had ever seen. With about 300,000 people, it was a larger urban area than any in Spain or the rest of Europe. Located on an island in a large lake, Tenochtitlan was not lacking clean water. Without motor vehicles or factories running on fossil fuels, the island city's air was pristine most of the time. Today, with 21 million people and the tendrils of urbanization lapping the hillsides of the valley, Mexico City has run out of both clean air and water.

Mexico City, still one of the largest urban areas on Earth, sits in a valley that caps air pollution, especially on winter days when the air is stagnant and cooler air drains into lower areas. It is also sinking as residents consume underground water. On top of all this, the entire area is seismically active, studded with volcanoes, and prone to earthquakes.

"A Stinking River of Sewage"

Because underground water has been pumped out much faster than rainfall replaces it, the Aztecs' lakes have long since disappeared. The ground under the city is sinking, "and Mexico City, a mile and a half above sea level, [is] sinking, collapsing in on itself. . . . It is a cycle made worse by climate change. More heat and drought mean more evaporation and yet more demand for water, adding pressure to tap distant reservoirs at staggering costs or further drain underground aquifers and hasten the city's collapse" (Kimmelman 2017). For many years, Mexico City grew on land reclaimed from surrounding swamps. Two three-mile-long aqueducts were built to carry fresh water from the mainland, each with two sluices, so one could be closed for cleaning without interrupting the water supply.

Today, what is left of the lake that once surrounded Tenochtitlan has been diverted into industrial canals. One of these, the 29-mile-long Grand Canal, which was built to ferry wastewater out of the city, is laden with waste so noxious that it can be smelled a mile away, "a stinking river of sewage belching methane and sulfuric acid" (Kimmelman 2017). Roughly 100 miles of canals remain from the immense lake that surrounded Tenochtitlan, used by farmers and tour-boat operators. Even these are often polluted, as they are sucked dry to feed water pipes in other parts of the city.

During January of 2017, a 20-foot-deep hole suddenly opened in one of the canals, draining it. "This is a warning," said Sergio Raúl Rodríguez Elizarrarás, a geologist at the National Autonomous University of Mexico. "We are driving the canals toward their extinction" (Burnett 2017, A-6).

Tap Water Is Scarce as City Sinks

Only 20 percent of Mexico City has reliable household tap water. Many people wait in line at neighborhood pipes (*pipas*) or buy expensive water from trucks. In Iztapalapa, one of many large parts of Mexico City (Iztapalapa alone houses 2 million people), tap water is scarce, flowing only a few hours a day, if at all. It is also polluted and likely to provoke bacterial infections that result in digestive problems ("Montezuma's revenge"). In Iztapalapa, some people have access to wells, but they provide only water laced with minerals and chemicals.

"Water becomes the center of women's lives in places where there is a serious problem," noted Mireya Imaz, a program director at the National Autonomous University of Mexico. "Women in Iztapalapa can spend all night waiting for the *pipas*, then they have to be home for the trucks, and sometimes they will ride with the drivers to make sure the drivers deliver the water, which is not always a safe thing to do. . . . It becomes impossible for many poor women to work outside the home," she said. "The whole system is made worse by corruption" (Kimmelman 2017).

Mexico City, 30 square miles in 1950, sprawled over more than 3,000 square miles by 2017 as engineering works sucked local aquifers dry. More and more of the city's water (40 percent by 2017) has been imported from outside the valley. As its underground water has been drained, Mexico City has sunken substantially over a century. A *New York Times* report observed that

> at a roundabout along the Paseo de la Reforma, the city's wide downtown boulevard, the gilded Angel of Independence, a symbol of Mexican pride, looks over a sea of traffic from the top of a tall Corinthian column. Tourists snap pictures without realizing that when Mexico's president cut the ribbon for the column in 1910, the monument sat on a sculptured base reached by climbing nine shallow steps. But over the decades, the whole neighborhood around the monument sank, like a receding ocean at low tide, gradually marooning the Angel. Fourteen large steps eventually had to be added to the base so that the monument still connected to the street. (Kimmelman 2017)

"Deeper in the city's historic center," the same report said, "the rear of the National Palace now tilts over the sidewalk like a sea captain leaning into a strong headwind. Buildings here can resemble Cubist drawings, with slanting windows, wavy cornices, and doors that no longer align with their frames" (Kimmelman 2017).

Part of the area's water shortage has been induced by urban sprawl. Before Mexico City sprawled across the entire valley, rainfall reached aquifers through its porous volcanic soil. As buildings and roads spread across the valley floor, more rainfall flowed into sewers and away from aquifers. Instead of replenishing the aquifers, rainfall causes floods (Kimmelman 2017).

With climate change, the situation will only get much worse, Claudia Shein-baum, a former environment minister who developed Mexico City's first climate-change program, told the *New York Times* (Kimmelman 2017). A warming climate will only increase the city's problems with pollution, specifically ozone. Heat waves mean health crises and rising costs for health care in a city where air-conditioning in poor neighborhoods is scarce (Kimmelman 2017).

Mexico City's Iconic Air Pollution

As Mexico City's water is depleted, its air pollution has become iconic. In 1992, the United Nations designated the city's air as the world's worst, a combination of automobile emissions, factory effluent, and a valley location that, especially in winter, produces classic thermal inversions. That title that has since passed on (by the number of very dangerous days per year) to New Delhi and Beijing. Mexico still has a high number of bad-air days. The legendary nature of the city's squalid air was enhanced by accounts describing air pollution that "was so poor, birds would fall out of the sky—dead. Locals said living there was like smoking two packs of cigarettes a day" ("Plan to Reduce" 2017).

A number of efforts have been made to ameliorate the pollution, but none have done much good. For a time, automobiles were banned on Saturdays, a step that was supposed to have cut pollution 15 percent. After nearly a decade of ineffectively enforced bans, little had changed, according to an analysis by Lucas W. Davis, an associate professor at UC Berkeley's Haas School of Business. "Saturday driving restrictions are a flawed policy. It's a big hassle for people and does not improve air quality," said Davis ("Plan to Reduce" 2017).

Driving has been curbed and industrial activity restricted during especially noxious periods (such as mid-March 2016), with little long-term effect. Some days, especially in mid-summer, air circulation improves and pollution disperses. During the March 2016 emergency, about 1.1 million cars in the Valley of Mexico, including nearly 450,000 in the capital, were ordered off the streets under the restrictions ("Choked Mexico" 2016). Mexico City's air quality is still the worst in the Western Hemisphere, with particulate levels three to four times higher on average than in New York City, São Paulo, Buenos Aires, or Los Angeles. Davis's solution is elegantly simple: "Test every car, test every year. If you have a car that's polluting the air, you can't drive it. Period" ("Plan to Reduce" 2017).

Beginning in 1989, motorists were banned from driving one day a week based on license numbers, but the affluent gamed the system by buying extra cars with different numbers. Others were transported by family members using their own cars or took taxis. A small number switched to public transport (buses or the subway), both of which are chronically overcrowded. Mexico City also enhanced bicycle sharing, which grew until the program was the largest in North America.

The city government also made tepid attempt to ban cars with diesel engines, but it was of little help because political jockeying postponed the ban's start date to 2025.

Matt McGrath of BBC World News observed that "massive growth in the use of cars coupled with a geographic location that trapped a toxic blanket of dirty air over the city saw tens of thousands of people hospitalized every year" (McGrath 2017).

Further Reading

Brandon, William. 1961. *The American Heritage Book of Indians*. New York: Dell.

Burnett, Victoria. 2017. "Mexico's Aquatic Paradise, Pushed to the Edge of Extinction." *New York Times*, February 23: A-6.

"Choked Mexico City Bans 1M [Million] Cars in Air Pollution Alert." 2016. *The Guardian* (United Kingdom), March 16. Accessed October 7, 2018. https://www.theguardian .com/world/2016/mar/16/choked-mexico-city-bans-1m-cars-in-air-pollution-alert.

Kimmelman, Michael. 2017. "Mexico City, Parched and Sinking, Faces a Water Crisis." *New York Times*, February 17. Accessed October 7, 2018. https://www.nytimes.com /interactive/2017/02/17/world/americas/mexico-city-sinking.html.

McDowell, Bart. 1980. "The Aztecs." *National Geographic*, December: 704–752.

McGrath, Matt. 2017. "Car Ban Fails to Curb Air Pollution in Mexico City." British Broad-casting Corp. (BBC) World News, February 2. Accessed October 7, 2018. http://www .bbc.com/news/science-environment-38840076.

Molina Montes, Augusto F. 1980. "The Building of Tenochtitlan." *National Geographic*, December: 753–766.

"Plan to Reduce Air Pollution Chokes in Mexico City." 2017. Phys.org, February 2. Accessed October 7, 2018. https://phys.org/news/2017-02-air-pollution-mexico-city.html.

RUSSIA

The Environmental Legacy of Soviet-Era Policies

According to one observer, environmental degradation across the former Soviet Union is a result of decisions "taken decades ago by the leaders of the Communist Party to give priority in the use of raw materials to the military-industrial complex and heavy industry without regard to the effect on human life [that has] turned many of Russia's cities into gas chambers under the open skies" (Yablokov et al. 1991, C-3).

Environmental destruction is said by Russian environmentalists to have been an inevitable result of command-economy decisions based on an ideology of gigantism pursued in the absence of any system for environmental protection. Ironically, Marxism (or "state socialism," as it was sometimes called) turned out to be as growth-oriented and heedless of environmental consequences as its ideological opposite, industrial capitalism.

A Heavy Emphasis on Industrialization

During the Soviet era, a heavy emphasis was placed on industrialization regardless of environmental consequences. Many very dirty industrial plants were constructed across Siberia. As a result, most of Siberia's large rivers were heavily contaminated.

Marine mammals (polar bears, walrus, bearded seals, ringed seals, white whales, and gray whales) have been found to be polluted by industrial waste bearing heavy metals, DDT, and PCBs. Some samples of plankton taken in the Arctic Ocean contain high levels of organochloric pollutants.

Following the fall of state socialism, little has changed in Russia during the last quarter century. The heedless development of Marxist doctrine has ended, but an emphasis on development of oil and coal without regard to environmental consequences continues. It's partially state-run, along with standard capitalism, with a 19th-century "mother lode" mentality. Officially, climate change is not regarded as much of a problem. Nor is disrespect for indigenous peoples' lands or rights.

During the 1930s and 1940s, indigenous lands were seized by Soviet state-run industries. Reindeer pastures and fishing sites were disrupted, "depriving Siberian indigenous peoples of food and severely disrupting their way of life" ("Peoples of

---

Trashing the Arctic

As the ice cover melts, the volume of human-generated garbage on the floor of the Arctic Ocean is rising with astounding speed, despite its location hundreds of miles from human urban areas. Shards of glass, plastic bags, and pieces of discarded fishing nets have become common, said Mine Tekman, of the Alfred Wegener Institute for Polar and Marine Research (AWI) in Germany.

Scientists photographed the Arctic Ocean floor in the Fram Strait between Greenland and Svalbard at 2,500 meters, taking 7,058 pictures over 10 years. Over a decade, the volume of garbage increased 23-fold, from 346 to about 8,800 pieces per square kilometer.

Tekman and colleagues wrote:

> Plastic accounted for the highest proportion (47 percent) of litter recorded at Hausgarten for the whole study period. When the most southern station was considered separately, the proportion of plastic items was even higher (65 percent). Increasing quantities of small plastics raise concerns about fragmentation and future micro-plastic contamination. . . . Litter densities were positively correlated with the counts of ship entering harbor at Longyearbyen, the number of active fishing vessels, and extent of summer sea ice. Sea ice may act as a transport vehicle for entrained litter, being released during periods of melting. The receding sea ice coverage associated with global change has opened hitherto largely inaccessible environments to humans and the impacts of tourism, industrial activities including shipping, and fisheries, all of which are potential sources of marine litter. (Tekman et al. 2017)

Further Reading

Tekman, Mine B., Thomas Krumpen, and Melanie Bergmann. 2017. "Marine Litter on Deep Arctic Seafloor Continues to Increase and Spreads to the North at the Hausgarten Observatory." *Deep Sea Research Part I: Oceanographic Research Papers,* February. Accessed October 11, 2018. http://www.sciencedirect.com/science/article/pii/S096706371630200X.

the Frozen North," n.d.). With industrialization came non-indigenous Russians from other parts of the Soviet Union.

What was true of indigenous lands became more and more typical of the country as a whole as years passed under the aegis of state socialism. Lake Baikal, for example, became one of the most polluted bodies of water in Asia. Pollution from logging, nickel mining, and gold mining became established in large areas of Siberia between 1930 and 1950, during Stalin's regime. In the 1950s, as logging operations increased rapidly, polluting factories' smokestacks were used as symbols of progress. Large swaths of Siberia, up to 40 percent of some areas, were cleared. Because of inefficient planning and transport, large amounts of harvested timber were left to rot ("Survival's Campaign" 1999). Exploration for oil and gas and the building of roads and railroads soon followed, accelerating during the 1960s.

Indigenous Peoples Crowded Out

Siberian indigenous peoples suddenly found themselves becoming minorities in their own homelands. Until the mid-1980s, the Soviet state pursued aggressive assimilative policies that have been summarized by Survival International, an international group based in England that advocates for indigenous peoples worldwide:

> From 1950 to the mid-1980s, the authorities in Moscow attempted to suppress all ethnic, linguistic, and cultural differences. In schools, indigenous languages ceased to be taught, and children were punished for speaking their own languages. This led to a breakdown between the generations; it is now common for the older people in a community to speak only their own languages, and the youngest to speak only Russian. In the same period, many indigenous villages were closed down, and the people were forced to move to larger ones. Villages of different peoples were amalgamated in an attempt to turn the country into a homogeneous Soviet state. Nomads were forced to settle in areas that were impractical for hunting or grazing. Bans on hunting and fishing led to a breakdown of indigenous economies and forced dependence on the state for subsidies and salaries. This loss of the indigenous peoples' ways of life resulted in unemployment, alcoholism, and high suicide rates, problems that still plague the peoples of the north. ("Peoples of the Frozen North," n.d.)

Following the collapse of the Soviet Union, much of the assimilative infrastructure dissolved. Many indigenous peoples returned to the land to herd and hunt for sustenance.

Native Peoples' Rights Given No Value

Under the Soviet state, indigenous peoples in Siberia had possessed no legal communal title to their lands. In fact, according to a law passed in 1968, immigrants' communal farms and industries were granted land (much of it originally used by indigenous peoples) free of charge. Under the prevailing legal system, land was regarded as being without value until it was utilized for economic improvement as defined by the Soviet state. Indigenous sustenance was given no value under the Soviet system. Compensation for environmental damage also was unknown under this system. Following the collapse of the Soviet Union, some indigenous groups demanded legal guarantees of land ownership and compensation for industrial damage to their homelands.

An analysis by the Slavic Research Center of Hokkaido University, Japan, reported, "The [Russian] government has tried resettling indigenous peoples as compensation, attempting to provide a 'civilized' mode of living with heated houses and electricity. Such measures have had certain benefits in industrialization, but they have accelerated the decline of indigenous cultures" ("Survival's Campaign" 1999).

Disregard for the environment in Russia did not end with the collapse of the Soviet Union. Russian Federation president Vladimir Putin, for example, during 2000 disbanded the State Committee for Environmental Protection soon after assuming office. This agency, which previously had functioned as an independent regulatory body, was merged with the Natural Resources Ministry; environmental protection was subsumed by that agency's pro-development mission.

The Oil Industry Expands with Warming

As the Russian oil industry continued to expand into the lands of indigenous peoples in Siberia, some of the same companies are following retreating polar ice caps provoked by global warming north into areas of the Arctic that earlier had been inaccessible most of the year because of ice and snow. A report in *Drillbits and Tailings* noted that "further melting of Arctic ice promises to be a tragic humanitarian and environmental disaster but could result in a huge windfall of natural gas for the Russian petroleum industry" ("Russian Gas" 2000).

The Russian Arctic shelf could hold as much as 2.3 trillion cubic feet (65 trillion cubic meters) of natural gas, more than present known reserves on the Russian mainland. This amount of gas could fulfill natural gas needs for Russia, Europe, and Turkey for as many as 20 years. Already, roughly 25 oil fields have been established on the northern Yamal Peninsula, most of which extend offshore into the Kara Sea. "The greater potential," according to this *Drillbits and Tailings* report, "lies in the currently inaccessible but rapidly melting Arctic Ocean" ("Russian Gas" 2000).

"It [Arctic sea ice] lost an average of 34,300 square kilometers—an area lager than the Netherlands—each year," said Lisa Mastny, a researcher at the Worldwatch Institute, which has compiled scientific reports on the melting of ice on a world-wide scale. Not only has ice coverage shrunken, but it also has thinned. "Between this period [the 1960s] and the mid-1990s, the average thickness dropped from 3.1 meters to 1.8 meters—a decline of nearly 40 percent in less than 30 years," said Mastny ("Russian Gas" 2000). With pack ice in retreat, an opportunity arises for Gazprom and other Russian petroleum firms to drill in the relatively shallow—and increasingly ice-free—waters. Gazprom has already built the German, Polish, and

Belarussian sections of a pipeline that stretches from Yamal to Europe. This links into Russia's existing pipeline network.

"There's a dangerous pathology at work here," said Rory Cox, communications director at U.S.-based Pacific Environment and Resources Center.

> By drilling for oil in the Far North and building these pipelines across pristine Siberian landscapes, Russian oil companies are creating a suicidal feedback loop. Global warming provides access to more fuel; burning it brings on more warming. The receding Arctic underscores the need to develop alternatives, not revel in the profits that Gazprom will make. ("Russian Gas" 2000)

Further Reading

McCauley, Martin. 2014. *The Soviet Union 1917–1991.* New York: Routledge.
"Peoples of the Frozen North." 1998. Survival International. Accessed November 30, 2018. https://assets.survivalinternational.org/static/files/related_material/39_21_81_siberi abg.pdf.
"Russian Gas Companies Follow Receding Arctic Ice." 2000. *Drillbits and Tailings* 5, no. 15 (September 19).
"Survival's Campaign for the Khanty. Economic Development and the Environment: The Sakhalin Offshore Oil and Gas Fields." 1999. Slavic Research Center, Hokkaido University, Japan. Accessed October 7, 2018. http://src-h.slav.hokudai.ac.jp/sakhalin/eng /71/kitagawa3.html.
Yablokov, Alexi, Sviatoslav Zabelin, Mikhail Lemeshev, Svetlana Revina, Galina Flerova, and Maria Cherkasova. 1991. "Russia: Gasping for Breath, Choking in Waste, Dying Young." *Washington Post*, August 18: C-3.

UNITED KINGDOM

Choking in London

The United Kingdom gave us the word "smog" (an amalgamation of "smoke" and "fog") from its first coal-burning years, when combustion of the first fossil fuels for industrial and home heating during the mid-19th century filled the air in London and other industrial centers with noxious smoke. This effluent combined with stagnant air and the island's frequent winter fogs to produce some of the dirtiest air in the world. The smog that sullied English air reach its worst level in the "Great Smog" of 1952, when stagnant high pressure combined with coal combustion to kill 12,000 people in a few days. After that, the United Kingdom undertook a program to cleanse its air.

Smog Returns

By 2017, however, smog had returned as population increased, along with traffic and emissions from diesel fuel that powers half of British motor vehicles. In addition, increasing numbers of people in an urban area that now envelops much of southwestern England have been using wood for home heating, which produces

Coal-mining pollution hangs like a fog over London, December 1952. Thousands died from the trapped, stagnant air. (Phil Dye/Mirrorpix/Getty Images)

smoke. By 2017, more than 1 million British homes had wood-burning stoves, with 175,000 more being installed annually. Many homeowners were burning wood to save on energy costs, despite the fact that "experts at the University of Southampton warned that wood burners 'liberate significant amounts of particulate pollution into the outdoor air' and said they risked undoing the good work of the Clean Air Act, which was brought in following the Great Smog of 1952" (Knapton 2017).

During mid-winter 2016–17, stagnant high-pressure air capped the atmosphere and had some Londoners comparing their air quality (or lack thereof) to that of contemporary Beijing and New Delhi. London was placed on a "very high" pollution alert for the first time, with air pollution at record levels. Even London's mayor, Sadiq Khan, has adult-onset asthma. In January 2017, particulate levels in London rose higher on some days than those in Beijing.

Illness from toxic air has become chronic in recent years. According to the *New York Times*:

Every winter, as if on cue, the coughing begins. As soon as the weather turns cold, Tara Carey, an international aid worker living in London, ritually places cough syrup on her bedside table because she knows her sleep will be punctuated by hacking coughs. She also coughs at work. And she coughs while cycling to her office, on a road so toxic that for a brief period last month the air pollution there was greater than in infamously smoggy Beijing. With her cough persisting winter after winter, Ms. Carey, 43, became worried she might have contracted tuberculosis during a work trip to Africa and sought medical help. She was shocked by the doctor's eventual diagnosis: asthma. In Ms. Carey's view, she said the only reasonable explanation for her illness was the pollution to which she was exposed over the last six years cycling through thick traffic on Brixton Road, one of London's busiest and most noxious routes. "You get a massive mouthful of fumes," she said, noting that asthma does not run in her family. "But we don't really realize how much toxic air we breathe in because we're acclimatized to it. It's pernicious." (Freytas-Tamura 2017)

Gas Masks for Children

Toxic air was dense enough (and frequent enough) in London by 2017 that face masks had been designed especially to protect children. The *Telegraph* reported that "Doodle Masks come in colorful designs, including Disney film themes, designed to attract children, and are sold for £7.50 each online" (Fuller 2017).

Today's London smog is not as thick as the infamous "pea soup" of earlier years, but it is laced with nitrogen dioxide, a toxic emission produced by diesel combustion. This type of pollution contributes to about 23,000 premature deaths in Britain during an average year, according to the U.K. Department for Environment, Food, and Rural Affairs (Freytas-Tamura 2017).

In the meantime, several schools refused to let children play outdoors, as public-health officials urged everyone to limit outdoor exercise. Schools were moving exercise areas away from roads with heavy polluting traffic. More than 440 London schools distributed gas masks to pupils because their lungs may be endangered by the dirty air.

Mayor Khan said: "Every child deserves the right to breathe clean air in London, and it is a shameful fact that more than 360 of our primary schools are in areas breaching legal pollution limits. London's filthy air is a health crisis, and our children are particularly vulnerable to the toxic effects of air pollution" (Knapton, 2017). In children, excess pollution may cause or aggravate coughs, bronchitis, and asthma, as it impedes development of their lungs and brains.

London has encouraged bicycle riding, but setting aside lanes for that purpose squeezes cars and trucks into fewer lanes and causes them to slow and stall, increasing pollution. In addition, bike riders are exposed to the worst of the pollution at tail-pipe level. The city also has announced a $12-a-day tax for the worst polluting vehicles.

Debate over Diesel Fuel

Ironically, the British government at one time urged British drivers to adopt diesel fuel to *reduce* pollution, following tests that turned out to be faulty, under-reporting toxicity of emissions by a factor of five. The government then provided financial incentives for purchase of vehicles with diesel engines. "No one at the time thought of the consequences of increased nitrogen-dioxide emissions from diesel, and the policy of incentivizing diesel was so successful that an awful lot of people bought diesel cars," said Anna Heslop, a lawyer at ClientEarth, an environmental law firm that last year forced the British government to produce a better plan to improve air quality (Freytas-Tamura 2017). The number of diesel vehicles licensed in London rose from 601,456 in 2012 to 774,513 in 2015, a 29 percent increase (Freytas-Tamura 2017).

In March of 2017, the British Broadcasting Corporation reflected upon "How Bad is Air Pollution in the UK?" (Harrabin 2017). The BBC also ran a series of television programs titled *So I Can Breathe* and reported that air pollution has recently been shortening the lives of 40,000 people a year in the United Kingdom, most of them with preexisting heart and lung problems. Dirty air also makes the lives of many people with ailments such as asthma and hay fever miserable without outright killing them. "Experts in air pollution argue that it has been under-reported for decades," the BBC reported, "but the issue has been thrust into the news because the U.K. government lost court cases over illegally dirty air and because car makers were found to be cheating tests on car emissions" (Harrabin 2017). In 2016, Britain's High Court ruled that central government measures meant to combat air pollution had been failing to comply with rules on nitrogen-dioxide limits passed by the European Union.

Solving the Problem

What should the British do to cleanse their air? The BBC series concluded:

> Solving air pollution needs a many-sided approach. The best value for money comes from targeting the really big individual polluters—that's old buses and lorries in cities. Most big cities are already doing that, although critics say not fast enough. Insulating homes so they don't burn as much gas, would save pollution, cash, and carbon emissions in the long term—but critics say the government appears to have no strategy for this. Stopping the spread of wood-burning stoves in cities might help a bit. Cutting pollution from ships would be good in port cities. (Harrabin 2017)

In 2017, government ministries were preparing to offer owners of older diesel cars as much as £3,500 (about US$4,200 at that time) to scrap them. However, according to the BBC, "The Green Party says it would be perverse to reward car makers with increased sales when they caused the problem in the first place by failing on their promises to government to make diesel engines clean" (Harrabin 2017). The government also was charging a £10 (US$12.43 in March 2017) daily pollution tax on diesel and gasoline-powered vehicles registered before 2006 that did not meet the European Union vehicle-emissions regulations.

The *Telegraph* reported:

Experts at King's College London said the recent spell of unhealthy pollution was the worst since April 2011 in the capital and was being caused by cold, calm, and settled conditions combined with "traffic pollution and air pollution from wood burning." Temperatures have fallen below zero [Celsius] overnight over the last few days, meaning householders are burning more fuel to keep warm. "This was the largest contribution from wood burning measured during the winter so far," said a spokesman for King's College. (Knapton 2017)

Recognizing the toll of worsening air pollution not only in London, but across the United Kingdom, the British government in July 2017 declared war on the internal-combustion engine, decreeing that sales of diesel and gasoline-powered cars will end by 2040. Months earlier, France had made a similar pledge, which, like Britain's, is also aimed at combatting greenhouse-gas emissions. The strategy aims to steer automotive and truck sales toward electric vehicles and assumes that increasing amounts of the power used to charge these cars will be generated by wind and solar power, instead of coal, oil, and natural gas. Chris Grayling, Britain's transport secretary, promised a "green revolution in transport" (Castle 2017, A-6).

Further Reading

Castle, Stephen. 2017. "To Fight Pollution, Britain Will Ban Sales of New Diesel and Gas Cars by 2040." *New York Times*, July 27: A-6.

Freytas-Tamura, Kimiko de. 2017. "A Push for Diesel Leaves London Gasping Amid Record Pollution." *New York Times*, February 17. Accessed October 7, 2018. https://www.nytimes.com/2017/02/17/world/europe/london-smog-air-pollution.html.

Fuller, George. 2017. "The Face Masks Helping Kids Beat London's Air Pollution." *The Telegraph*, March 6. Accessed October 7, 2018. http://www.telegraph.co.uk/news/2017/03/06/face-masks-kids-counter-london-air-pollution/.

Harrabin, Roger. 2017. "How Bad Is Air Pollution in the UK?" BBC News, March 6. Accessed October 7, 2018. http://www.bbc.com/news/science-environment-38979754.

Knapton, Sarah. 2017. "Air Pollution in London Passes Levels in Beijing . . . and Wood Burners Are Making Problem Worse." *The Telegraph*, January 25. Accessed October 7, 2018. http://www.telegraph.co.uk/science/2017/01/24/air-pollution-london-passes-levels-beijingand-wood-burners-making/.

Chapter 8: Renewable Energy

OVERVIEW

This chapter focuses on ways countries are reducing greenhouse-gas production with alternative energy, including solar and wind power, as well as ethanol. Many such efforts are underway—so much effort, on so many fronts, that an encyclopedia of solutions could be composed. What follows is merely a sampling.

More and more, coal is becoming yesterday's fuel. By 2015, coal mines were closing steadily in Kentucky, Ohio, Illinois, and West Virginia, and thousands of miners had been laid off, as U.S. coal production had declined 15 percent in six years (Krauss 2015). The proportion of U.S. electrical power generated by coal fell from 50 percent in 2000, to 40 percent in 2012, and 30 percent in 2015, as the U.S. Energy Department anticipated further declines. By 2017, wrote former New York City mayor Michael Bloomberg, "the average cost of wind power has dropped to $20 per megawatt, compared with the more than $30 cost per megawatt for electricity from many coal plants in the [Midwest] region. Why would consumers pay more for a power source that may kill them?" (Bloomberg 2017).

The Kentucky Coal Mining Museum in Benham (part of Southeast Kentucky Community and Technical College) in 2017 announced plans to switch from grid-supplied coal-generated electricity to solar power, because it is less expensive. In 2011, the United States had 514 coal-fired power plants. Half closed or announced plans to shut down within six years, and the Sierra Club expects two-thirds to get off the grid by 2022 (Friedman 2017, A-25).

China, once a seemingly limitless market for U.S. coal, was curtailing its consumption. Coal's share of U.S. electricity generation was being eroded not only by cheaper wind and solar power, but also by natural gas supplied by hydraulic fracturing ("fracking").

After 15 gigawatts of coal power were retired from service in the United States during 2015, another 7 gigawatts were taken off-line in 2016. At the same time, the solar-power workforce grew by 25 percent, accounting for one in 50 new jobs in the country. By 2016, 260,000 people were working in the solar power industry. Total U.S. greenhouse-gas emissions in the United States were falling, and household spending on energy had dropped to 3.9 percent of U.S. personal consumption, down from a peak of 9.1 percent during the 1980s and 7.3 percent in 1960 (Loveless 2017, 6).

Until very recently, wind and solar power were very minor players in an energy-generation mix dominated by coal, oil, and natural gas. Within roughly 15 years, since the turn of the millennium, wind and solar have become much more efficient

Wind turbines at the Horse Hollow Wind Energy Center in Texas, a state that is among the leaders in this type of technology. (Neonriver/Dreamstime.com)

and cost-competitive. Between 2009 and 2015, the cost of generating electricity with wind declined 61 percent; solar power's cost dropped by 82 percent during the same period (Krugman 2016, A-21).

Some U.S. states (Iowa is an example) will have approached half-wind generation by the time this book is published. Iowa also will lay claim to the tallest wind turbine in the United States, at 554 feet, mounted on a concrete tower that raises the tip of its blade to a level equaling the top of the Washington Monument (555 feet). Customers of the Omaha Public Power District (OPPD) woke up one morning in 2017 to learn that their utility, which has been mainly reliant on coal and nuclear, had quickly switched to one-third wind-powered.

By 2015, Germany as a whole was approaching 30 percent of electrical generation by solar, wind, and other renewable sources, even as it phased out nuclear power. "Germany is the first country in the world to show they can uncouple growth from burning of fossil fuels," said Jim Yong Kim, president of the World Bank. "This is the main task of our generation" (Eddy 2015). German household consumers have borne much of the burden of this energy paradigm shift through higher rates.

Even so, given rising energy demand around the world, coal was still being burned in new power plants in places such as India, even as solar and wind technologies improve and costs decline. By 2017, however, planners in India were beginning to scale back the number of new coal-fired plants in favor of solar arrays

Doctors' Orders: Get Off Coal

The British medical journal *Lancet* in 2015 assembled internationally known doctors and other public-health experts to assess global warming's impact on Earth by the end of the 21st century and what will be necessary to hold warming within survivable limits while maintaining human health. Their number-one prescription was to substitute non-fossil-fuel energy for coal as quickly as possible. Calling coal combustion a "medical emergency," Dr. Anthony Costello, a pediatrician and director of the Global Health Institute at the University College of London, said that coal, the fuel that produces the most carbon pollution per unit of energy generated, also produces more health-damaging air pollution than any other source of energy (Borenstein 2015).

Further Reading

Borenstein, Seth. 2015. "Top Doctors' Prescription for Feverish Planet: Cut Out Coal." *St. Louis Post-Dispatch*, June 22. Accessed December 3, 2018. https://www.stltoday.com/business /local/top-doctors-prescription-for-feverish-planet-cut-out-coal/article_0f9dea74-2ebd -52f0-be60-d6abfae5affa.html.

Watts, Nick, W. Neil Adger, Paolo Agnolucci, Jason Blackstock, Peter Byass, Wenjia Cai, . . . Anthony Costello. 2015. "Health and Climate Change: Policy Responses to Protect Public Health." *The Lancet*, June 23. Accessed October 11, 2018. http://www.thelancet.com /commissions/climate-change.

as relative costs fell and coal-provoked pollution increased. India's changing power mix is one example (and an important one) of why the rise in worldwide greenhouse-gas emissions (and the proportion of carbon dioxide in the atmosphere) may soon peak for the first time since coal use became widespread in the mid-19th century, due to "undisputed worldwide technological progress and expansion of renewable technologies," according to one analysis (Edenhofer 2015, 1286). In 2013, for the first time, electric utilities worldwide installed more new capacity from renewable sources (143 gigawatts, mainly wind power) than fossil fuels (141 gigawatts), according to an analysis presented April 14, 2014 at the Bloomberg New Energy Finance annual summit in New York City. Analysts with BNEF expect that "the shift will continue to accelerate, and by 2030, more than four times as much renewable capacity will be added" (Randall 2014).

Even as wind and solar power boomed worldwide, the International Energy Agency warned that the change was too slow to keep temperatures from rising to levels that will seriously distort climate. India, where 25 percent of the population (300 million people) have no electricity at all (and the per capita carbon footprint is 4 percent that of the United States), occupies an especially important position in the world energy equation, as its leaders seek to increase power by all means—coal to solar. India recognizes a growing toll of coal combustion; half the schoolchildren in New Delhi have lungs that have been scarred by the worst air pollution on Earth, as asthma rates rise rapidly, but production of power has taken priority, and in 2015, the government announced plans to double its use of domestic coal by 2019 (Barry 2015, A-19).

At the same time, small villages in India are installing localized solar power, a few light bulbs at a time, where no one heretofore has had any electricity at all. In laboratories, scientists are testing an "artificial leaf" that produces solar power as plants do it—a revolutionary step that, once it is commercialized, will hasten the day when fossil fuels are burned mainly in memory. The question is how quickly. Will the solar revolution arrive before we have burned enough fossil fuels to ruin large parts of the global ecosphere? Stay tuned.

This chapter will focus on alternative energy initiatives in Brazil, China, Denmark, Germany, India, Kenya, Morocco, and the United States.

Further Reading

Barry, Ellen. 2015. "For Indians, Smog and Poverty Are Higher Priorities Than Talks in Paris." *New York Times*, December 10: A-19.

Bloomberg, Michael R. 2017. "Climate Progress, with or without Trump." *New York Times*, March 31. Accessed October 8, 2018. https://www.nytimes.com/2017/03/31/opinion/climate-progress-with-or-without-trump.html.

Eddy, Melissa. 2015. "Germany May Offer Model for Reining in Fossil Fuel Use." *New York Times*, December 3. Accessed October 8, 2018. http://www.nytimes.com/2015/12/04/world/europe/germany-may-offer-model-for-reining-in-fossil-fuel-use.html.

Edenhofer, Ottmar. 2015. "King Coal and the Queen of Subsidies." *Science* 349 (September 18): 1286–1287.

Friedman, Thomas L. 2017. "Coal Museum Sees Future; Trump Doesn't." *New York Times*, April 19: A-25.

Krauss, Clifford. 2015. "Coal Miners Struggle to Survive in an Industry Battered by Layoffs and Bankruptcy." *New York Times*, July 17. Accessed October 8, 2018. http://www.nytimes.com/2015/07/18/business/energy-environment/coal-miners-struggle-to-survive-in-an-industry-battered-by-layoffs-and-bankruptcy.html.

Krugman, Paul. 2016. "Wind, Sun, and Fire." *New York Times*, February 1: A-21.

Loveless, Bill. 2017. "Trump Might Not Be Able to Reverse Coal Industry Slump." *USA Today*, February 14: 6-B.

Randall, Tom. 2015. "Fossil Fuels Just Lost the Race against Renewables." *Bloomberg News*, April 14. Accessed November 30, 2018. https://www.bloomberg.com/news/articles/2015-04-14/fossil-fuels-just-lost-the-race-against-renewables.

BRAZIL

Sugarcane Ethanol

Ethanol can be produced through fermentation of many natural substances, but sugarcane offers advantages over most others, including corn. For each unit of energy expended to turn sugarcane into ethanol, 8.3 times as much energy is created, compared with a maximum of 1.3 times for corn, according to scientists at the Center for Sugarcane Technology and other Brazilian research institutes. "There's no reason why we shouldn't be able to improve that ratio to 10 to one," said Suani

Teixeira Coelho, director of the National Center for Biomass at the University of São Paulo. "It's no miracle. Our energy balance is so favorable not just because we have high yields, but also because we don't use any fossil fuels to process the cane, which is not the case with corn" (Rohter 2006). Sugarcane ethanol is generally more economical than oil, with the per-barrel price under $30.

Ethanol Use Expands

Use of ethanol in Brazil accelerated after 2003, following the introduction of "flex-fuel" engines, designed to run on ethanol, gasoline, or any mixture of the two. By 2013, more than half of Brazil's transportation fuel was supplied by sugarcane ethanol. By 2013, more than 80 percent of the 1.3 million automobiles sold in Brazil had flex-fuel engines, compared to 20 percent in the United States. Most fueling stations offered E85 fuel, compared to 1 percent in the United States.

Brazil, using ethanol from sugarcane, became energy self-sufficient in 2006, even as demand for fuel grew. Brazil's full-court press on ethanol was three decades old by that time. During the 1970s, Brazil's government began developing the ethanol industry by subsidizing the sugarcane industry and requiring its use in government vehicles. By the late 1990s, however, the subsidies were phased out as the cost of producing ethanol dropped to 80 cents a gallon, less than $1.50, the worldwide average for producing gasoline (Luhnow and Samor 2006, A-1, A-8). Many gas stations in Brazil have two sets of pumps, marked "A" for alcohol and "G" for gasoline. "Renewable fuel has been a fantastic solution for us," Brazil's minister of agriculture, Roberto Rodrigues, said. "And it offers a way out of the fossil fuel trap for others as well" (Rohter 2006).

In the past, the residue remaining when sugarcane stalks were compressed to squeeze out juice was discarded. Today, Brazilian sugar mills use that residue to generate electricity to process cane into ethanol and use other by-products to fertilize the fields where cane is planted. Some mills are now producing so much electricity that they sell their excess to the national grid. Brazil's government taxes ethanol at 9 cents per gallon, compared to 42 cents for gasoline, and all gasoline is legally required to contain at least 10 percent ethanol. Researchers in Brazil have decoded the genetics of sugarcane and used the knowledge to breed varieties with higher sugar content. Brazil has increased the per-acre productivity of sugarcane threefold since 1975 (Luhnow and Samor 2006, A-1, A-8).

Victims of the Ethanol Boom

Sugarcane ethanol may be an energy winner, but local Native people and sugarcane field workers in Brazil often have become victims of the boom. *Birdwatchers,* an Italian film, describes the Brazilian Guarani-Kaiowá Indians' struggle against the factory farming that is crowding them off their land. *Birdwatchers* parses the issues (land invasion, suicides, and rebellion) against the background of a love story involving the daughter of a wealthy landowner and a young Guarani who has become a shaman apprentice in the Brazilian state of Mato Grosso do Sul. The Guarani in Brazil have lost much of their land to sugarcane cultivation.

Mato Grosso means "thick forest" in Portuguese. Today, however, most of the trees in the area have been felled. During the last 70 years, the Guarani and neighboring peoples have been evicted from their land by cattle ranchers, as well as sugarcane and soya planters. Many now work on the ranches or farms for subsistence wages, in perpetual peonage. Over a period of 20 years, more than 500 Native people (some as young as nine years of age) have killed themselves. Others have been shot to death while trying to reoccupy alienated land.

Why Not Use Sugar Beets?

If sugarcane is such a good source of biofuel, why not use sugar beets for ethanol? The processing equipment for sugar beets and sugarcane is somewhat similar. The problem with sugar beets lies in their harvest cycle in temperate regions. While sugarcane is grown nearly year-round (and thus can supply a processing plant almost all the time), sugar beets are grown on an annual cycle and harvested in the fall. Thus, according to Kenneth P. Vogel, a professor in the University of Nebraska at Lincoln's Agronomy Department, a processing plant that costs several million dollars would run only for a few months a year. This hurdle could perhaps be overcome with technology (not yet designed) to switch from sugar beets to other types of ethanol sources. Instead, ethanol interests on the Great Plains are looking at switchgrass (which is handled like hay), corn stover, and other plant biomass. The manufacturing technology for these needs to be developed as well ("No Sugar-Beet" 2007, 6-B).

Market Forces Trump "Green" Rationales

While Brazil's sugarcane ethanol may be "greener" than the corn-based fuel produced in the United States, politics and economics can manipulate markets in the opposite direction. By 2016, according to a report in Reuters, the British-based news service, "U.S. ethanol plants and traders [were] rushing to sell biofuel to Brazil as tightening supplies and logistics in the South American market gave the world's top producer a rare opportunity to ship south during the peak sugarcane harvest" (Prentice and Ewing 2016). While the price of corn had fallen because of a record harvest in the United States, sugar prices rose in Brazil, making imports of U.S. ethanol to Brazil relatively less expensive. Because of the difference in prices, Raizen, the largest sugar and ethanol producer in the world, purchased 14.5 million gallons of biofuel from the United States in July and August of 2016 for shipment to Brazil, the largest amount since 2012.

The ethanol industries in both the United States and Brazil remained generally profitable in 2016, despite a decline in oil prices to an average of about $50 per barrel. "Several years ago, if you had suggested to a room full of ethanol producers, 'we're going to have $45 crude oil and you guys are going to still make money,' you would have found some doubters," said Geoff Cooper, senior vice president at the Renewable Fuels Association. "These are decent margins" (Prentice and Ewing 2016).

With a severe deficit in public accounts, the Brazilian government at the end of 2016 ended a tax exemption for domestic ethanol sales, a move that, according to a report in Reuters, "should make the biofuel less attractive for mills and motorists and prompt Brazil to produce and export more sugar while importing more gasoline" (Soto and Teixeira 2016).

The tax exemption, which had started in 2013, gave ethanol a tax break compared to gasoline. It had aimed to offset low sugar prices on international markets (prices rose sharply in 2016, eroding that rationale) and to match subsidies for gasoline, the price of which was being kept relatively low to act as a check on inflation in Brazil. The gasoline subsidy also was reduced to remove a drain on government spending, amid an economic crisis that included inflation as high as 10.7 percent, at the same time that unemployment was rising and Brazil's economy was contracting at the fastest pace in at least 100 years.

According to Reuters, "The Brazilian sugar industry wants the government to maintain the ethanol exemption in recognition of the environmental benefit of the biofuel and to avoid discouraging its production" (Soto and Teixeira 2016). Many Brazilian drivers chose ethanol if it cost less than 70 percent of gasoline because it yields 30 percent less energy per gallon. "Over the past 12 months [in 2016], ethanol sales climbed 36 percent to a record 17.8 billion liters as the biofuel became a more attractive option following a rise in gasoline taxes a year ago. Ethanol fell to as low as 60 percent of the gasoline price during the production peak in August, a move that led consumers to wipe out the nation's inventories" (Freitas 2016).

Further Reading

Freitas, Gerson, Jr. 2016. "Sugarcane Fuel Wins in Brazil as Cheap Ethanol Beats Gasoline." Bloomberg News, February 29, Accessed November 30, 2018. https://www.bloomberg.com/news/articles/2016-03-01/sugar-cane-fuel-wins-in-brazil-as-cheap-ethanol-beats-gasoline.

Luhnow, David, and Geraldo Samor. 2006. "As Brazil Fills Up on Ethanol, It Weans Off Energy Imports." *Wall Street Journal*, January 9: A-1, A-8.

"No Sugar-Beet Answer." 2007. *Omaha World-Herald* (Editorial), April 7: 6-B.

Prentice, Chris, and Reese Ewing. 2016. "U.S. Ethanol Industry Finds Sweet Deals in Brazil as Sugar Prices Soar." Reuters, August 4. Accessed October 8, 2018. http://www.reuters.com/article/us-usa-ethanol-brazil-idUSKCN10F1WY.

Rohter, Larry. 2006. "With Big Boost from Sugarcane, Brazil Is Satisfying Its Fuel Needs." *New York Times*, April 10. Accessed October 8, 2018. http://www.nytimes.com/2006/04/10/world/americas/10brazil.html.

Soto, Alonso, and Marcelo Teixeira. 2016. "Brazil to Let Ethanol Tax Break Expire in December." Reuters, August 25. Accessed October 8, 2018. http://www.reuters.com/article/us-brazil-ethanol-tax-exclusive-idUSKCN1100PB.

CHINA

Solar Power's Workshop

In less than a decade after 2008, China has become the world's predominant producer and user of solar power on a mass scale, a trend that accelerated after Donald J. Trump was elected president of the United States. Trump's description of global warming as a Chinese "hoax" has been highly resented in China, which has been using solar power as part of a global propaganda offensive to show the world that it is a leader in combatting climate change, under the explicit direction of its president, Xi Jinping. By 2017, China was manufacturing and using more than half of the world's solar panels.

World's Solar Leader

China's solar-power surge in 2017 included the world's largest array, which, with 4 million panels across 27 square kilometers on the Tibetan plateau, is the size of a small city. By 2015, China was adding 15 to 20 gigawatts of solar capacity per year, having surpassed Germany as the largest solar-power market in the world. By early 2016, China had 43.2 GW of solar capacity, compared to 38.4 GW in Germany and 27.8 in the United States. World capacity reached 200 GW in 2015 and 320 GW in at the end of 2016 (Martin 2016). Even so, Chinese solar plants generated 66.2 billion kilowatt-hours in 2016, only 1 percent of the country's total power generation (Woo 2017).

Even as solar expands, China is still the world's largest consumer of coal (as much as the rest of the world combined) and by far the planet's number-one producer of greenhouse gases. Hundreds of millions of people in China suffer from some of the world's dirtiest air, largely caused by carbon combustion. According to Richard Martin, writing in the *MIT Technology Review*, "China's leaders are desperate to reduce the coal-fired air pollution that renders the air in big cities like Shanghai and Beijing virtually unbreathable" (Martin 2016).

Growing Chinese dominance of solar-panel manufacturing was a major reason why world prices fell by 80 percent between 2008 and 2013. John Fialka, writing in *Scientific American*, said that "China had leapfrogged from nursing a tiny, rural-oriented solar program in the 1990s to become the globe's leader in what may soon be the world's largest renewable energy source" (Fialka 2016).

Changing Solar's Economics Worldwide

The Chinese have "fundamentally changed the economics of solar all over the world," said Amit Ronen, director of the Solar Institute of George Washington University, one of many scholars following the intense competition in the emerging $100 billion industry that supports the world's growing solar-energy demands (Fialka 2016). The U.S. solar industry may have invented the technology and holds most of the patents, but China by 2015 had assumed world leadership in production and use. Two-thirds of the solar panels installed in the United States were being made in China by 2016. For example, JinkoSolar Holding Co, Ltd. (New York Stock Exchange Stock Code: JKS), which employs more than 10,000 people in eastern China, maintains offices and warehouses in Europe and the United States.

China began its use of solar in rural areas that were not served by the conventional power grid, much like the way solar is developing in India, Kenya, and other developing countries today. Soon thereafter, China developed manufacturing on a mass scale. China became intrigued by development of a solar-panel industry that would export to Germany, the United States, and other affluent countries. The Chinese central government encouraged this trend with generous tax credits. By 2016, more than two-thirds of the world's solar hot water heaters were being installed in China.

"The Chinese took it and basically ran with it," said Donald Chung, one of the authors of a U.S. Department of Energy report, who studies the solar industry for DOE's National Renewable Energy Laboratory in Golden, Colorado (Fialka 2016). China bought solar companies and invited others to move to China, where skilled labor was supplied at relatively low wages. Solar was designated as one of seven favored industries in the national government's five-year plans. China thus found itself with a glut of solar panels (with twice as much production as overseas sales), which was absorbed within the country by policies that favored installation of rooftop solar collectors. Within two years, by 2015, China had become not only the world leader in solar panel manufacturing, but also its leading domestic market.

Wyatt Metzger, a principal scientist at the U.S. Department of Energy National Renewable Energy Laboratory (NREL), said, "They have a centralized government and terrible pollution problems. They understand the need to get away from coal and to invest in clean energy" (Fialka 2016). Chinese activity created jobs in the United States, where, by 2016, 250,000 people were employed in aspects of the industry that could not be outsourced, such as installation (Fialka 2016).

An International Solar Grid?

China's advanced thinking regarding solar power includes anticipation of an international grid in which it will be the main advocate and supplier, as a backward-looking United States fades into irrelevance under President Donald Trump, with his reliance on outdated fossil-fuel technology. According to a report in *Scientific American* (Fialka 2016), "In October, Liu Zhenya, former chairman of China's state-owned power company, State Grid Corp., came to the United Nations to shed more light on his nation's evolving solar ambitions, which he said are part of a plan aimed at organizing a global power grid [which he calls the Global Energy Interconnection] that could transmit 80 percent renewable energy by 2050." The Chinese have invited the nations of the world to support a Global Energy Interconnection Development and Cooperation Organization (GEIDCO), which would be chaired by Liu and create "a global grid that would transmit solar, wind, and hydroelectric-generated power from places on Earth where they are abundant to major population centers" (Fialka 2016).

> He [Liu] gave three reasons for his new mission. Expanding energy demands will exhaust coal, oil, and natural gas supplies over the next 110 years. Environmental pollution from fossil fuels will exacerbate serious pollution and health problems. And world leaders need a mechanism to cut the world's greenhouse-gas emissions by half to prevent a potential 4°C rise in the Earth's average temperature, a possibility that Liu called "seriously threatening human survival." (Fialka 2016)

A Rebuke to Trump

Tom Phillips reported from the Longyangxia Dam Solar Park in the western province of Qinghai, Tibet, illustrating just how closely China's leadership identifies with solar power's potential for international influence:

> High on the Tibetan plateau, a giant poster of the Chinese president, Xi Jinping, has been hung near the entrance to one of the greatest monuments to Beijing's quest to become a clean energy colossus. To Xi's right, on the road leading to what is reputedly the biggest solar farm on earth, a billboard greets visitors with the slogan: "Promote green development! Develop clean energy!" Behind him, a sea of nearly 4 million deep-blue panels flows towards a spectacular horizon of snow-capped mountains— mile after mile of silicon cells tilting skyward from what was once a barren, windswept cattle ranch. (Phillips 2017)

According to *The Guardian* (London), "Xi said that unlike Donald Trump, a climate denier whose election as U.S. president has alarmed scientists and campaigners, he was convinced global warming was a real and present danger that would wreak havoc on the world unless urgent action was taken" (Phillips 2017). Xi, who was born in Qinghai, said: "When I was a child, rivers usually froze over during the winter; heavy snowfall hit the area every year, so we could go skiing and skating . . . people weren't very rich, and nobody had a fridge, but you could still store your meat outside. We cannot do that anymore" (Phillips 2017).

"Our response to climate change bears on the future of our people and the well-being of mankind," Xi said, vowing to "unwaveringly pursue sustainable development" (Phillips 2017). Sam Geall, the executive editor of China Dialogue, a bilingual website on the environment, said that Beijing sees Trump's election and his ignorant denial of climate change "as a rare and unexpected opportunity to boost Chinese soft power by positioning itself as the world's premier climate-change fighter" (Phillips 2017). China also anticipates selling low-carbon technology to developing nations in Africa, south Asia, and Latin America during coming decades.

The Guardian reported:

> The Chinese president scoffed at Trump's suggestion that climate change was a Chinese hoax and said such claims would do nothing to dampen his country's

enthusiasm for a low-carbon future. "Even if President Trump doesn't care about the climate, that's America's point of view," he said. "The Chinese government will carry out and fulfill its international commitments as they always have done in the past, and as they are doing now in order to try to tackle climate change." (Phillips 2017)

Further Reading

Fialka, John. 2016. "Why China Is Dominating the Solar Industry." *Scientific American*, December 19. Accessed October 8, 2018. https://www.scientificamerican.com/article/why-china-is-dominating-the-solar-industry/.

Martin, Richard. 2016. "China Is on an Epic Solar Power Binge." *MIT Technology Review*, March 22. Accessed October 8, 2018. https://www.technologyreview.com/s/601093/china-is-on-an-epic-solar-power-binge/.

Phillips, Tom. 2017. "China Builds World's Biggest Solar Farm in Journey to Become Green Superpower." *The Guardian* (U.K.), January 19. Accessed October 8, 2018. https://www.theguardian.com/environment/2017/jan/19/china-builds-worlds-biggest-solar-farm-in-journey-to-become-green-superpower.

Woo, Ryan. 2017. "China's Solar Power Capacity More Than Doubles in 2016." Reuters, February 4. Accessed October 8, 2018. http://www.reuters.com/article/us-china-solar-idUSKBN15J0G7.

DENMARK

Wind Power Becomes Basic

Denmark was dependent on imported oil during the 1970s and made an enduring commitment to achieve energy independence when supplies were embargoed and its economy devastated. Wind power is an important part of this strategy, and as a result, Denmark has become a world leader in wind-turbine technology. Work on Danish turbines is a major reason why the technology today generates electricity that competes in price with oil, coal, and nuclear power. In the meantime, Denmark has built infrastructure that provides several thousand jobs. Wind power is a large part of Denmark's strategy to become the world's first carbon-neutral nation by 2050.

An Early Pioneer in Wind Energy

In matters of advanced technology, Denmark dominated the worldwide wind-power industry until about 2010, when China took the lead. Until then, Danish companies supplied more than half the wind turbines in use worldwide, making wind-energy technology one of the country's largest exports. Some Danish wind turbines have blades almost 300 feet wide—the length of a football field. During January 2007, a very stormy month, Denmark harvested 36 percent of its electricity from wind, almost double the usual. By 2016, Danish wind power routinely supplied half of the nation's electricity.

In 2007, a center-right coalition took power in Denmark and reduced subsidies for wind-power development, as its growth slowed but did not stop. Suddenly, the tax environment for new wind-power development was better in Texas than in Denmark. The Danes, however, had a very long head start, and momentum toward use of wind power continued even on reduced subsidies.

Danish wind energy also has experienced technical setbacks, as Danish wind operators, hoping to bypass local objections and take advantage of stronger, steadier air currents, have tried to build giant turbines at sea. In one case, in 2004, turbines at Horns Reef, about 10 miles off the Danish coast, broke down, their equipment damaged by storms and salt water. Vestas, a Danish manufacturer, fixed the problem by replacing the equipment at a cost of €38 million. Peter Kruse, the head of investor relations for Vestas, warned that the lesson from Horns Reef was that wind farms at sea would remain far more expensive than those on land. "Offshore wind farms don't destroy your landscape," Kruse said, but the added installation and maintenance costs are "going to be very disappointing for many politicians across the world" (Kanter 2007).

Wind Power Profit Sharing

The Danish wind-power industry also provides 150,000 families with shares of profits from the national electricity grid. Thousands of rural Danish residents have

Sheep graze under solar panels at the Marup local district heating plant in Samsø, Denmark, which is fueled by wood chips and solar power. (Melanie Stetson Freeman/The Christian Science Monitor via Getty Images)

joined wind-power cooperatives, buying turbines and leasing sites to build them, often on members' land (Woodard 2001, 7). At the same time, power companies in Denmark are now taxed for each ton of carbon dioxide emitted above a low (and, over time, gradually declining) limit.

By 2014, Denmark's 5.6 million people were drawing more than 40 percent of their electric power from renewables, mainly wind, with plans to be off fossil fuels by 2060—in any form, including transport. Denmark's was the most ambitious renewable energy commitment on Earth, but it wasn't being accomplished without technical political friction.

One problem that might be regarded as a benefit by consumers: renewable energy was causing power prices to crash during high-demand periods. However, this is when conventional plants make their money, and Denmark needs to keep them running in case the sun doesn't shine and the wind doesn't blow. "We are really worried about this situation," Anders Stouge, the deputy director general of the Danish Energy Association, said. "If we don't do something, we will in the future face higher and higher risks of blackouts" (Gillis 2014). The Danes have been experimenting with attaching a premium value to standby capacity, which would reward old-source plants when their power is needed most.

The biggest obstacle to going completely off fossil fuels, however, is transportation. The Danes, who are already using bicycles more often than anyone else for short-range transport, still need cars for longer trips, as well to transport any load that will not fit into a backpack or a small trailer. Electric cars have been tried but did not do well with consumers, with their high cost, limited power, and range. Transportation technology without gasoline still requires considerable work.

Samsø Island as Energy Example

The 4,000 residents of largely agricultural Samsø Island, in Denmark, accepted a challenge from the Danish government during the 1990s to convert to a carbon-neutral style of life and, by 2007, had largely accomplished the task, with no notable sacrifice of comfort. Farmers grow rapeseed and use the oil to power their machinery; homegrown straw is used to power centralized home-heating plants; solar panels heat water and store it for use on cloudy days (of which the 40-square-mile island has many); and wind power provides electricity from turbines in which most families on the island have a share of ownership. The island is now building more wind turbines to export power.

During the late 1990s, most of the 4,300 Samsingers' homes were heated with oil imported on tankers, and their electricity was generated from coal and imported from the Danish mainland on cables. The average resident of the island produced 11 tons of carbon dioxide a year (Kolbert 2008). In 1997, the Danish Ministry of Environment and Energy sponsored a renewable-energy contest that the city government entered after an engineer who did not live on the island filed a proposal after having asked the mayor. Quite to everyone's surprise, Samsø was picked as Denmark's "renewable-energy island." The designation carried no prize money, so nothing happened for a few months. After that, residents established energy cooperatives and began to tutor themselves in such things as insulation and wind power

According to Elizabeth Kolbert, writing in the *New Yorker*, "They removed their furnaces and replaced them with heat pumps. By 2001, fossil-fuel use on Samsø had been cut in half. By 2003, instead of importing electricity, the island was exporting it, and by 2005, it was producing from renewable sources more energy than it was using" (Kolbert 2008). "'People on Samsø started thinking about energy,' Ingvar Jørgensen, a farmer who heats his house with solar hot water and a straw-burning furnace, told me," Kolbert wrote. "'It became a kind of sport'" (Kolbert 2008).

Ten years later, Samsø found itself a subject of study by energy tourists, as researchers traveled great distances, burning copious fossil fuels, to observe Samsø's experiment, with its ranks of wind turbines, eleven in all, spinning in a relatively constant wind that makes an ideal wind-power site.

Samsø also built district heating plants that run on biomass, mainly bales of locally grown straw, burned to warm water that circulates into local homes, where it provides both heat (the houses are very tightly insulated, and the climate usually mild) and hot water. One district plant, in Nordby, burns wood chips, with solar panels providing hot water when the sun is shining. Kolbert explained that "burning straw or wood . . . produces CO_2, but while fossil fuels release carbon that would have remained sequestered, biomass releases carbon that would have entered the atmosphere through decomposition" (Kolbert 2008).

A few Samsø farmers also have converted their cars and tractors to operate on canola oil. Some of them grow their own canola seeds, which also can be used as an ingredient in salad dressing.

Further Reading

Gillis, Justin. 2014. "A Tricky Transition from Fossil Fuel: Denmark Aims for 100 Percent Renewable Energy." *New York Times*, November 10. Accessed October 8, 2018. http://www.nytimes.com/2014/11/11/science/earth/denmark-aims-for-100-percent-renewable-energy.html.

Kanter, James. 2007. "Across the Atlantic, Slowing Breezes." *New York Times*, March 7. Accessed October 8, 2018. http://www.nytimes.com/2007/03/07/business/businessspecial2/07europe.html.

Kolbert, Elizabeth. 2008. "The Island in the Wind: A Danish Community's Victory over Carbon Emissions." *New Yorker*, July 7. Accessed October 8, 2018. http://www.newyorker.com/reporting/2008/07/07/080707fa_fact_kolbert/.

Woodard, Colin. 2001. "Wind Turbines Sprout from Europe to U.S." *Christian Science Monitor*, March 14: 7.

GERMANY

Wind and Solar Power Rising

By 2017, Germany was obtaining one-third of its power from renewable sources, more than any other large nation on Earth. While some jurisdictions within nations have approached that level (the state of Iowa in the United States produced almost half of its electricity from wind in 2017, for example), Germany leads as a large nation-state, with a population of more than 80 million people. Occasionally, for very short periods (one example was May 15, 2016, at 2:00 p.m.), renewable energy sources (biomass, hydroelectric, wind, solar, and geothermal) have supplied nearly all of Germany's electricity. By 2016, Germany was drawing power from more than 23,000 wind turbines and 1.4 million photovoltaic solar systems. The nations of the European Union in 1997 pledged to achieve 12 percent renewable electricity production by 2010, but Germany passed that target in 2007 and reached one-third in 2016, with more than 80 percent anticipated by 2050.

Solar Power in a Cloudy Place

Germany, a land of fog and weeping skies, now uses solar cells so sensitive that a hot sun-aided shower on a cloudy day was no longer a problem. In 2006, half the world's new solar capacity was installed in Germany, despite the fact that the country has only half as many sunny days as Portugal, a more obvious solar success story (Whitlock 2007). The German solar panels work on drizzly days, although they generate only a quarter to half the electricity as on a sunny day.

Coal mining was Espenhain's largest employer under the East German regime that collapsed during 1989, providing 8,000 jobs. After German reunification, the mining jobs vanished. "'This region was known as the dirtiest in all of Europe,' said Jürgen Frisch, mayor of Espenhain. 'The solar plant came at a very good time for Espenhain. It's helped to change our image.' Unlike the coal mines, the solar plant makes almost no noise, save for the low thrum of a few outdoor air-conditioning units that cool its electrical transformers. The plant, with 33,500 solar panels, sits on a 37-acre site in a field off a rural road and requires scarcely any maintenance" (Whitlock 2007).

Enacted in 2000, a German law requires the country's utility companies to subsidize new solar installations by buying their electricity and using it on their grids at marked-up rates that allow small solar enterprises to earn profits. Wind and biofuels also have preferred status under German law. As the world's sixth-biggest producer of carbon-dioxide emissions, Germany is trying to slash its output of greenhouse gases and wants renewable sources to supply a quarter of its energy needs by 2020 (Whitlock 2007). Germany also has decided to phase out all nuclear power plants by 2020, a reaction to Japan's Fukushima nuclear disaster in 2011.

In northwestern Germany, Freiburg has been working to become the world's first "solar city," with a "solar-powered train station, energy-efficient row houses, innumerable rooftop photovoltaic systems, and, high on a hill overlooking the vineyards, the world-famous Heliotrop, a high-tech cylindrical house that rotates to follow the sun" (Roberts 2004, 192). Freiberg is home to the Fraunhofer Institute for Solar Energy Systems, where scientists have been seeking breakthrough research that will reduce the cost of photovoltaic solar energy to competitive levels with fossil fuels and nuclear power.

The small German island of Heligoland, off the coast in the North Sea, has become a wind-power showpiece, with "wind turbines, standing as far as 60 miles from the mainland, stretching as high as 60-story buildings and costing up to $30 million apiece. On some of these giant machines, [the size of] a single blade roughly equals the wingspan of the largest airliner" (Gillis 2014). Long-range plans call for wind farms as far as 125 miles offshore, costing as much as $340 million per turbine for the machine, power cables, installation, and other costs.

Also in the North Sea, the Netherlands in 2015 was building a large array of wind turbines 50 miles offshore at a cost of $3 billion that provided power for 1.5 million homes beginning in 2017. Wind power from sea-based platforms costs as much as three times that of land-based wind farms because of relatively high construction, maintenance, and transmission costs, but government subsidies have been used to support the Dutch program.

Energy Transition as National Policy

The Germans have adopted a phrase, *Energiewende*, meaning "the energy transition," as national policy as the country takes the initiative on renewable energy (Gillis 2014). However, because renewable power is intermittent, German utilities retain their conventional power plants and have had to engage in a complex dance in which they are used sometimes—and quickly—when the sun doesn't shine, wind does not blow, and storage does not compensate.

Germany's demand for wind turbines and solar panels has helped inculcate manufacturing of both (a large proportion of which are now imported from China) and has driven down costs. Wind power in Germany provided an early template for several of the United States, which by 2016 generated 25 to 50 percent of their electricity from wind. "It's pretty amazing what's happening, really," said Gerard Reid, an Irish financier who worked in Berlin on energy projects. "The Germans call it a transformation, but to me it's a revolution" (Gillis 2014).

Germany's price tag for *Energiewende* by 2014 was more than $140 billion, which includes:

> guaranteed returns for farmers, homeowners, businesses, and local cooperatives willing to install solar panels, wind turbines, biogas plants, and other sources of renewable energy. The plan is paid for through surcharges on electricity bills that cost the typical German family roughly $280 a year, though some of that has been offset as renewables have pushed down wholesale electricity prices. The program has expanded the renewables market and created huge economies of scale, with worldwide sales of solar panels doubling about every 21 months over the past decade and prices falling roughly 20 percent with each doubling. (Gillis 2014)

The *Energiewende* by 2017 had been very successful, but it also has been expensive because wind and solar power have been heavily subsidized. With the costs of both now declining rapidly, the transition is paying off. "At the center of the transformation has been a slate of renewable-energy subsidies that have dramatically scaled up once-niche solar and wind technologies and in the process have slashed their cost, making them competitive in some cases with fossil fuels," wrote Jeffrey Ball in *Fortune* magazine (2017).

The cost has been mind-boggling: €25 billion (US$28 billion) spent in 2016 alone, most of it collected as surcharges on electricity bills, which increased 50 percent between 2007 and 2016. The cost also has been shouldered by stockholders in traditional utilities, who have seen the market values of their shares plummet.

Energy Leadership on a Global Scale

Germany can now point with pride to its role as international energy pioneer as China, India, the United States, and others develop solar and wind energy on commercial scales. Germany has done all of this while shutting down its nuclear-power industry (which provided 13 percent of power in 2016) because of concerns for safety. Reacting to the 2011 Fukushima disaster in Japan, Germany has pledged to shut down all of its nuclear-power plants within a decade. Coal, which provided 40 percent of power in 2016, is in decline as well in a planned campaign to bring down emissions of carbon dioxide.

"If renewable energy ends up significantly helping curb climate change, then history may judge the *Energiewende* as a remarkable example of global leadership," Jeffrey Ball wrote (2017). "Germany has demolished one of the most fundamental reservations about alternative energy: that wind and solar power are too flaky to be relied on." Germany's energy grid remains an exemplar of reliability, as "blackouts remain as rare in the world's fourth-largest economy as late trains or bad beer. . . . Germany has solidified the start of an epic shift toward renewable energy" (Ball 2017).

Germany is trying to reduce its carbon-dioxide emissions 40 percent below 1990 levels by 2020, a very tall order—and one that it probably will not achieve. Even with heavy subsidies of renewable energy, greenhouse-gas emissions increased in 2015 and 2016. The reduction from 1990 levels has been 27 percent by 2016, still making Germany a world leader. At the same time, wind and solar energy are booming, as is the use of lignite, Germany's cheap and dirty domestic coal.

In January 2012, Jürgen Grossmann, until 2016 the chief executive officer of RWE AG, a large German power producer, told a conference that solar power in Germany "makes as much sense as growing pineapples in Alaska" (Ball 2017). Regardless, the shift to renewables in Germany is "an irreversible process now," Thomas Birr of RWE told Jeffrey Ball. Purely on profitability, he said, solar projects "are completely outcompeting fossil" ones. "You cannot build a coal plant anymore for this price. And we don't talk about nuclear; nuclear is totally outpriced. . . . We will be pioneers" (Ball 2017).

Further Reading

Ball, Jeffrey. 2017. "Germany's High-Priced Energy Revolution." *Fortune*, March 14. Accessed October 8, 2018. http://fortune.com/2017/03/14/germany-renewable-clean-energy-solar/.

Gillis, Justin. 2014. "Sun and Wind Alter Global Landscape, Leaving Utilities Behind." *New York Times*, September 13. Accessed October 8, 2018. http://www.nytimes.com/2014/09/14/science/earth/sun-and-wind-alter-german-landscape-leaving-utilities-behind.html.

Roberts, Paul. 2004. *The End of Oil: The Edge of a Perilous New World.* Boston: Houghton Mifflin.

Whitlock, Craig. 2007. "Cloudy Germany a Powerhouse in Solar Energy." *Washington Post,* May 5: A-1. Accessed October 8, 2018. http://www.washingtonpost.com/wp-dyn/content/article/2007/05/04/AR2007050402466_pf.html.

INDIA

Suddenly Going Solar

In India, solar power has been developing rapidly. Very suddenly, in 2016 and 2017, the declining cost of solar power provoked India to switch its generation strategy to renewable energy from coal. Planners also became convinced that a drive toward solar power would help alleviate choking air pollution. By 2017, with existing coal-fired power plants running at less than 60 percent of capacity, India pledged to quit building new ones, as solar panels were being installed across the country, in rural and urban areas, providing rudimentary electric power to people for the first time. Coal consumption may have peaked in 2016. This historic turnabout occurred at the point where projected production costs for solar power declined to less than the cost of coal. Just as importantly, the cost of lithium ion batteries, which are used to store solar power, declined as well. India canceled 13.7 gigawatts of proposed coal-fired power plants in one month (May 2017) alone (Anand 2017).

Village Solar Grids Spreading

In some instances, villages that had lacked electricity entirely are skipping the fossil-fuel age and building solar-based power grids. The solar villages are providing an example for the 300 million people in India (almost one quarter of the country's 1.3 billion people) who—even in 2017—had not become part of a centralized power grid. Many of these "solar microgrids" have been developed by nonprofit organizations, but not without problems. In a microgrid, an array of photovoltaic solar cells serves a few dozen homes and businesses connected by a cable. Villages usually manage the microgrids through a council of local residents.

India's Solar Power Goals

During December 2015, India's prime minister, Narendra Modi, said the nation as a whole plans to generate 40 percent of its electricity from sustainable sources by 2030, including 100 gigawatts of solar power by 2022. That would be more than 10 times the amount of solar power online in India as of March 31, 2016. The goal sounds ambitious, but India's growth in solar power has been astonishing, considering that the first sizable solar array was not installed until May 2011.

With 300 sunny days a year and a population growing in both size and affluence, India has been installing solar power as quickly as possible, as it maximizes all sources (including coal-fired plants) to provide power to everyone. In one year's time (March 31, 2015, to March 15, 2016), India's solar power capacity nearly doubled, from 3.7 to 6.7 gigawatts, according to the central government's Ministry of New and Renewable Energy. By December 31, 2016, solar capacity had passed nine gigawatts. In rural areas, only half the population has grid-supplied power and many villagers still cook with cow dung, a major contributor to air pollution.

Electrification of several thousand villages has been a popular political issue, and Modi will probably be on the ballot in 2019. Rural solar power has become one way to extend the grid without expensive infrastructure. Even larger cities that are connected to the grid contain shantytowns populated by poor migrants from rural areas that have no electricity.

The first Indian village, Dharnai, in Bihar province (northeastern India), was electrified with 100 kilowatts of solar power in 2014. Only a fraction of India's villages (a number in the hundreds of thousands) had gone solar as of 2017 (no one knew exactly how many), but those with solar power have received copious attention in the country and around the world. One such village, Kannauj, in Utter Pradesh province (on and near the Ganges River), electrified 450 homes (RYOT Studios 2016). Dharnai also had electrified 450 homes housing about 2,400 people, along with two schools, a health center, a farmer training center, an *anganwadi* (child development center), and 50 shops, as well as several streetlights and water pumps. In several other villages, solar was being used to electrify water pumps and provide relief from drought, a matter of vital importance in a country that is subject to very hot summers that are relieved only by periodic monsoon rains (Weber 2014). The micropower grid in Dharnai was constructed by Greenpeace in cooperation with two Indian groups, BASIX and CEED.

Solar microgrids also have been expanding in other parts of the tropics, most notably in Africa, which (like India) has large amounts of direct sunlight and high levels of rural poverty. The microgrids could change the lives of 1.2 billion people worldwide who lacked electricity in 2016, allowing many people to skip fossil-fueled power. Solar often offers the only practical alternative in rural areas.

An Example of Village Solar

Fred Pearce, writing in *Yale Environment 360,* published at the Yale School of Forestry & Environmental Studies, described one such village, Rajanga, in the Kandhara forest reserve of eastern Odisha state:

> Standing outside his mud hut, illuminated by a solar-powered light bulb dangling from the straw roof, the president of Rajanga village's energy committee, Suresh Pradhan, described the best thing about having this new source of light. His children could play outside in the evening, he said. And his wife could carry on sewing. But

being free of elephants was the biggest gain. "They used to come into our village at night. We were in fear, especially my children, and they did damage. Now there is light, they don't come." (Pearce 2016)

Rajanga's 550 residents were chosen as candidates for a microgrid because of their remote location, "surprisingly cut off, with no tarred road, no power grid, no water supply mains, no shops, no lodging houses or cafes, and only an intermittent and patchy mobile phone signal" (Pearce 2016). Because the forest is a protected forest reserve, residents of the village may not gather firewood, bamboo, or any other forest products. The village also sits on an elephant migration route, so roads or power cables may not be constructed or laid though the forest.

Residents of Rajanga subsist on rice, organic vegetables, and red chilies, as well as goats and chickens. According to Pearce, "In a booklet on its work in the villages, IRADA somewhat brutally summarizes the problems of the villagers as 'poverty, ignorance, and alcoholism.' If microgrids can work here, the NGO reasons, they can work almost anywhere" (Pearce 2016).

Many of India's solar microgrids begin with investments by private entrepreneurs, of which two examples are Mera Gao Power in Uttar Pradesh and the Mlinda Foundation in West Bengal. India is densely populated, and the grid (which is heavily subsidized by government, so its power is very inexpensive) is also being expanded. Investors may lose their money. While India is investing in solar, it is also expanding its network of coal-fired power plants. Because businesses have been reluctant to invest in solar, nonprofits have introduced most of the microgrids, which often depend on foreign aid.

Size and Scope of Village Solar

The generating capacity of solar panels in many villages in India is about 12 kilowatts, the power requirements of one average family home in the United States, so power is rationed, according to Pearce.

> Every household gets sockets for two three-watt LED bulbs and a mobile-phone charging point, for which they pay a flat fee of just 50 rupees (75 cents) a month. But no TVs or fans—much less refrigerators or sound systems—are allowed. When someone in Kanaka bought a TV, the energy committee said he would have to put it in the community center, where I found it plugged into a DVD player containing the village's only DVD—a copy of *Bandit Queen*, a 1994 biopic about an abused lower caste girl who leads a Robin Hood-style bandit gang. (Pearce 2016)

Villages in India were celebrated in the media as being the first of a given province or region to electrify with solar power.

As with many such projects, the amount of power provided was very modest by American standards. This project, which was co-funded by ECCO Electronics, which manufactures solar products, and Jakson Group, called "a diversified power solutions provider" by the *Times of India*, installed two lamps in each of the village's 61 households. Other accounts pointed out that lamps and cell-phone chargers were being powered. A central one-kilowatt unit "powers eight street lamps, and an LED television set and a TV set-top box for the community center" (Mazumdar 2017). The central unit also can be used to power an irrigation pump. The microgrid was powered by eight solar panels that were built to be folded and stored in the event of cyclonic storms (called hurricanes in the Atlantic Ocean), a constant threat along the coast of the Bay of Bengal.

Further Reading

Anand, Geeta. 2017. "India, Once a Coal Goliath, Is Fast Turning Green." *New York Times*, June 2. Accessed October 8, 2018. https://www.nytimes.com/2017/06/02/world/asia/india-coal-green-energy-climate.html.

Mazumdar, Jaideep. 2017. "Odisha Gets Its 1st 100% Solar-Powered Village." *Times of India*, October 4. Accessed October 8, 2018. http://timesofindia.indiatimes.com/india/Odisha-gets-its-1st-100-solar-powered-village/articleshow/49212485.cms.

Pearce, Fred. 2016. "In Rural India, Solar-Powered Microgrids Show Mixed Success." *Yale Environment 360*, January 14. Accessed October 8, 2018. http://e360.yale.edu/features/in_rural_india_solar-powered_microgrids_show_mixed_success.

RYOT Studio. 2016. "7 Villages of India That Are an Example for the World." *Huffington Post,* October 31. Accessed November 30, 2018. https://www.huffingtonpost.in/2016/04/28/7-villages-of-india-that-_n_9794264.html.

Weber, Peter. 2014. "A Village in India Now Runs Entirely on Its Own Solar Power Grid." *The Week*, July 22. Accessed October 8, 2018. http://theweek.com/speedreads/449594/village-india-now-runs-entirely-solar-power-grid.

KENYA

Solar Power Lights Up the Night

Solar-power usage has been rising rapidly in Kenya, a world leader in the number of solar power systems installed per capita. Kenyans, especially in rural areas outside the standard electrical grid, have been buying into solar microgrids in numbers that would have been inconceivable a few years ago. In Kenya by late 2015, "private entrepreneurs are solving the complex problems of collecting revenues from poor and remote communities by using mobile phone-style, top-up credit linked to village smart meters. The meters cut off power when the credit runs out. The system is brutally effective, and the certainty of payment encourages investors" (Pearce 2016). The direct, hot African sun supplies power in ways that recently had resided only in the realms of science fiction.

Several private solar-power distributors (the most notable being M-Kopa, a Canadian company founded in 2012) have made purchase of solar power affordable through pay-as-you-go and microfinancing in places that are seeing the benefits of electrification for the first time. In so doing, these distributors are enabling many Kenyans to skip the fossil-fueled stage of electrification. A Lighting Africa initiative has been underwritten by the World Bank and the International Finance Corporation (IFC).

Climate activist Bill McKibben described the ways in which everyday life is being changed across Africa by the swift spread of solar microgrids. For the first time, cold water is available in hot places, along with international television. One villager in Ivory Coast told McKibben: "'I like the National Geographic Channel'— that is, the broadcast arm of the institution that became famous showing Western-ers pictures of remote parts of Africa" (McKibben 2017, 46).

A Large Market

Nearly two-thirds of Kenya's population had no access to the conventional elec-trical grid in 2016, providing a large market for solar microgrids. As the cost of solar technology declines, its reach is expected to grow not only in Kenya, but across much of the tropical Third World. Kenya, like many tropical countries, has a plentiful supply of direct sunlight year-round. By 2016, close to one-sixth of Kenyan households had a solar connection, and numbers were rising rapidly.

Cell phones are so pervasive in Kenya that as of 2016, 85 percent of households, including all of M-Kopa's customers, used them to pay bills (as "mobile money"). The system is exacting and updated every day. According to one account, "In each solar kit is a SIM card that customers transfer money onto every day using their mobile phones. If the customers don't pay, the system doesn't work. But the pay-as-you-go system also gives customers more control—if customers are not using the solar energy system or if it breaks, they won't pay" (Lynch 2015).

"For a deposit of $35, buyers get the system, then make 365 daily payments of $0.43 through mobile money system M-Pesa. When it is all paid off, the sys-tem belongs to the buyer outright" (Shapshak 2016). By 2017, M-Kopa had sold about 500,000 units in Kenya, Tanzania, Uganda, and Ghana, with sales aver-aging 800 units per day (McKibben 2017, 48). M-Kopa's 650 employees and 1,000 field agents anticipate eventually having a million African households on solar microgrids (Shapshak 2016). According to M-Kopa, "each home in its program will save $750 over a four-year period compared with using kerosene. That's significant in a part of the world where $15 feeds a family for a month" (Onyulu 2016).

"After two years of paying the daily fee, [the user] will take full ownership of [an] eight-watt solar home system" (Onyulo 2016). The owner of an M-Kopa solar system can pay it off in a few months, after which customers are offered upgrades, such as a fuel-efficient refrigerator or stove. As prices of solar technology decline, M-Kopa's solar systems are becoming less expensive than the grid. By 2015, solar energy used for lighting in Kenya already was cheaper than the national grid and 10 times cheaper than kerosene wick lamps. Kenya in 2014 began to upgrade its conventional grid with nine solar power stations, with an eventual goal of obtain-ing half its power from the sun in a partnership that will be half government and half private ("Kenya to Generate" 2014).

Grow by Day, Sell by Night

With solar lights in the market, farmers can grow by day and sell by night.

> Violet Karimi . . . takes fruit and vegetables harvested on her farm to sell in Embu's open-air market. "I collect my stock and head to the market, where I trade until late in the evening," said the 36-year-old. "This is possible because the solar lights in the market and the rest of Embu town are switched on the whole night." On a good day, she said, she can bring home as much as $30—at least three times as much as she could sell in a shorter evening, before the lights made longer trading hours possible. "Customers want to shop in the evening because that is when they leave work," she said. "The solar lighting has encouraged them to stay and buy as long as they like." (Njagi 2016)

Maina Waruru reported for Reuters from Kenya:

> Fuel-seller Nancy Kaisa has for years used solar power to light her premises at the Entasopia shopping center in Kajiado, in southern Kenya. But recently she's started using energy from a solar minigrid system to operate her fuel pumps and meet her other energy needs, ditching her diesel generator. "I now serve more customers a day than before because (the pump) dispenses fuel faster than it used to before, when it was hand-operated," Kaisa said. (Waruru 2016)

In Kenya, "everyone wanted a refrigerator and a TV, and bars were opening with sound systems blasting their music into the night" (Pearce 2016).

Changing Daily Life Fundamentally

In rural Kenya, reliable electricity, even in small amounts, changes daily life fundamentally. An account in *Newsweek* described just how fundamental these changes can be, describing Janet Nakhonwe, who had never before lived been in a place connected to an electrical grid, nor slept in a room with an electric light. With solar power, she had electricity throughout her home in Lumino, on the Ugandan border. The biggest impact, she says, has been on her six children's education: "My children can now study well and complete their assignments on time," she said, because they had light to read by. "Last term, they performed excellent in their exams" (Onyulo 2016).

The solar microgrids are very popular in Kenyan villages (as elsewhere in the developing world) because they replace smoky lamps lit by expensive and dangerous kerosene or candles. Both release toxic fumes and pose fire threats. Streets once shrouded in darkness at night are now illuminated by solar lights. "For us, investing in the solar project is a double win," said Embu County governor Martin Wambora. "Solar energy is cheaper to maintain in the long term and puts us in solidarity with the world's push for a global green economy" (Onyulo 2016).

One Embu villager, Joe Njiru, pays $10 a month for a solar microgrid and leverages it into $560 a month at his bar, which uses eight solar-powered light bulbs, with enough power remaining to light a television set that receives cable broadcasts. Customers at the bar now remain long past midnight, buying beer. In the pre-solar days, the bar's evenings rarely lasted past 8:00 p.m. It had a grid connection, but power often faded away in the evening due to system overloads from many people who tapped into the system illegally (Hanley 2016).

Microgrids' Capacity Increases

By 2016, the microgrids in the area could store some power in batteries, but only enough for a few LED lights and small television sets. M-Kopa's home solar energy unit—about the size of a thick computer tablet—provides enough solar power for three lights, five USB connections that will charge mobile phones, charging, and one portable radio. As microgrids grow and battery capacity increases, so will their applications. "This is where development partners should come in and support solar microgrids with finances so that they can be able to expand their generation and storage capacity," said Lois Gicheru, the chief executive officer of Solafrique Ltd., an enterprise that works with African communities to help them access renewable energy (Hanley 2016).

"Prices for renewable technologies, especially solar and wind power, are falling at an extraordinary rate to the point at which they are competitive with fossil fuels," the African Progress Report, which was formerly headed by the late former UN secretary general Kofi Annan, said in late 2015 (Lynch 2015).

Further Reading

Hanley, Steve. 2016. "Solar Power Helps Raise Income Levels in Kenya." Clean Technica, December 17. Accessed October 8, 2018. https://cleantechnica.com/2016/12/17/solar-power-helps-raise-income-levels-kenya/.

"Kenya to Generate over Half of Its Electricity through Solar Power by 2016." 2014. *The Guardian*, January 17. Accessed October 18, 2018. https://www.theguardian.com/environment/2014/jan/17/kenya-solar-power-plants.

Lynch, Justin. 2015. "Kenya's M-Kopa Is Set to Deliver Solar Power to a Million Homes." *Quartz Africa,"* December 9. Accessed November 30, 2018. https://qz.com/africa/569815/kenyas-m-kopa-is-set-to-deliver-solar-power-to-a-million-homes-in-east-africa/.

McKibben, Bill. 2017. "Power Brokers: Africa's Solar Boom Is Changing Life beyond the Grid." *New Yorker*, June 26: 46–55.

Njagi, Kagondu. 2016. "As Solar Power Expands the Work Day, Incomes Rise in Eastern Kenya." Reuters, December 8. Accessed October 8, 2018. http://allafrica.com/stories/201612080556.html.

Onyulo, Tonny. 2016. "Solar Power Brightens Residents' Prospects in East Africa." *Newsweek*, July 2. Accessed October 8, 2018. http://www.newsweek.com/2016/07/15/kenya-solar-power-clean-energy-476915.html.

Pearce, Fred. 2016. "In Rural India, Solar-Powered Microgrids Show Mixed Success." *Yale Environment 360*, January 14. Accessed October 8, 2018. http://e360.yale.edu/features/in_rural_india_solar-powered_microgrids_show_mixed_success.

Shapshak, Toby. 2016. "How Kenya's M-Kopa Brings Prepaid Solar Power to Rural Africa." *Forbes*, January 28. Accessed October 8, 2018. http://www.forbes.com/sites

/tobyshapshak/2016/01/28/how-kenyas-m-kopa-brings-prepaid-solar-power-to-rural-africa/#1c5fb33d70f4.

Waruru, Maina. 2016. "Taste of Solar Power Builds Appetite for More in Kenya, Survey Finds." Reuters, October 31. Accessed October 8, 2018. http://www.reuters.com/article/us-kenya-solar-grids-idUSKBN12V1UG.

MOROCCO

Solar in the Sahara

There's a race to build the world's largest solar array. Will it be Chinese, in Tibet? Or Morocco, in the Sahara Desert? Either way, the world is going solar in a hurry, from electric meters that run backward in California to microgrids in Kenya and Ghana, paid for in daily installments by back-country villagers on smart phones. In 2017, even U.S. president Donald Trump made an off-the-cuff remark at a "campaign rally" in Cedar Rapids, Iowa, that he wants to install solar panels on his prospective "great wall" on the Mexican border. Find a time machine, take it back 50 years, and describe today's world to people back when. Watch their jaws drop. It would be science fiction to them.

43,000 Square Miles to Supply the World?

Mehran Moalem, a former nuclear-power engineer who works to establish solar factories around the globe, made a case in *Forbes* that 43,000 square miles of solar panels operating at moderate efficiencies available today (such things have been improving rapidly during the past 20 years) could supply total world energy usage, which, in 2015, amounted to about 17.3 terawatts. That calculation brought Moalem to the Sahara Desert in Africa, which contains 3.6 million usually sunny, sparsely populated square miles and is prime real estate for solar power, usually for more than 12 hours per day.

Moalem's calculation indicated that covering 1.2 percent of the Sahara Desert with solar panels would produce enough power to supply the world's needs. "There is no way coal, oil, wind, geothermal, or nuclear can compete with this," he wrote. The cost of the project will be about $5 trillion, a one-time cost at today's prices without any economy-of-scale savings. That is less than the bail-out cost of banks by U.S. president Barack Obama in the last recession.

If this sounds like science fiction, consider Noor 1 (*noor* is Arabic for "sun"). In Morocco, with 3,000 hours of sunshine a year, Noor 1, at the "door of the desert" near the south-central Moroccan town of Ouarzazate, sits, as Chris Bentley of Public Radio International wrote, "on a dusty, red-earth plateau where the Atlas Mountains begin to descend into the Sahara Desert. Its dramatic landscape has made it a popular setting for movies and TV shows, from Middle Eastern epics like *Lawrence of Arabia* and *The Mummy* to HBO's *Game of Thrones*" (Bentley 2017).

An aerial view of 500,000 solar mirrors at the Noor 1 Concentrated Solar Power (CSP) plant, about 12.5 miles outside the central Moroccan town of Ouarzazate, 2016. (Fadel Senna/AFP/ Getty Images)

Noor's first phase of construction, with 500,000 solar panels, was completed in 2016 and began production with a power-generating capacity of 160 mega-watts. Two more phases of construction remain, which will make Noor the largest Concentrated Solar Power (CSP) plant in the world when it becomes completely operational in 2020, covering 2,500 hectares, or 6,178 acres (Dvorsky 2016). A Reuters dispatch published in *Fortune* magazine sketched the size of Noor 1 in everyday terms: "At Noor, curved mirrors totaling 1.5 million square meters (16 million square feet)—the size of about 200 soccer pitches—capture the sun's heat in the reddish desert" ("Vast Moroccan" 2016).

How the System Works

The first phase of Noor, which began operation in 2016, produces 580 mega-watts—enough to supply 2 million people—created by 500,000 12-meter-tall par-abolic mirrors that focus energy onto a pipeline filled with hot fluid (393°C/739°F), the heat source used to warm the water and create steam. By that time, accord-ing to NASA Earth Observatory, "almost half of Morocco's energy is expected to come from renewables, about one-third of which will be from solar" ("Solar in the Sahara" 2016).

The NASA Earth Observatory explained that

concentrated solar power plants use the sun's energy to heat water and produce steam that spins energy-generating turbines. The system at Ouarzazate uses 12-meter-tall parabolic mirrors to focus energy onto a fluid-filled pipeline. The pipeline's hot fluid—393°C (739°F)—is the heat source used to warm the water and make steam. The plant doesn't stop delivering energy at nighttime or when clouds obscure the sun; heat from the fluid can be stored in a tank of molten salts. ("Solar in the Sahara" 2016)

The plant was originally planned to supply power to Europe via cables through the Strait of Gibraltar, but after original sponsors withdrew, the African Development Bank and the Moroccan government assumed financing with expectations that the plant will be used to supply increasing internal Moroccan energy demands, with supply to Spain under sponsorship of the Spanish consortium TSK-Acciona-Sener. The World Bank, which also helped to finance Noor 1, points out that until it started going solar, 97 percent of Morocco's energy was imported, most of it oil. The price tag on the Noor 1 project is $2.45 billion; all three phases are expected to cost $9 billion ("Vast Moroccan" 2016).

A Regional Solar Grid?

Several very large arrays of solar thermal power plants across the Sahara Desert could supply most, if not all, of the power needs for Europe, the Middle East, and North Africa, where plans began in 2007 to establish a renewable energy "supergrid." Thermal power stores solar energy in a heat-retaining fluid and uses it to drive turbines. One early plan called for roughly 1,000 100-megawatt power plants' worth of solar and other renewable energies (Feresin 2007, 595). The technology for this system was available, said Gerhard Knies, a retired physicist based in Hamburg, Germany; the grid "could offer unlimited, cheap, and carbon-dioxide-free energy to Europe" (Feresin 2007, 595). Knies has joined with other advocates of the idea. While the energy will be cheap once infrastructure has been built, getting to that point could cost at least €400 billion (almost US$600 billion). Europeans also may be unwilling to place their energy fates in the hands of North African states. The idea of the Sahara as solar-energy bank actually goes back as far as Frank Shuman, an inventor who, working in Philadelphia during 1913, built the first prototype thermal solar plant in Egypt. Today's advocates propose that solar be the major component of an energy system that also will use wind and biomass fuels.

> "Whether they bet on the right technology, they will find out in the years to come," said Christian Breyer, a professor of solar economy at Lappeenranta University of Technology in Finland. Morocco is pumping up its solar capacity fast, but Breyer says it might not be fast enough to keep up with the speed of innovation in the industry. Photovoltaic panels and batteries to store their output have improved so quickly that they threaten to outshine concentrated solar plants like Noor. (Bentley 2017)

"By the way," noted Moalem, "the cost of a 1 GWe [gigawatt electric] nuclear plant is about three billion dollars. The cost of 17.3 terawatt nuclear power will be $52 trillion, or 10 times that of solar, even if all the other issues with safety and uranium supply are resolved" (Moalem 2016). He figured that large solar arrays in sunny equatorial regions (such as the Sahara) will produce solar power much more efficiently than rooftop units in places such as the United States and Europe, with variably cloudy weather and relatively low sun angles. The technology of long-distance transmission to population centers has been mastered. Moalem also asserts that a truly large-scale solar array should manufacture its own panels and other equipment, reducing costs. In the future, according to Moalem, solar also will be adapted to produce truly "clean" hydrogen fuel that eventually will replace gasoline to propel motor vehicles.

Furthermore, Moalem asserted that using 1.2 percent of the African Sahara to replace all fossil-fuel and nuclear energy production "will save vast tracts of land that are currently suffering from strip mining for coal and from contamination by acid rain, not to say anything about possible radioactive land regions in case of nuclear accidents" (Moalem 2016).

Tony Patt, professor of climate policy at the Swiss Federal Institute of Technology in Zurich, who leads the research for the European Research Council on whether the Saharan sun could power Europe, said:

> The technology is good. It's matured a lot in the last few years in terms of thermal storage. That allows you to take the heat that you capture from the sun and store it for, let's say, up to a day, and produce the power later. That means you can generate it around the clock. And the Sahara Desert is so big that if there is cloudy weather, it's localized, and with thermal storage, it can provide absolutely reliable power. ("Should We" 2015)

Daniel Egbe, an evaluator for the World Bank, who is also a chemist and founder of ANSOLE, a network of Africans for Africa with a focus on renewable energy, stressed that sub-Saharan Africa should have a seat at the table of solar energy, arguing that the new energy paradigm could be Africa's path out of poverty. "As an African," he said, "knowing the history about the exploitation of the continent, where there is a big gap when it comes to riches, and Africa is still poor due to the colonial past and the slave time, nobody can just come and do things as if we are still in the past" ("Should We" 2015). The new solar arrays in Morocco may be just the beginning.

Further Reading

Bentley, Chris. 2017. "Now Blooming in the Desert: Morocco's Grand Dream of Energy Independence." Public Radio International, February 21. Accessed October 8, 2018. https://www.pri.org/stories/2017-02-21/now-blooming-desert-moroccos-grand -dream-energy-independence.

Dvorsky, George. 2016. "A Massive Solar Power Plant Is Taking Shape in the Sahara Desert." Gizmodo, January 11. Accessed October 8, 2018. https://gizmodo.com/watch -a-massive-solar-power-plant-take-shape-in-the-sah-1752261396.

Feresin, Emiliano. 2007. "Europe Looks to Draw Power from Africa." Nature 450 (November 29): 595.

Moalem, Mehran. 2016. "We Could Power the Entire World by Harnessing Solar Energy from 1 Percent of the Sahara." *Forbes*, September 22. Accessed October 8, 2018. https://www.forbes.com/sites/quora/2016/09/22/we-could-power-the-entire-world -by-harnessing-solar-energy-from-1-of-the-sahara/#5da5c316d440.

"Should We Solar Panel the Sahara Desert?" 2015. BBC News, December 30. Accessed October 8, 2018. http://www.bbc.com/news/science-environment-34987467.

"Solar in the Sahara." 2016. NASA Earth Observatory, January 10. Accessed October 8, 2018. http://earthobservatory.nasa.gov/IOTD/view.php?id=87293&src=eoa-iotd.

"Vast Moroccan Solar Power Plant Is Hard Act for Africa to Follow." 2016. Reuters in *Fortune,* November 5. Accessed October 8, 2018. http://fortune.com/2016/11/05 /moroccan-solar-plant-africa/.

UNITED STATES

An "Artificial Leaf"

Solar technology is undergoing a revolution that may eventually allow power to be acquired from nearly any surface on which the sun shines. One such technology is the "artificial leaf." Richard Martin reported in the *MIT Technology Review* that Daniel G. Nocera and his colleague Pamela Silver

> have devised a system that completes the process of making liquid fuel from sunlight, carbon dioxide, and water. And they've done it at an efficiency of 10 percent, using pure carbon dioxide—in other words, one-tenth of the energy in sunlight is captured and turned into fuel. That is much higher than natural photosynthesis, which converts about 1 percent of solar energy into the carbohydrates used by plants, and it could be a milestone in the shift away from fossil fuels. . . . "Bill Gates [founder of Microsoft] has said that to solve our energy problems, someday we need to do what photosynthesis does, and that someday we might be able to do it even more efficiently than plants," says Nocera. "That someday has arrived." (Martin, 2016)

Developing Work on the "Artificial Leaf"

Several teams of scientists have been working on "artificial leaf" technology. At Caltech's Jorgensen Laboratory, a team of more than 190 people have been using silicon, nickel, iron, and other materials in the Joint Center for Artificial Photosynthesis (JCAP), a $16-million, five-year program funded by the U.S. Department of Energy that is attempting to replicate photosynthesis as an energy source.

In about 2010, for the first time, a team of researchers at the Massachusetts Institute of Technology (MIT) led by Daniel Nocera (in 2016 he was a professor of energy science at Harvard) created the prototype of an "artificial leaf" that can replicate the way plants use sunlight to knit chemical bonds, possibly (with further development) providing a practical, inexpensive source of solar energy. The device was described by Robert F. Service in *Science* as "a silicon wafer about the size and shape of a playing card. Different catalysts coat each side of the wafer. The silicon absorbs sunlight and passes that energy to the catalysts to split water (H_2O) into molecules of hydrogen (H_2) and oxygen (O_2)" (Service 2011, 25). The hydrogen then may be used in a fuel cell, reemerging as water. This technology makes water a cheap, clean, and available form of energy.

Seeking the Best Catalysts

Although practical application was regarded as years away, "it's spectacular," said Robert Grubbs, a chemist at the California Institute of Technology in Pasadena, after watching a demonstration of the process at a meeting of the American Chemical Society (Service 2011, 25). The MIT team previously had developed a catalyst that used cobalt and phosphorous to break water at the molecular level and knit oxygen atoms into O_2 molecules. Hydrogen-forming catalysts also had been developed, but all of this was very expensive. Nocera by early 2011 had developed an inexpensive catalyst that uses three different metals. He did not name the metals, pending patents and publication. Nocera said in 2011 that he hoped to have the technology operable on a commercial scale within two to three years with the aid of his company, Sun Catalytix. Nocera has plans to join with Ratan Tata, chair of Tata Group, of India, to produce a power plant the size of a refrigerator that will manufacture electric power from sunlight and water to provide inexpensive energy to large numbers of people who presently have no access to power.

"For years the efficiency of polymer-based solar cells scraped along at a feeble 3 percent to 5 percent. But things have improved markedly. In early April, Mitsubishi Chemical reportedly set a new efficiency record, producing organic solar cells with a 9.2 percent conversion efficiency," Robert Service reported in *Nature*. "Meanwhile, three other companies—Konarka Technologies, Solarmer Energy Inc., and Heliatek—are now reporting cells with efficiencies greater than 8 percent. Many researchers in the field are confident that the figure could soon top 10 percent and possibly reach 15 percent" (Service 2011, 293).

Jessica Marshall explained the process in *Nature*:

> At the heart of JCAP's artificial-leaf design are two electrodes immersed in an aqueous solution. Typically, each electrode is made of a semiconductor material chosen to capture light energy from a particular part of the solar spectrum and coated with a catalyst that will help to generate hydrogen or oxygen at useful speeds. Like many other artificial-photosynthesis devices, JCAP's system is divided by a membrane to keep the resulting gases apart and reduce the risk of an explosive reaction. Once the water has been split, the hydrogen is harvested. It can be used as a fuel by itself—perhaps in hydrogen-powered cars such as those already making their way into showrooms in California—or be reacted with carbon monoxide to make liquid-hydrocarbon fuels. (Marshall 2014, 22)

The challenge comes in finding materials that will work. Silicon, for example, works as an electrode that produces hydrogen gas, but it is stable only in an acidic solution. The best photoanodes (electrodes), which produce oxygen, work best only in a basic solution. Good oxygen-producing electrodes, such as iridium,

---

When Pig Fat Flies

Will pigs fly? In the world of biomass fuel, they might. By 2007, the U.S. Department of Defense and the National Aeronautics and Space Administration were funding exploratory projects into biofuels for jet airplanes. In 2015, United Airlines was investing in biofuels and had tested it, mixed with jet fuel, on a handful of its commercial flights. The airline believes that it can produce biomass fuel from farm waste and animal fats at less than half the cost of conventional jet fuel (a large part of any airline's costs), with lower carbon-dioxide emissions. United Airlines on July 1, 2015 announced a $30 million investment in Fulcrum BioEnergy. Cathay Pacific, which is based in Hong Kong, also has invested in the same company.

Further Reading

Mouawad, Jad, and Diane Cardwell. 2015. "Farm Waste and Animal Fats Will Help Power a United Jet." *New York Times*, June 30. Accessed October 11, 2018. http://www.nytimes .com/2015/06/30/business/energy-environment/farm-waste-and-animal-fats-will-help -power-a-united-jet.html.

---

are rare and too expensive for commercial development. These experiments may remind historians of science of the many substances that Thomas Edison tested to find the best filament for the first practical electric light bulbs, but on a much more massive scale.

In one experiment to find the best proportions of nickel, iron, cobalt, and cerium oxides to generate oxygen from water, *Nature* reported, the team screened nearly 5,500 combinations for stability and function using a miniaturized chemical lab that glided over the glass plates tirelessly. The best-performing combination is not the most effective catalyst ever found for this reaction, but it is transparent, allowing light to pass through to the photo-absorber, and it has good chemical compatibility with that material (Marshall 2014, 23).

Thus far, the best combination of economy and stability (with 10 to 20 percent efficiency) has eluded the researchers (Reece et al. 2011, 645). A coating of titanium dioxide on silicon may be the answer. "That's basically the last piece of the puzzle to create the first-generation prototype," said a spokesman for JCAP (Marshall 2014, 124). A less expensive fabrication method for silicon must be found for single-crystal silicon, however. The United States' effort is being spurred by similar work in Japan, so a race of sorts has developed.

Solar-to-Chemical Conversion Rates Soar

By 2016, scientists were studying a form of artificial photosynthesis that, it was reported in *Science*, "when combined with solar photovoltaic cells, solar-to-chemical conversion rates should become nearly an order of magnitude more efficient than natural photosynthesis" (Liu et al. 2016, 1210). Liu and colleagues wrote that "on a previous artificial photosynthesis design, [we] combined the hydrogen-oxidizing bacterium *Ralstonia eutropha* with a cobalt-phosphorus water-splitting catalyst. This biocompatible self-healing electrode circumvented the toxicity challenges of previous designs and allowed it to operate aerobically" (Liu et al. 2016, 1210).

In other words, according to Richard Martin:

> Solar energy, water, and carbon dioxide [have now been used] to produce energy-dense liquid fuels. Nocera and Silver's system uses a pair of catalysts to split water into oxygen and hydrogen and feeds the hydrogen to bacteria along with carbon dioxide. The bacteria, a microorganism that has been bioengineered to specific characteristics, converts the carbon dioxide and hydrogen into liquid fuels. (Martin 2016)

This new work "is really quite amazing," said Peidong Yang of the University of California, Berkeley, who had developed a similar system with much lower efficiency. "The high performance of this system is unparalleled" in any other artificial photosynthesis system reported to date" (Martin 2016).

Further Reading

Liu, Chong, Brendan C. Colón, Marika Ziesack, Pamela A. Silver, and Daniel G. Nocera. 2016. "Water Splitting–Biosynthetic System with CO_2 Reduction Efficiencies Exceeding Photosynthesis." *Science* 352, no. 6290 (June 3): 1210–1213.

Marshall, Jessica. 2014. "Solar Energy: Springtime for the Artificial Leaf." 2014. *Nature* 510 (June 4): 22–24. Accessed October 8, 2018. http://www.nature.com/news/solar-energy-springtime-for-the-artificial-leaf-1.15341.

Martin, Richard. 2016. "A Big Leap for an Artificial Leaf: A New System for Making Liquid Fuel from Sunlight, Water, and Air Is a Promising Step for Solar Fuels." *MIT Technology Review*, June 7. Accessed October 8, 2018. https://www.technologyreview.com/s/601641/a-big-leap-for-an-artificial-leaf/.

Reece, Steven Y., Jonathan A. Hamel, Kimberly Sung, Thomas D. Jarvi, Arthur J. Esswein, Joep J. H. Pijpers, and Daniel G. Nocera. 2011. "Wireless Solar Water Splitting Using Silicon-Based Semiconductors and Earth-Abundant Catalysts." *Science* 334, no. 6065 (November 4): 645–648.

Service, Robert F. 2011. "Artificial Leaf Turns Sunlight into a Cheap Energy Source." *Science* 332, no. 6025 (April 1): 25.

Service, Robert F. 2011. "Outlook Brightens for Plastic Solar Cells." *Science* 332, no. 6027 (April 15): 293.

Wigginton, Nicholas S. 2016. "Artificial Photosynthesis Steps Up." *Science* 352, no. 6290 (June 3): 1185–1186.

Chapter 9: Oceanic Issues

OVERVIEW

This work is keyed to a format that describes the effects of environmental and energy-related issues within specific national borders. This section aims to augment this format to provide a more complete, global perspective. First, many energy-related and environmental issues are global in scope and do not stop at artificial, human-drawn lines on maps. One example is the decline of biodiversity. While extinction of a single species may be treated within one nation (the golden toad in Costa Rica is an example, see chapter 4), the problem as a whole cannot be contained that way. Secondly, two-thirds of the Earth's surface is covered by oceans, outside land-based national boundaries. Therefore, a global perspective requires attention to environmental issues that play out (often with human provocation) in the maritime world. The state of coral reefs and rising levels of acidity in the oceans are two examples of issues that should be included in any global survey of environmental and energy-related issues.

Most of these issues occur in the oceans. Our focus, when considering the implications of global warming, usually is fixed on the third of the planet that comprises dry land. While the land warms, the other two-thirds of the Earth has been warming as well, with profound implications for the species that inhabit it. Some of these issues are very important, such as the rise in ocean acidity caused by rising levels of carbon dioxide in the oceans. This increase occurs gradually and is only rarely discussed outside of a few scientific journals. It threatens everything in the oceans that is clothed in calcium, including phytoplankton, the basis of the oceanic food chain.

On a practical level that human beings experience, rising seas provoked by melting ice and thermal expansion of seawater will become the most notable challenge related to global warming (ranging from inconvenience to disaster) for many millions of people around the world. Human beings have an affinity for the open sea. Thus, many major population centers have been built within a mere meter or two of mean sea level. From Mumbai (Bombay) to London to New York City, many millions of people will find warming seawater lapping at their heels during coming years. Sea levels have been rising very slowly for a century or more, and the pace will increase in the coming years.

Melt all of Greenland's ice, and sea level worldwide rises an average of 20 feet; the West Antarctic ice sheet adds 16 feet; the East Antarctic Ice sheet adds 164 feet; and all of the planet's mountain glaciers add one to two feet. Total sea-level rise from melting ice could add up to about 200 feet, maximum (Pilkey et al. 2016, 17).

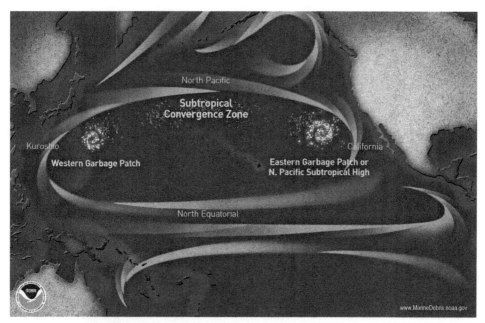

Ocean currents that steer the Great Pacific Garbage Patch as it moves between Hawaii and California. (NOAA)

Oceans Absorb Excess Heat

After setting record highs by large margins during 1998, global surface temperatures on land appeared to back and fill for the next decade and a half. Global warming contrarians asserted that warming's back had been broken, even as indicators of climate change, such as rises in sea level, Arctic ice cover, and atmospheric carbon dioxide levels, indicated continued warming. By 2015, scientists had reached widespread consensus on what had caused the "hiatus" in air temperatures. The excess heat had been absorbed, for the most part, by the oceans, according to research by Veronica Nieves and colleagues, who compiled 20 years of ocean temperature records from underwater floats and other instruments.

Further Reading

Nieves, Veronica, Josh K. Willis, and William C. Patzert. 2015. "Recent Hiatus Caused by Decadal Shift in Indo-Pacific Heating." *Science* 349, no. 6247 (July): 532–535. Accessed October 11, 2018. http://www.doi.org/10.1126/science.aaa4521.

The last time that temperatures were 3°C higher than today worldwide was during periods of the Pliocene, 2 to 3 million years ago. The seas at that time were about 50 to 115 feet higher than today. The level of carbon dioxide in the atmosphere at that time peaked at about 425 parts per million. The level as of 2015 has breached 400 ppm and has been increasing two to three ppm per year. Thus, by the end of this century, if not before, we probably will have enough warming "in the pipeline" to raise sea levels as much as 80 feet within 150 to 200 years. One billion people today live within 25 meters (80 feet) of sea level. Because of thermal inertia, a century or two will be required for our Pliocene-level carbon-dioxide level to melt enough ice to raise sea levels several dozen feet.

Scientists at the Lawrence Livermore National Laboratory and NOAA measured the amount of heat absorbed by the oceans since 1865 and learned that about half had occurred within the past two decades. Peter J. Gleckler and colleagues reported in *Nature Climate Change,* "Our model-based analysis suggests that nearly half of the industrial-era increases . . . have occurred in recent decades, with over a third of the accumulated heat occurring below 700 meters and steadily rising" ("Livermore" 2016). "In recent decades the ocean has continued to warm substantially, and with time the warming signal is reaching deeper into the ocean," said Gleckler ("Livermore" 2016). The oceans absorb more than 90 percent of the Earth's excess heat increase associated with global warming.

On land, we are experiencing a major decline in biodiversity that threatens many species, from the bees that pollinate many of our fruits and vegetables, to frogs and toads, to many large primates. Humankind is the major provocation of these looming extinctions, another set of important issues that cannot be contained within a single set of national borders. Solutions also are a worldwide matter, so we will survey the spread of solar and wind power worldwide. The nationalistic focus of this volume also obscures the fact that humankind needs to pursue solutions by overcoming the political narrowness that accompanies a focus purely on national interests. We need to think in those terms as well to construct a sustainable future. The good of individuals and their own cultures must mesh with a realization that, to paraphrase Benjamin Franklin on the eve of the American Revolution, we must hang together, or we will most assuredly hang separately.

Each oceanic issue that follows is keyed to a national case study. To illustrate the state of corals, we consider the state of Australia's Great Barrier Reef. The toll of radiation comes into play with a description of Japan's Fukushima nuclear accident. Sea-level rise's focus is on the United States' East and Gulf of Mexico coastlines. The United Kingdom stands here as an example of an island nation that is losing many of its well-known fish, especially the cod that has long been a fixture of English fish and chips, due to warming waters. Dead zones, caused mainly by an overload of nitrogen-based fertilizers, are keyed to examples off the coasts of China and Bangladesh. Methane bubbles in the oceans near Norway are linked to global warming and increasing ocean acidification. Finally, the worldwide issue of ocean acidity is the subject of a profile describing Canada's west coast. Each of these nation-based profiles is couched in a global context.

Further Reading

"Livermore Scientists Find Global Ocean Warming Has Doubled in Recent Decades." 2016. Lawrence Livermore National Laboratory, January 18. Accessed October 8, 2018. https://www.llnl.gov/news/livermore-scientists-find-global-ocean-warming-has-dou bled-recent-decades.

Pilkey, Orrin H., Linda Pilkey-Jarvis, and Keith C. Pilkey. 2016. *Retreat from a Rising Sea: Hard Choices in an Age of Climate Change.* New York: Columbia University Press.

AUSTRALIA

The Great Barrier Reef Is Dying

When aquatic environments nurture coral reefs, they are among the most productive, diverse ecosystems on earth. Coral reefs are sometimes called the rain forests of the sea. Coral reefs cover much less than 1 percent of the world's ocean floors, while at the same time hosting more than a third of the marine species presently described by science, with many species remaining undocumented. Some of these organisms may provide new sources of anti-cancer compounds and other medicines. Coral reefs also protect shorelines from erosion by acting as breakwaters that, if healthy, can repair themselves.

Coral bleaches when heat kills its living substance, resulting in a lifeless skeleton that turns white. According to NOAA, coral bleaching was reported throughout the Indian Ocean and Caribbean in 1998 and throughout the Pacific, including Mexico, Panama, Galapagos, Papua New Guinea, American Samoa, and Australia's Great Barrier Reef starting in 1997. The severity and extent of coral bleaching in 1997–1998 was widely acknowledged among coral reef scientists as unprecedented in recorded history to that time (Mathews-Amos and Berntson 1999). The main contributor to the bleaching was hotter-than-usual ocean water provoked by El Niño conditions that would be exceeded in 2014–2016.

The International Society for Reef Studies concluded that 1997 and 1998 had witnessed the most geographically widespread coral bleaching in the recorded histories of at least 32 countries and island nations. Reports of bleaching came from sites in all the major tropical oceans of the world, some for the first time in recorded history. The 1997–1998 bleaching episode was exceptionally severe, as a large number of corals died. According to one study, parts of Australia's Great Barrier Reef have been so severely affected that many of the usually robust corals, including one dated over 700 years of age, were badly damaged or had died during 1997 and 1998 (Mathews-Amos and Berntson 1999).

Corals support fish that feed more than a billion people, including many millions in island nations such as the Philippines and Indonesia. The largest bleaching was occurring in Australia's Great Barrier Reef, where two-thirds of northern reefs sustained significant bleaching (Innis 2016). Damage to reefs worldwide in 2016 exceeded that of El Niños in 1998 and 2010. "We are currently experiencing the longest global coral bleaching event ever observed," said C. Mark Eakin, the Coral Reef Watch coordinator at the National Oceanic and Atmospheric Administration in Maryland. "We are going to lose a lot of the world's reefs during this event" (Innis 2016). Reefs that had required several centuries to grow were being destroyed in a matter of weeks.

Acidification Disorients Fish

A study led by Danielle Dixson of Australia's James Cook University in Queensland (published in *Ecology Letters*) reports that increasing ocean acidification leaves some fish (orange clownfish in this study) unable to sense chemical signals given off by predators (Dixson et al. 2009). Acidification also disorients fish behavior by confusing sensory cues (hearing, smell, sight), according to Dixson's study. Dixson found that clownfish were drawn back to particular spots on reefs adjacent to New Guinea's island rain forests by the scents of the trees' leaves that had fallen into the water. Elevated carbon dioxide destroyed the fishes' ability to use smell as an environmental cue. They failed to be able tell one smell from another—including the scent, in the case of clownfish, of predators such as dottybacks and rock cod. The clownfish soon swam directly toward predators and were eaten.

Further Reading

Dixson, Danielle L., Philip L. Munday, and Geoffrey P. Jones. 2009. "Ocean Acidification Disrupts the Innate Ability of Fish to Detect Predator Olfactory Cues." *Ecology Letters*, November 16. Accessed October 11, 2018. http://www.doi.org/10.1111/j.1461-0248.2009.01400.x.

2016 El Niño Intensifies Coral Devastation

Record-high temperatures on both land and in the world's oceans from 2014 to 2017 intensified coral damage, threatening a third or more of reefs with eradication. Kim Cobb, a marine scientist at the Georgia Institute of Technology, was stunned by damage off Kiritimati Island, near the center of the Pacific Ocean. "The entire reef is covered with a red-brown fuzz," Cobb said. "It is otherworldly. It is algae that has grown over dead coral. It was devastating" (Innis 2016). Around Kiritimati, within a broad swath at the center of the El Niño (near the equator in the Pacific Ocean), water temperatures rose from the usual 78°F to 88°F.

By 2017, according to an account by Damien Cave and Justin Gillis in the *New York Times*:

> Huge sections of the Great Barrier Reef, stretching across hundreds of miles of its most pristine northern sector, were recently found to be dead, killed last year by overheated seawater. More southerly sections around the middle of the reef that barely escaped then are bleaching now, a potential precursor to another die-off that could rob some of the reef's most visited areas of color and life. "We didn't expect to see this level of destruction to the Great Barrier Reef for another 30 years," said [Professor]Terry P. Hughes… "Climate change is not a future threat," Professor Hughes said. "On the Great Barrier Reef, it's been happening for 18 years." (Cave and Gillis 2017).

Surveys of the Great Barrier Reef late in 2016 indicated that 67 percent of the Northern Barrier Reef had died; in some areas, mortality had reached 83 percent.

Damage in southern regions was much less, ranging down to 6 percent near Townsville. The Australian government has tried to restrict port development and agricultural runoff, as well as other risks that can be locally controlled, but it cannot prevent temperature rises that are bleaching large parts of the reef. "The reefs in muddy water were just as fried as those in pristine water," Professor Terry Hughes of James Cook University in Townsville, Australia, said. "That's not good news in terms of what you can do locally to prevent bleaching—the answer to that is not very much at all. You have to address climate change directly" (Cave and Gillis 2017).

Can Corals Cope?

Hughes and colleagues hold out some hope that while climate change will alter species composition of many coral reefs, it will not completely wipe them out. Hughes and his team collected samples at 132 sites in Australia's Great Barrier Reef. "Of the 12 coral taxa sampled," they reported, "11 showed significant differences in abundance across the reef, regardless of how susceptible they were to thermal stress and bleaching. These differences in abundance did not follow changes in latitude or temperature. This flexibility may enable coral reefs to continue functioning as the environment alters with climate change" ("Can Coral Cope" 2012).

Corals can adapt to a certain amount of gradual warming that precedes bleaching. However, a report in *Science* published in 2016 found that massive, sudden warming of their aquatic habitat overwhelms these defenses, killing corals on a massive scale. Tracy D. Ainsworth and colleagues, having studied Australia's Great Barrier Reef, found that the 2015–2016 El Niño, which produced record-high temperatures in the oceans as well as atmosphere, disabled corals' natural bleaching mechanisms. Furthermore, they wrote, "Near-future increases in local temperature of as little as 0.5°C result in this protective mechanism being lost, which may increase the rate of degradation of the GBR" (Ainsworth et al. 2016, 338).

"Designer" Corals?

Some biologists are now seeking to develop "designer corals" that may be able to survive warmer and more acidic, polluted ocean waters. They're propagating reefs that thrive in hot water so that they may become the new global normal as the rest fall victim to humankind's appetite for fossil fuels. Scientists now range the world seeking candidate corals, such as the *Acropora hyacinthus* near Ofu Island in American Samoa, which can withstand water temperatures of 35°C (95°F) in shallow pools under direct sunlight. These corals can be interbred with more fragile species that are being killed by warming waters. A fifth of corals have been lost since 1950 worldwide, and another 35 percent are in critical condition (Mascarelli 2014).

Large-scale diffusion of "super" coral would involve problems of expense and scale that have not been confronted, much less solved. Test projects conducted thus far involve an infinitesimally small portion of the coral reefs that face destruction as oceans warm and become more acidic. Hughes, director of the Australia Research Council's Centre of Excellence for Coral Reef Studies, studied all ongoing restoration projects (about 250 of them as of 2015) and their cost (about a quarter of a billion dollars). The total area covered by all of them was about the size of two football fields. "When you consider just the Great Barrier Reef, which is a tiny fraction of the world's reefs, it has the area of Finland," he said. "So going from a test tube or an aquarium to millions of football fields is hugely expensive, obviously" (Kolbert 2016, 26).

Some varieties of coral (notably *Porites cylindrica* in Australia's Heron Island Lagoon and the Great Barrier Reef) resist ocean acidification, according to a study published during 2015 in the United States *Proceedings of the National Academy of Sciences* by researchers at the University of Queensland and the University of Western Australia. This type of coral contains a calcifying fluid at a constant pH level, even as the acid level changes in water around it. "The regulatory mechanism allows the coral to grow at a relatively constant rate, suggesting it may be more resilient to the effects of ocean acidification than previously thought," said lead author Lucy Georgiou, of UWA's ARC Centre of Excellence for Coral Reef Studies ("Self-Regulating" 2015).

"This newly discovered phenomenon of pH homeostasis during calcification indicates that coral living in highly dynamic environments exert strong physiological controls on the carbonate chemistry of their calcifying fluid, implying a high degree of resilience to ocean acidification within the investigated ranges," the scientists wrote (Georgiou et al. 2015). This study supports the conclusions of earlier work, also conducted in Australia. Research published in *Nature Climate Change* during 2012 "identified a powerful internal mechanism that could enable some corals and their symbiotic algae to counter the adverse impact of a more acidic ocean" ("Corals 'Could Survive'" 2012).

Further Reading

Ainsworth, Tracy D., Scott F. Heron, Juan Carlos Ortiz, Peter J. Mumby, Alana Grech, Daisie Ogawa, . . . William Leggat. 2016. "Climate Change Disables Coral Bleaching Protection on the Great Barrier Reef." *Science* 352, no. 6283 (April 15): 338–342.

"Can Coral Cope with Climate Change?" 2012. *Nature* 484 (April 19): 290.

Cave, Damien, and Justin Gillis. 2017. "Large Sections of Australia's Great Reef Are Now Dead, Scientists Find." *New York Times*, March 15. Accessed October 9, 2018. https://www.nytimes.com/2017/03/15/science/great-barrier-reef-coral-climate-change-dieoff.html.

"Corals 'Could Survive a More Acidic Ocean.'" 2015. University News. University of Western Australia, April 2. Accessed October 9, 2018. http://www.news.uwa.edu.au/201204024487/international/corals-could-survive-more-acidic-ocean.

Georgiou, Lucy, James Falter, Julie Trotter, David I. Kline, Michael Holcomb, Sophie G. Dove, . . . Malcolm McCulloch. 2015. "PH Homeostasis during Coral Calcification in a Free Ocean CO_2 Enrichment (FOCE) Experiment, Heron Island Reef Flat, Great Barrier Reef." *Proceedings of the National Academy of Sciences*, October 6. Accessed October 9, 2018. http://www.pnas.org/content/early/2015/10/01/1505586112.full.pdf.

Innis, Michelle. 2016. "Climate-Related Death of Coral Around World Alarms Scientists." *New York Times*, April 9. Accessed October 9, 2018. http://www.nytimes.com/2016/04/10/world/asia/climate-related-death-of-coral-around-world-alarms-scientists.html.

Kolbert, Elizabeth. 2016. "Unnatural Selection: What Will It Take to Save the World's Reefs and Forests?" *New Yorker,* April 18: 22–28.

Mascarelli, Amanda. 2014. "Climate-Change Adaptation: Designer Reefs." *Nature* 508, no. 7497 (April 24): 444–446. Accessed October 9, 2018. http://www.nature.com/news/climate-change-adaptation-designer-reefs-1.15073.

Mathews-Amos, Amy, and Ewann A. Berntson. *Turning Up the Heat: How Global Warming Threatens Life in the Sea.* Redmond, WA: Marine Conservation Biology Institute, 1999.

Perry, Michael. 1998. "Global Warming Devastates World's Coral Reefs." Reuters, November 26.

Ryan, Siobhain. 2002. "National Icons Feel the Heat." *Courier Mail* (Australia), February 4: 1.

"Self-Regulating Coral Protect Themselves Against Ocean Acidification." 2015. *University News*, University of Western Australia, October 6. Accessed October 9, 2018. http://www.news.uwa.edu.au/201510068035/research/self-regulating-coral-protect-themselves-against-ocean-acidification.

BANGLADESH

The World's Worst "Dead Zone"

The Ganges River system, which traverses most of northern India and laces Bangladesh before emptying into the Bay of Bengal, is the most densely populated area on Earth. One quarter of Earth's population lives in eight countries that border the bay (one of them is Bangladesh); 200 million live along its coastlines.

The Bay of Bengal dead zone (so-called because the zone is nearly bereft of oxygen, killing most animate life) is one of the most severe in the world, but it is far from alone. Others have been found nearly anywhere that agricultural chemicals have been poured into an ocean, including locations off the coasts of Namibia, the West Coast of North America, the United States Gulf of Mexico coast south of the Mississippi River delta, and the west coast of India opposite Mumbai ("Huge 'Dead Zone'" 2016). Dead zones also exist in the eastern tropical Pacific near Peru's border with Chile and in the Pacific off Mexico, as well as in the Arabian Sea.

The Ganges River system is also one of Earth's most polluted regions. Its urban areas, including Delhi, Kolkata, Dhaka, and many other cities, together with farmers that number in the tens of millions, produce the world's dirtiest air, much of it from cow-dung-fueled cooking fires as well as coal-fired power plants, motor vehicles, and other sources. The area is also home to the planet's thickest blanket of nitrogen-based fertilizers, which (as part of the "Green Revolution") have fed growing numbers of people. The growing number of farmers also are drawing down underground water reserves (aquifers) across much of the region, which are important because most moisture falls during a monsoon that usually begins in June and ends in August.

An Intense Dead Zone

At the same time, the Ganges conducts to the Bay of Bengal huge amounts of nitrogen that has fed one of the planet's most intense "dead zones," south of the Ganges' mouth in Bangladesh, which has been killing many of the fishing stocks that for many centuries have provided protein for that region's people. It was not until 2016 that teams of scientists inspected the growing dead zone and realized just how severe conditions had become. Already, more than two-thirds of Bangladeshis who fish for a living do not earn enough to rise out of poverty.

Bangladesh thus has become something of a postcard from the future—its land slowly sinking as the Ganges delta erodes into the Bay of Bengal, the ocean slowly rising as the atmosphere warms. Never a rich nor a well-fed region, its people now find their fish supplies imperiled as the population continues to rise. At the same time, erosion is reducing the amount of land available for farming in Bangladesh, forcing more people to chase a declining number of smaller fish. Many once-abundant species have all but disappeared. "Particularly badly affected are the species at the top of the food chain," according to one report. "The bay was once feared by sailors for its man-eating sharks; they are now rare in these waters. Other apex predators like grouper, croaker, and rays have also been badly hit. Catches now consist mainly of species like sardines, which are at the bottom of the marine food web" (Bhathal and Pauly 2008, 26).

The "Pink Gold Rush"

Western aid agencies not only encouraged heavy use of nitrogen fertilizers (which have played a major part in creating the dead zone), but also taught fishing people to use trawlers, which depleted what fish runs the anoxic waters didn't. For a short period, Bangladeshi and other fishermen sold fish to foreign markets, according to an account in *The Guardian*:

> This led to a "pink gold rush," in which prawns were trawled with fine mesh nets that were dragged along the seafloor. But along with hauls of "pink gold," these nets also scooped up whole seafloor ecosystems as well as vulnerable species like turtles, dolphins, sea snakes, rays, and sharks. These were once called bycatch and were largely discarded. Today the collateral damage of the trawling industry is processed and sold to the fast-growing poultry and aquaculture industries of the region. In effect, the processes that sustain the Bay of Bengal's fisheries are being destroyed in order to produce dirt-cheap chicken feed and fish feed. (Ghosh and Lobo 2017)

This boom was soon followed by a bust, as fish stocks fell due to overfishing and the intensifying anoxic nature of the dead zone. Questions of sustainability were ignored at every turn for short-term gains. By the late 1990s, fish catches were falling as trawlers were forced to range over wider areas. Many fishermen found themselves going into debt. Some fish were poisoned with cyanide as coral reefs were destroyed by use of dynamite by some fishermen. The collusion of all of these factors devastated areas that an English fisheries officer had described in the late

19th century as being "literally alive with fish" (Ghosh and Lobo 2017). Even as fish stocks dwindled, wrote Amitav Ghosh and Aaron Savio Lobo in *The Guardian*, quoting S. Schonhardt in the *Wall Street Journal*:

> At night, specially equipped, long-armed boats materialize around the islands and shine high-powered green lights into the water to attract plankton and the squid that follow in their wake. After nightfall, a glow that is bright enough to be visible from outer space hangs above the archipelago, like a miasmic fog. These squid boats, some of which are probably crewed by men who have been trafficked like slaves, help to make Thailand the world's largest exporter of squid—at least for the time being. (Ghosh and Lobo 2017; Schonhardt 2014)

The Dead Zone Grows

The Ganges and its tributaries also carry untreated industrial waste, sewage, and plastic, as well as chemicals from agriculture and aquaculture, all combining with nitrogen runoff to feed massive plankton blooms that consume oxygen. By 2017, the dead zone had grown to 60,000 square kilometers.

Local fishermen had noticed the depletion of animal life in the Bay of Bengal for several years before it was confirmed by scientists from India's National Institute of Oceanography (NIO), of Goa, India, with aid from the University of Southern Denmark and the Max Planck Institute for Marine Microbiology in Bremen, Germany. "Oxygen depletion in the oceans occurs both due to natural causes and human activities," Wajih Naqvi, former director of NIO and a coauthor of the study, told *Nature India* (Jayaraman 2016). Lack of oxygen makes the ecosystem shift to anaerobic metabolism. This pushes microorganisms living within these oxygen minimum zones (OMZ) to get energy by degrading organic matter in a process that removes vast amounts of nitrogen—a key nutrient for life—from the oceans. This upsets the nitrogen balance of the planet.

There have been reports that the OMZs of the world have been expanding and intensifying over the past few decades, although a clear cause for this is yet to be established. Naqvi said that "should a similar global trend apply to the BoB [Bay of Bengal], its OMZ will trip to anaerobic mode, like in the Arabian Sea" (Jayaraman 2016). Naqvi continued, "If the BoB does turn completely anoxic, and I suspect it will happen in future, global biogeochemical fluxes and the nitrogen cycle will be substantially affected and change the community structure at depth. However, it will probably not have any implications for India because we know that Arabian Sea OMZ operates anaerobically but has little effect on surface processes" (Jayaraman, 2016).

> Unlike other oxygen minimum zones, our measurements using a highly sensitive oxygen sensor demonstrate that the Bay of Bengal has persistent concentrations of oxygen. . . . We propose that this oxygen supports nitrite oxidation, thereby restricting the nitrite available for anammox or denitrification. If these traces of oxygen were removed, nitrogen loss in the Bay of Bengal oxygen minimum zone waters could accelerate to global significance. (Bristow et al. 2017)

Further Reading

Amrith, Sunil. 2013. *Crossing the Bay of Bengal: The Furies of Nature and the Fortunes of Migrants*. Cambridge, MA: Harvard University Press.

Bhathal, B., and D. Pauly. 2008. "'Fishing Down Marine Food Webs' and Spatial Expansion of Coastal Fisheries in India, 1950–2000." *Fisheries Research* 91, no. 1 (May): 26–34.

Bristow, L. A., C. M. Callbeck, M. Larsen, M. A. Altabet, J. Dekaezemacker, M. Forth, . . . D. E. Canfield. 2017. "N_2 Production Rates Limited by Nitrite Availability in the Bay of Bengal Oxygen Minimum Zone." *Nature Geoscience* 10: 24–29. Accessed October 9, 2018. http://www.nature.com/ngeo/journal/v10/n1/abs/ngeo2847.html.

Ghosh, Amitav, and Aaron Savio Lobo. 2017. "Bay of Bengal: Depleted Fish Stocks and Huge Dead Zone Signal Tipping Point." *The Guardian*, January 31. Accessed October 9, 2018. https://www.theguardian.com/environment/2017/jan/31/bay-bengal-depleted-fish-stocks-pollution-climate-change-migration.

"Huge 'Dead Zone' Discovered in Bay of Bengal." 2016. *The Hindu,* December 20. Accessed October 9, 2018. http://www.thehindu.com/todays-paper/tp-national/Huge-%E2%80%98dead-zone%E2%80%99-discovered-in-Bay-of-Bengal/article16908928.ece.

Jayaraman, K. S. 2016. "'Dead Zone' Found in Bay of Bengal." *NatureIndia*, December 12. Accessed October 9, 2018. http://www.natureasia.com/en/nindia/article/10.1038/nindia.2016.163.

Lobo, A. S., A. Balmford, R. Arthur, and A. Manica. 2010. "Commercializing Bycatch Can Push a Fishery Beyond Economic Extinction." *Conservation Letters* 3 (August): 277–285.

Schonhardt, S. 2014. "What's the One Thing in Thailand Visible from Space?" *Wall Street Journal*, February 28. Accessed October 9, 2018. https://blogs.wsj.com/indonesiarealtime/2014/02/28/whats-the-one-thing-in-thailand-visible-from-space/.

CANADA

Sea Life Surrenders to Acidity

Generally rising atmospheric levels of carbon dioxide are combining with industrial discharges, septic runoff, and motor-vehicle traffic to accelerate acidity of waters in urban areas of British Columbia, Canada, as well as along the neighboring Washington State coast and in Puget Sound.

The coastal waters have become so acidic that wild oysters have not reproduced since 2005. Acidity problems are worst in Hood Canal and in Canada's Strait of Georgia, sites where most of the area's commercial shellfish are harvested. Scientists from the University of Washington and NOAA warned on July 12, 2010 that the entire area had become a hot spot for marine acidity. Similar warnings have been sounded by scientists and shellfish growers in British Columbia.

Scientists at Seattle's NOAA Northwest Fishery Science Center by 2016 had learned that in addition to eroding calcium shells, rising acidity harms animals with chiton shells (in this instance, Dungeness crabs, in the Pacific Northwest and British Columbia), because it alters their metabolism. Rising acidity also reduces the survival rate of the crabs' larvae, an important food source for many other sea creatures, including salmon.

Acidity Spike in the Strait of Georgia

Rob Saunders of Islands Scallops in Qualicum Beach, inland from British Columbia's Baynes Sound, regards his shellfish as "canaries in a coal mine" for ocean acidity. Until 2011, his scallops were disease-resistant and meaty. Then they began to die. Between January and August of 2013, nearly 10 million scallops in British Columbia's Georgia Strait died en masse. Saunders's 40,000 scallop larvae, housed in 40,000-liter tanks, also died, as he was forced to lay off three-quarters of his employees. At first Saunders suspected disease, but a water test revealed that its pH level had dropped sharply, increasing acidity of the water.

"As a biologist I was always taught that this ocean doesn't change much, not drastically at least, so it came as a shock when we saw the pH drop," he said (lower pH equals higher acidity). "It became clear to us that the issue was carbon dioxide in the ocean" (Pollon 2015). Saunders then read studies relating climate change with Canada's Pacific ecosystems and learned that the water in which his scallops were dying was more acid than it had been in 20 million years.

The Georgia Strait, like Puget Sound to its south, is, according to many reports, "particularly vulnerable to ocean acidification, because it already has some of the planet's lowest pH waters. That's because the northeast Pacific Ocean resides at the terminus of a vast oceanic conveyor belt that delivers oxygen-poor, CO_2-rich water to our corner of the Pacific through a natural process known as upwelling" (Pollon 2015).

"Upwelling zones," along the U.S. and Canadian west coasts, where prevailing winds drive water toward the shoreline, are notably vulnerable to amplification of severe ocean acidification. "Near-shore waters of the California current system already have a low carbonate saturation rate, wrote Nicolas Gruber and his colleagues. They used eddy-resolving model simulations in their study, which projects acidification along the western coastal areas through the year 2050, when "habitats along the seafloor will become exposed to year-round undersaturation within the next 20 to 30 years. These projected events have potentially major implications for the rich and diverse ecosystem" (Gruber et al. 2012, 220).

Acidic water was preventing shellfish from fully developing their calcium carbonate shells and "forcing the animals to expend so much energy trying to build a shell [in] the acidic waters that it is weakened and vulnerable to disease and problems it could normally withstand. While Saunders can "sweeten" the water in his tanks by adding sodium carbonate to raise its pH level, any shellfish swimming in the open ocean is on its own.

Kenneth Denman, an adjunct professor at the University of Victoria's School of Earth and Ocean Sciences and pioneer in studying how climate change affects the marine environment, said: "We can say definitively that there has been a reduction in the pH over the last 25 years, and we can say with virtual certainty that it's caused by the increase in carbon dioxide in the atmosphere and the amount that gets into the ocean. The carbon dioxide levels in the ocean are tracking the carbon dioxide levels in the atmosphere immediately above the surface" (Pollon 2015).

Acidic water also kills oyster larvae. On the west coasts of both countries, the highest acidity is being found near the coast, the area richest in marine life. Coastal acidity is being carried by ocean currents into the enclosed waters of Hood Canal and Puget Sound, raising levels from industrial sources there. Hood Canal is becoming a dead zone as pollution and runoff from septic tanks kills fish, octopus, crab, and other sea life. pH in southern Puget Sound was as low as 7.4 by 2010 (Welch 2013, B-1, B-2).

Ocean acidification is damaging the crab fishery along British Columbia's west coast, as well as in Alaska, most notably affecting the huge Bristol Bay red king crab. With arms the size of baseball bats, this crab has become the star of restaurant "surf and turf" feasts around the world, a $100 million a year industry, and "the showboat of the Northwest's billion-dollar fishing industry. . . . a television sensation and a marketer's dream, its image emblazoned on bumper stickers, mugs, caps, and T-shirts throughout the Pacific Northwest and Alaska" (Welch 2013a).

King Crabs Endangered

Snow crab, another major West Coast harvest, also may be damaged. "With red king crab, it's all doom and gloom," said Robert Foy, who oversaw the crab research for the National Oceanic and Atmospheric Administration (NOAA) in Kodiak, Alaska. "With snow crab, there's so little known we just can't say. But we don't see anything from our experience that's good for any of these crabs. Some is just not as bad as others" (Welch 2013a). Rising ocean acidification is damaging the crabs' shells.

The problems in the crab industry also illustrate the paradoxes of ocean acidification on sea life. Similar species in the same habitat react differently. "The real issue here is unpredictability," said Richard Aronson, a Florida-based marine scientist who has tracked king crab in Antarctica. "There are all these unanticipated collateral impacts. The problem is, most of them are nasty surprises" (Welch 2013a)

"Many crab species appear hardy in the face of souring seas, or at least not so frail," reported Craig Welch (2013a). He continued, "Exceedingly corrosive waters actually pump up some blue crab to three times their size and turn them into voracious predators. Sour waters kill Dungeness crab, but far less often than Alaska red king crab" (Welch 2013 a). Laboratory tests with acidity levels expected at mid-century intensify the damage, especially to Alaska red king crab. The models were developed with the future in mind, but in some areas, acidity had reached danger levels by 2015.

Crabs react differently depending on the water chemistry of their habitat. Very little has been proven, but some species may quickly develop tolerance to some level of acidity. Golden king crabs, which live in very deep water with high levels of carbon dioxide, may be evolving a tolerance, but acidity levels may be rising more quickly than they can change. "It's almost like an arms race," said Gretchen Hofmann, a marine biologist at the University of California, Santa Barbara. "We can see that the potential for rapid evolution is there. The question is, will the changes be so rapid and extreme that it will outstrip what they're capable of?" (Welch 2013b).

Golden king crab, for example, live extremely deep, below 1,000 feet, where waters already are naturally rich in CO_2, making them highly tolerant of sea-chemistry changes. Crabs that live near the water's surface and close to shore may evolve tolerance from swings in acidity experienced from rising and falling tides and even photosynthesis. Eventually, crabs that cannot survive acidic water caused by human emissions of carbon dioxide may be raised in aquatic farms under managed conditions. One such farm, the Alutiiq Pride Shellfish Hatchery, already exists in Seward, Alaska, providing seed stock for new species and providing for shellfish enhancement and other research.

Commercial Catch Imperiled

By 2008, scientists surveying waters near the West Coast of North America had found rising levels of acidified ocean water within 20 miles of the shoreline, raising concern for marine ecosystems from Canada to Mexico. Researchers on the *Wecoma*, an Oregon State University research vessel, discovered that the acidified upwelling from the deeper ocean is probably 50 years old. Future ocean acidification levels probably will rise as atmospheric levels of carbon dioxide increase (Feely et al. 2008, 1490; "Pacific Coast" 2008).

"When the upwelled water was last at the surface, it was exposed to an atmosphere with much lower CO_2 levels than today's," said Burke Hales, an associate professor in the College of Oceanic and Atmospheric Sciences at Oregon State University and a coauthor of the study. "The water that will upwell off the coast in future years already is making its undersea trek toward us, with ever-increasing levels of carbon dioxide and acidity." The researchers found that the 50-year-old upwelled water had CO_2 levels of 900 to 1,000 parts per million, placing it "right on the edge of solubility" for calcium carbonate-shelled aragonites, Hales said ("Pacific Coast" 2008). Continued carbon-dioxide overload in the oceans could make ocean water more acidic than it has been "for tens of millions of years and, critically, at a rate of change 100 times greater than at any time over this period" (Riebesell et al. 2007, 545).

Isaac Kaplan, a computer modeler at NOAA in Seattle, anticipates declines in flounder, sole sharks, skates, Pacific whiting (hake), and rays. "Some species will go up, some species will go down," said Phil Levin, ecosystems leader for NOAA's Northwest Fisheries Science Center in Seattle. "On balance, it looks to us like most of the commercially caught fish species will go down" (Welch 2013b). The North Pacific pollock catch (which supplies a number of products from imitation crab to fish sticks, is hauled in by ocean-going fishermen and factory trawlers, and by itself is a $1 billion annual industry) has problems recognizing their food sources in water at enhanced carbon-dioxide levels. However, over several generations they might evolve a tolerance.

Both British Columbia and Washington State have been among the first to monitor ocean acidity—to tally levels in water and emissions sources. This is acknowledged as just a first step: "The only long-term solution to acidification is for the world to reduce industrial emissions of carbon dioxide, allowing the ocean to reach a less acidic equilibrium" ("Marine Life" 2012). Local officials next "will seek to reduce carbon pollution from land-based sources, including agricultural and urban runoff. There will also be practical, site-based steps to offset carbon, like planting sea grasses (which themselves are endangered globally) in shellfish hatcheries. And there will be an extensive campaign to educate the public, business leaders, and policy makers about the risks of increasing acidification" ("Marine Life" 2012).

Further Reading

Feely, Richard A., Christopher L. Sabine, J. Martin Hernandez-Ayon, Debby Ianson, and Burke Hales. 2008. "Evidence for Upwelling of Corrosive 'Acidified' Water onto the Continental Shelf." *Science* 320, no. 5882 (June 13): 1490–1492.

Gruber, Nicolas, Claudine Hauri, Zouhair Lachkar, Damian Loher, Thomas L. Frölicher, and Gian-Kasper Plattner. 2012. "Rapid Progression of Ocean Acidification in the California Current System." *Science* 337, no. 6091 (July 13): 220–223.

"Marine Life on a Warming Planet." 2012. Editorial in the *New York Times,* December 2. Accessed October 9, 2018. http://www.nytimes.com/2012/12/03/opinion/marine-life -on-a-warming-planet.html.

Oregon State University. "U.S. Pacific Coast Waters Turning More Acidic." ScienceDaily. Accessed November 30, 2018. www.sciencedaily.com/releases/2008/05/080522181511 .htm.

Pollon, Christopher. 2015. "The B.C. Scallop Farmer's Acid Test." Huffington Post, November 19. Accessed October 9, 2018. http://www.huffingtonpost.ca/2015/11/19/climate -change-scallops_n_8597502.html.

Riebesell, U., K. G. Schulz, R. G. J. Bellerby, M. Botros, P. Fritsche, M. Meyerhöfer, . . . E. Zöllner. 2007. "Enhanced Biological Carbon Consumption in a High CO2 Ocean." *Nature* 450 (November 22): 545–548.

Welch, Craig. 2013a. "Sea Change: Scientists Fear Ocean Acidification Will Drive the Collapse of Alaska's Iconic Crab Fishery." *Seattle Times*, September 11. Accessed October 9, 2018. http://apps.seattletimes.com/reports/sea-change/2013/sep/11/alaska-crab -industry/.

Welch, Craig. 2013b. "Sea Change: The Pacific's Perilous Turn." *Seattle Times*, September 11. Accessed October 9, 2018. http://apps.seattletimes.com/reports/sea-change/2013/sep /11/pacific-ocean-perilous-turn-overview/.

CHINA

Shanghai's Many Dead Zones

China has paid a steep environmental price for economic prosperity, evident by the fact that the phrase "dead zone" has meant many things in Shanghai. The term has been applied to an anoxic or "hypoxic" (without oxygen) region off the coast,

A chemical plant visible on the horizon of the Chinese Yellow River estuary in the Bohai Bay fills the air with pollution as women collect seashells. (China Photos/Getty Images)

fed by a polluted Yangtze River, as well as seepage from a city-sized, toxic plastics recycling zone that slowly kills its residents.

Shanghai's oceanic dead zone in the East China Sea is one of about 400 such areas of unusually low dissolved oxygen content worldwide. They may form through natural processes, but that is rare. In today's world, anoxic areas are usually a result of runoff from land that has been heavily bathed in nitrogen-based fertilizers. Excess phosphorus runoff from cities can play a role, as well. Nearly all of these "dead zones" are growing in lakes and ocean waters as global warming raises temperatures. Nitrogen and phosphorus feed abnormally lush algae blooms that sink when they die, sucking up oxygen, leaving little or nothing for other marine life. Dead zones have doubled in number each 10 years since the 1960s as nutrient-rich runoff has risen.

Most dead zones have developed in the oceans, but lakes and rivers loaded with large deposits of sewage may be similarly afflicted. Emissions from motor vehicles and industrial plants also may contribute to dead zones. Scientists began to notice such areas during the 1970s, nearly always near sites of heavily polluted water.

Chinese Rivers' Dead Zones

In 2006, several estuaries of China's two longest rivers, the Yellow and Yangtze, were classified as dead zones by the United Nations due to high amounts of industrial pollutants. "Experts warn that these areas are fast becoming major threats to fish stocks and to people who depend upon fisheries for food and livelihoods," the *China Daily* reported, citing a recent study by the U.N. Environmental Program ("Estuaries of China" 2006). The report noted that the river systems, afflicted with rising levels of sewage and industrial pollution, were becoming anoxic, threatening fish and other aquatic life.

Another report, by China's State Environmental Protection Administration, documented a rise in the number of "red tides," in which algae blooms deprive water of oxygen, along China's coasts. In 2005, 82 such tides occurred along China's coastlines. The report said that "large-scale red tides have become an annual occurrence in waters off eastern China's Zhejiang province, where the Yangtze River flows into the sea, and farther north in the Bohai Sea near the Yellow River estuary" ("Estuaries of China" 2006). Both reports identified rising use of nitrogen and phosphate-based fertilizers as an important reason why both rivers and coastal areas are becoming uninhabitable. In June 2006, a red tide that spread over 1,000 square kilometers (620 square miles) in the Yangtze River killed more than 12 million fish. People in Shanghai were warned not to eat local seafood.

In 2013, an outbreak of green algae (*hutia* in Chinese) frothed up in seawaters off eastern Shandong. Newspapers carried photos of people on the beaches playing in the green mass as it suffocated sea life below. About the same time that the Yellow River was declared hazardous to fish, newspapers in China carried reports that people were earning money hauling corpses out of it.

> Wei Jinpeng pointed to a fisherman's cove below and began counting his latest catch [in Changpo Village]. He stopped after six and guessed that perhaps a dozen human corpses were bobbing in the murky waters. The bodies were floating facedown and tethered by ropes to the shore, their mud-covered limbs and rumps protruding from the water. Wei is a fisher of dead people. He scans the river for cadavers, drags them to shore with a small boat, and then charges grieving families to recover their relatives' corpses. (Lasseter 2006)

Wei does not investigate causes of the murders (suicides, in some cases). Some may be victims of floods; others may be suicides by people who can't pay their debts. In seven years of corpse hauling, he has recovered about 50 bodies and been paid an average of $500 each by relatives.

The City of Toxic Plastic

The term "dead zone" also has acquired another meaning near Shanghai. For his book, *Junkyard Planet: Travels in the Billion-Dollar Trash Trade,* Adam Minter visited a gigantic dump in Wen'an, near Shanghai, which at that time (in 2010) was a center of the global scrap plastic trade. A year later, the Chinese government shut it down. Adam Minter, *The Guardian's* Shanghai correspondent, wrote:

> It's a bustling, crowded, and incomprehensibly dirty main street, crossed by the occasional stray dog, partly blocked by a broken-down truck, and frequently scarred by

black spots where, I'm later told, unrecyclable plastics were burned in the night. Above me, plastic bags are captured by the wind, floating on the breeze. But what I find most striking about Wen'an is this: there's nothing green. It's a dead zone. (Minter 2014)

Thirty years ago, Wen'an was an agricultural region known for its relatively clean streams and peach trees, fragrant soil, and fishing, before "the plastics recycling trade plasticized the lungs of men in their 20s" (Minter 2014). Instead, Wen'an had become a network of about 50 villages built on recycled plastic: 20,000 small-scale, family-owned businesses, processing more waste plastic than any other location on Earth, set along a river choking on algae, in an atmosphere thick with toxic stench from plastics being melted and strained in a boiling, smoking mass as workers inhaled "the visible fumes that fill the room with a chemical choke" (Minter 2014). A doctor who had served people in the area since 1968 spoke to Minter:

"Since the '80s, high blood pressure has exploded," he explained. "In the past nobody had it. Now 40 percent of the adults in this village have it. Back in the [19]80s, you'd only see it in people in their 40s. In the [19]90s, we started seeing it in people over 30. Now we're seeing it in people age 28 and up. And it comes with pulmonary problems that restrict movement. People have it in their 30s so badly that they can't move anymore. They're paralyzed." (Minter 2014)

Young villagers are developing pulmonary fibrosis and paralyzing strokes. "Back in the [19]70s and [19]80s, you didn't die from high blood pressure. Now you die from it," another local doctor told Minter (2014).

Further Reading

"Estuaries of China Rivers Named 'Dead Zones.'" 2006. *China Post,* October 21.
Lasseter, Tom. 2006. "China's Dark Side: On Yellow River, Corpses Mean Cash." McClatchy Newspapers in *China Post,* October 21. https://www.mcclatchydc.com/news/nation -world/world/article24593404.html.
Minter, Adam. 2014. "Plastic, Poverty, and Pollution in China's Recycling Dead Zone." *The Guardian,* July 16. Accessed October 9, 2018. https://www.theguardian.com/life andstyle/2014/jul/16/plastic-poverty-pollution-china-recycling-dead-zone.

JAPAN

Fukushima, Radiation, and Fishing

Japan's Fukushima nuclear disaster changed the environment of northern Japan immeasurably, including the sea life of the adjacent Pacific Ocean. Traces of radio-activity even reached the United States via ocean currents, although not enough to do much harm. Japan, which suffered as a target of the only nuclear weapons to be used in war (at Hiroshima and Nagasaki in 1945), compiled a detailed record of the damage done by the Fukushima accident. While a large body of ocean water dilutes the intensity of radioactivity, the incident is still a matter of concern, as it resulted in a widespread shutdown of fishing in the Fukushima area. Several countries quickly banned Japanese seafood imports. The United States for a time banned imports from areas in and close to Fukushima.

The accident on March 11, 2011, was provoked by a 9.0 magnitude earthquake near the coast that was followed by a major tsunami (tidal wave) that devastated the city of Fukushima as well as the nuclear power plant. Thousands of tons of radioactive water poured from the damaged Daiichi nuclear power plant from its oceanfront site directly into the Pacific Ocean. Quickly, high levels of radioactivity were detected 25 miles off-shore, spreading on oceanic currents. Contamination also was ejected into the sky and spread eastward on prevailing winds.

One report said that "the Tokyo Electric Power Company (TEPCO) said that seawater containing radioactive iodine-131 at 5 million times the legal limit was detected near the plant. According to the Japanese news service, NHK, another sample contained 1.1 million times the legal level of radioactive cesium-137" (Grossman 2011).

A Shot of Radiation for the Food Chain

"Given that the Fukushima nuclear power plant is on the ocean (the water was used to cool its reactors, four of which were damaged), with leaks and runoff directly into it, the impacts on the ocean exceeded those of Chernobyl [then in the Soviet Union, now in Ukraine], which was hundreds of miles from any sea," said Ken Buesseler, senior scientist in marine chemistry at the Woods Hole Oceanographic Institution in Massachusetts (Grossman 2011). As it was being absorbed by phytoplankton and zooplankton, the radiation also contaminated kelp, as well as other marine life. It then was transmitted up the food chain to marine mammals, fish, and humans. Radioactive wastewater continued to leak into the ocean for several weeks after the initial surge.

Within a few days, the ruins of the Fukushima plant bathed the land and the food chain of the ocean around it in a pulse of radioisotopes, including cesium. David Pacchioli, writing in *Oceanus Magazine*, published by the Woods Hole Oceanographic Institution, offered a synopsis by Scott Fowler, who helped pioneer marine radioecology for more than 30 years at the International Atomic Energy Agency's Marine Environment Laboratories. Fowler spoke in Tokyo at the Fukushima and the Ocean Conference in November 2012.

The damage permeated the food chain, beginning with marine phytoplankton, the tiny plants that sustain life at its base, which began to soak up radioactivity within hours of the accident. These phytoplankton then formed a basic conduit, as described by Fowler:

> As the phytoplankton are eaten by larger zooplankton, small fish, and larger animals up the food chain, some of the contaminants end up in fecal pellets or other detrital particles that settle to the seafloor. These particles accumulate in sediments, and some radioisotopes contained within them may be remobilized back into the

overlying waters through microbial and chemical processes. How much radioactivity gets into marine life depends on a host of factors: How long the organisms are exposed to radioactivity is certainly important, but so too are the sizes and species of the organisms, the radioisotopes involved, the temperature and salinity of the water, how much oxygen is in it, and many other factors such as the life stage of the organisms. (Pacchioli 2013)

Different Types of Radiation

In measuring the effects of Fukushima's radiation, Fowler said that it is important to screen out natural background radiation, such as polonium-210 and potassium-40, both of which are omnipresent in the ocean. Although potassium-40 occurs most often, polonium-210 accumulates in marine organisms most often and most intensely. "Polonium is responsible for the majority of the radiation dose that fish and other marine organisms receive," he said (Pacchioli 2013).

The most dangerous radiation introduced into the maritime environment by the Fukushima accident was plutonium, which was absorbed from seawater by a large variety of marine life. Each species reacts to plutonium differently. For example, said Fowler, "Phytoplankton accumulated roughly 10 times as much plutonium as microzooplankton, which took up 100 times more than clams. Octopi and crabs took up about half as much plutonium as clams but about 100 times more than bottom-dwelling fish" (Pacchioli 2013).

In addition to absorption through the food chain, radiation also may be absorbed from sediments that have been contaminated by radioactivity. Once again, species-specific response varies by type of animal, as well as by location. For example, Fowler described one experiment that measured uptake of americium, a specific type of radioactivity:

> Worms exposed to contaminated sediments took up significantly more of the radio-isotope than clams did. But both worms and clams took up much more of the radio-isotopes from Pacific sediments, which contain relatively high amounts of silica minerals, than they did from Atlantic sediments, which contain more carbon minerals. (Pacchioli 2013)

Food remained the most important conduit of radioactivity, however, "a far more efficient route than if they are absorbed externally from the environment. Marine invertebrates, such as bottom-dwelling starfish and sea urchins, are particularly proficient at absorbing a wide range of ingested radioisotopes, he said, but fortunately, they lose that incorporated radioactivity over time, via excretion" (Pacchioli 2013). The radioactivity eventually would reach porpoises and seals that had eaten contaminated fish.

Tom Hei, professor of environmental sciences and vice-chairman of radiation oncology at Columbia University, explained that the mechanisms that determine how an animal takes in radiation are the same for fish as they are for humans.

> Once in the body—whether inhaled or absorbed through gills or other organs— radiation can make its way into the bloodstream, lungs, and bony structures,

potentially causing death, cancer, or genetic damage. Larger animals tend to be more sensitive to radiation than smaller ones. Yet small fish, mollusks, and crustaceans, as well as plankton and phytoplankton, can absorb radiation.... How the radiation accumulates depends on the degree of exposure—dose and duration—and the half-life of the element, said Hei. (Grossman 2011)

Radioactive iodine may be absorbed by the thyroid gland in marine mammals and humans, as well as in fish species' thyroid tissue. Cesium is absorbed by muscles. "Cesium behaves like potassium, so would end up in all marine life," said Arjun Makhijani, president of the Institute for Energy and Environmental Research in Maryland. "It certainly will have an effect" (Grossman 2011).

Nicholas Fisher, a marine biogeochemist at Stony Brook University (New York), who studies the effects of radiation on the marine environment, described the results of a research trip sponsored by the Woods Hole Oceanographic Institution, led by Buesseler, along the Japanese coast astride Fukushima during June of 2011. They concentrated on the ecological toll of cesium-137 because of its enduring effects (the Pacific Ocean still bears the impact of cesium-137 from U.S. atomic-bomb tests as late as the 1960s). Other types of radiation (such as iodine-131, which was emitted at Fukushima in large quantities) has a half-life of only eight days. The radioactive half-life of cesium-137, by contrast, is 30 years. The scientists also studied the role of radiation in the food chain as it biomagnifies—that is, its impact increases with each step up. "Cesium shows only modest biomagnification in marine food chains—much less than mercury, a toxic metal, or many other harmful organic compounds such the insecticide DDT and polychlorinated biphenyls (PCBs)," Fisher said (Pacchioli 2013).

Diluted Radiation Reaches the United States

By 2016, salmon with minute traces of cesium-134, which the press was calling "the fingerprint of the Fukushima," were caught near Tillamook Bay and Gold Beach, Oregon, in the United States, 6,000 miles from the site of the accident. By that time, the radioactivity, while still detectable by instruments, was not considered harmful. Buesseler told *USA Today*: "To put it in context, if you were to swim every day for six hours a day in those waters for a year, that additional radiation from . . . Japan . . . [would be] one thousand times smaller than one dental X-ray" (Baynes 2016). Cesium-134 was also was detected in Canadian salmon, according to a report by University of Victoria chemical oceanographer Jay Cullen in the Fukushima InFORM project. "In Japan, at its peak [celsium-134 levels] it was 10 million times higher than what we are seeing today on the West Coast," he said (Bowerman and Loew 2016).

Further Reading

Baynes, Chris. 2016. "Radioactive Fish Contaminated by Fukushima Nuclear Disaster Found Off U.S. Coast 6,000 Miles Away." *The Mirror*, December 9. Accessed October 9, 2018. http://www.mirror.co.uk/news/world-news/radioactive-fish-contaminated-fukushima-nuclear-9427679.

Bowerman, Mary, and Tracy Loew. 2016. "Should We Be Worried about Fukushima Radiation?" *USA Today*, December 9. Accessed October 9, 2018. https://www.usa today.com/story/news/nation-now/2016/12/09/should-we-worried-fukushima-radia tion/95196156/.

Grossman, Elizabeth. 2011. "Fukushima: Radioactivity in the Ocean: Diluted, But Far from Harmless." *Yale Environment 360*, April 7. Accessed October 9, 2018. http://e360.yale .edu/features/radioactivity_in_the_ocean_diluted_but_far_from_harmless.

Pacchioli, David. 2013. "How Is Fukushima's Fallout Affecting Marine Life?" *Oceanus Magazine*, May 2. Accessed October 9, 2018. http://www.whoi.edu/oceanus/feature /how-is-fukushimas-fallout-affecting-marine-life.

NORWAY

Methane Hydrates Bubbling Up

Warming oceans eventually could cause intense eruptions of methane from the seafloor, accelerating global warming caused by humankind's industries and transportation. About 10,000 billion tons of solid methane clathrates, described as "a lattice of ice crystals rather like a honeycomb," are stored beneath the ocean and on the continents (Pearce 2007, 91). By comparison, the contribution of humans to the atmosphere's inventory of greenhouse gases via burning of fossil fuels has

---

A Methane "Burp" and Noah's Ark

Gregory Ryskin, a geologist at Northwestern University, has asserted that a small-scale methane "burp" may help to explain the flood that was navigated in the Bible by Noah's Ark. The biblical flood, according to Ryskin, may be attributable to an eruption from Europe's Black Sea. Some geological evidence suggests an event of this type 7,000 to 8,000 years ago. According to an account by Tom Clarke in *Nature* Online (Clarke 2003; Ryskin 2003, 737), "Ryskin contends that methane from bacterial decay or from frozen methane hydrates . . . began to be released. Under the enormous pressure from water above, the gas dissolved . . . [and] was trapped there as its concentration grew." A single disturbance, according to this account, "a small meteorite impact or even a fast-moving mammal, could then have brought the gas-saturated water closer to the surface. Here it would have bubbled out of solution under the reduced pressure. Thereafter the process would have been unstoppable: a huge overturning of the water layers would have released a vast belch of methane" (Clarke 2003).

Further Reading

Clarke, Tom. 2003. "Boiling Seas Linked to Mass Extinction; Methane Belches May Have Catastrophic Consequences." *Nature* Online, August 22. Accessed October 11, 2018. https:// www.nature.com/news/2003/030822/full/news030818-16.html.

Ryskin, Gregory. 2003. "Methane-Driven Oceanic Eruptions and Mass Extinctions." *Geology* 31 (September): 737–740.

amounted to about 200 billion tons of carbon. If even a small portion of the oceans' stored methane were to escape into the atmosphere, greenhouse warming could greatly accelerate. Among scientists, this mechanism has come to be called the "methane burp" or "clathrate gun hypothesis."

Submarine hydrates total several thousand petagrams (Pg) worldwide. A petagram is a billion metric tons, and several thousand would equal at least a thousand times the methane now in the atmosphere. In addition, the top three meters of Arctic permafrost contains about 1,000 Pg of carbon as organic matter that could be converted to methane that would equal 300 times the methane in the atmosphere (Kerr 2010, 620).

Whether—and if so, then when—methane hydrates pump up the atmosphere's greenhouse-gas load has become a matter of intense debate in scientific circles. Most of the hydrates are several hundred meters deep in the oceans and under varying depths of mud in areas that probably will not enter the atmosphere as gas for centuries after much of the world's surface ice has melted. However, once warmed, they would react at a temperature similar to that of water ice, so any reaction could be rapid and massive.

In the Arctic, writes David Archer in *The Long Thaw: How Humans Are Changing the Next 100,000 Years of Earth's Climate,* methane hydrates have formed within 200 meters of the surface. Since Arctic warming is intense, some of them could melt in decades to centuries (Archer 2009, 132). The potential is complicated by the complicated nature of hydrate chemistry (methane turns to carbon dioxide as it liquefies), as well as by the very rapid speed of anthropogenic warming compared to natural changes observed in the paleoclimate. However, ultimately, writes Archer, "Methane hydrates could release as much carbon as the CO_2 released from fossil fuels, doubling the long-term impact of global warming" (Archer 2009, 136).

Methane Bubbles in Norway and Elsewhere

A group of British and German research scientists using sonar has found that at least 250 sources of methane bubbles have been rising from hydrates in the seabed off Norway, an early indication, perhaps, of the "methane gun." The group reported in *Geophysical Research Letters* that the methane was rising from between 150 and 4500 meters near West Spitsbergen (Westbrook et al. 2009). Temperature records indicate that this area has warmed by about 1°C during the last 30 years, destabilizing some of the hydrates. Professor Tim Minshull of the National Oceanography Centre at Southampton told BBC News:

> We already knew there was some methane hydrate in the ocean off Spitsbergen, and that's an area where climate change is happening rather faster than just about anywhere else in the world. . . . Our survey was designed to work out how much methane might be released by future ocean warming; we did not expect to discover such strong evidence that this process has already started. (Burns 2009)

This research indicates that the methane release may be part of a long-term pattern that dates, in some cases, to the last years of the most recent ice age. Most of the methane ejected from the hydrates reacts with oxygen in the ocean, forming carbon dioxide as carbonic acid, contributing to ocean acidification. "If this process becomes widespread along Arctic continental margins, tens of teragrams of methane per year could be released into the ocean," the scientists said (Westbrook et al. 2009).

About 8 million tons of methane a year has been bubbling from a frozen seabed north of Siberia, according to a March 5, 2010, article in *Science*. Investigators are not sure whether this is a sign of global warming or merely the first detection of methane from stores trapped under permafrost that has been going on for decades. "Subsea permafrost is losing its ability to be an impermeable cap," said study coauthor Natalia Shakhova, a scientist at the University of Fairbanks, Alaska. Shakhova said that the seafloor was believed to be an impermeable barrier. The scientists measured methane levels at 5,000 sites on the East Siberian Arctic Shelf between 2003 and 2008 (Doyle 2010; Heimann 2010, 1211).

Whether the venting was caused by global warming or by natural factors, Shakhova said "no one can answer this question." A future rise in temperatures could accelerate the thaw, however. "It's good that these emissions are documented. But you cannot say they're increasing," said Professor Martin Heimann, an expert at the Max Planck Institute for Biogeochemistry in Germany. "These leaks could have been occurring all the time" since the last Ice Age ended 10,000 years ago, he said (Doyle 2010). Monitoring could resolve whether the venting was "a steadily ongoing phenomenon or signals the start of a more massive release period," according to the scientists, based at United States, Russian, and Swedish research institutions. The release of just a "small fraction of the methane held in [the] East Siberian Arctic Shelf sediments could trigger abrupt climate warming," they wrote (Shakhova et al. 2010).

In the December 8, 2009 issue of the *Proceedings of the National Academy of Sciences,* David Archer of the University of Chicago and colleagues reported a model that suggests that a 3°C warming of the oceans could melt half of existing hydrates. Thus far, the deep ocean has warmed only a tiny fraction of that amount in most places, although one location near Norway's Svalbard Archipelago has warmed 1°C in 30 years. Methane plumes have been reported there. Calculations indicate that rises in atmospheric methane levels thus far are stemming mainly from other sources, such as wetlands in northern latitudes, the Amazon Valley, and Indonesia, as well as human emissions. However, according to Archer, tipping points for submarine hydrates and permafrost will come, but slowly, driven by thermal inertia that will take time to transfer ocean or earth-bound warming to the air (Kerr 2010, 621).

Methane "Hot Spots" Today

Even today, melting permafrost is producing "hot spots" across the Arctic where ejections of methane bubble out so strongly that ice never forms, even in the coldest winter weather. "It could be 10 or 30 liters of methane per day, from one little hole," said Katey Walter Anthony, an ecologist at the University of Alaska Fairbanks, who has observed this phenomenon in Alaska, Greenland, and the Siberian regions of Russia. "And it does that all year. And then you realize there are hundreds of spots like that, and millions of lakes" (Lavelle 2012, 94). Some of this methane comes not only from near-surface mud, but also from deeper geologic formations that were capped by permafrost "and that contain hundreds of times more methane than is in the atmosphere now" (Lavelle 2012, 95). These may be harbingers of the theorized "methane burp."

By 2010, scattered emissions of methane hydrates had been reported in parts of the oceans, but they did not seem to be adding up—yet—to the kind of steady increase that would raise worldwide atmospheric levels markedly. Scientists were detecting destabilization of some clathrates in the vicinity of the Gulf Stream off the eastern coast of the United States. "A changing Gulf Stream has the potential to thaw and convert hundreds of gigatonnes of frozen methane hydrate trapped below the seafloor into methane gas, increasing the risk of slope failure and methane release," wrote Benjamin J. Phrampus and Matthew J. Hornbach in *Nature* (2012). Using thermal models and seismic data, the researchers associated this destabilization with warming water and attempted to gauge what effect these changes may have on stability of methane hydrates. They found that:

> [R]ecent changes in intermediate-depth ocean temperature associated with the Gulf Stream are rapidly destabilizing methane hydrate along a broad swath of the North American margin. The area of active hydrate destabilization covers at least 10,000 square kilometers of the United States eastern margin and occurs in a region prone to kilometer-scale slope failures. Previous hypothetical studies postulated that an increase of 5°C in intermediate-depth ocean temperatures could release enough methane to explain extreme global warming events like the Palaeocene–Eocene thermal maximum (PETM) and trigger widespread ocean acidification. Our analysis suggests that changes in Gulf Stream flow or temperature within the past 5,000 years or so are warming the western North Atlantic margin by up to 8°C and are now triggering the destabilization of 2.5 gigatonnes of methane hydrate (about 0.2 percent of that required to cause the PETM). This destabilization . . . may continue for centuries. It is unlikely that the western North Atlantic margin is the only area experiencing changing ocean currents. Our estimate of 2.5 gigatonnes of destabilizing methane hydrate may therefore represent only a fraction of the methane hydrate currently destabilizing globally. The transport from ocean to atmosphere of any methane released—and thus its impact on climate—remains uncertain. (Phrampus and Hornbach 2012)

Linking Methane Releases and the Carbon Cycle

Writing in *Science*, Kai-Uwe Hinrichs and colleagues Laura Hmelo and Sean Sylva of the Woods Hole Oceanographic Institution provided a direct link between methane reservoirs in coastal marine sediments and the global carbon cycle, an indicator of global warming and cooling (Hinrichs et al. 2003, 1214–1217). Molecular fossils of methane-consuming bacteria found in the Santa Barbara Basin off California deposited during the last glacial period (70,000 to 12,000 years ago) indicate that large quantities of methane were emitted from the seafloor during warmer phases of Earth's climate in the recent past. Preserved molecular remnants found by the Woods Hole team result from bacteria that fed exclusively on methane and indicate that large quantities of this powerful greenhouse gas were present in coastal waters off California. The team studied samples that were deposited 37,000 to 44,000 years ago.

"For the first time, we are able to clearly establish a connection between distinct isotopic depletions in forams and high concentrations of methane in the fossil record," said Hinrichs, an assistant scientist in the Woods Hole Institution's Geology and Geophysics Department. "The large amounts of methane presumably released during one event about 44,000 years ago suggest a mechanism different from those underlying the emissions at warmer periods, i.e. slow decomposition of methane hydrate triggered by warming of bottom waters," Hinrichs continued. "The sudden release of these enormous quantities of methane was probably caused by landslides and melting of the methane hydrate" (Hinrichs et al. 2003, 1214–1217).

Further Reading

Archer, David. 2009. *The Long Thaw: How Humans Are Changing the Next 100,000 Years of Earth's Climate.* Princeton, NJ: Princeton University Press.

Burns, Judith. 2009. "Methane Seeps from Arctic Seabed." BBC News, August 18. Accessed October 9, 2018. http://news.bbc.co.uk/go/pr/fr/-/2/hi/science/nature/8205864.stm.

Doyle, Alister. 2010. "Study Finds Methane Bubbling from Arctic." Reuters in Australian Broadcasting Corp., March 5. Accessed October 9, 2018. http://www.abc.net.au/science/articles/2010/03/05/2837443.htm.

Heimann, Martin. 2010. "How Stable Is the Methane Cycle?" *Science* 327, no. 5970 (March 5): 1211–1212.

Hinrichs, Kai-Uwe, Laura R. Hmelo, and Sean P. Sylva. 2003. "Molecular Fossil Record of Elevated Methane Levels in Late Pleistocene Coastal Waters." *Science* 299, no. 5610 (February 21): 1214–1217.

Kerr, Richard A. 2010. "'Arctic Armageddon' Needs More Science, Less Hype." *Science* 329, no. 5992 (August 6): 620–621.

Lavelle, Marianne. 2012. "Good Gas, Bad Gas." *National Geographic*, December: 90–109.

Pearce, Fred. 2007. *With Speech and Violence: Why Scientists Fear Tipping Points in Climate Change.* Boston: Beacon Press.

Phrampus, Benjamin J., and Matthew J. Hornbach. 2012. "Recent Changes to the Gulf Stream Causing Widespread Gas Hydrate Destabilization." *Nature* 490 (October 25): 527–530.

Shakhova, Natalia, Igor Semiletov, Anatoly Salyuk, Vladimir Yusupov, Denis Kosmach, and Örjan Gustafsson. 2010. "Extensive Methane Venting to the Atmosphere from Sediments of the East Siberian Arctic Shelf." *Science* 327, no. 5970 (March 5): 1246–1250.

Westbrook, Graham K., Kate E. Thatcher, Eelco J. Rohling, Alexander M. Piotrowski, Heiko Pälike, Anne H. Osborne, . . . Alfred Aquilina. 2009. "Escape of Methane Gas from the Seabed along the West Spitsbergen Continental Margin." *Geophysical Research Letters* 36, L15608 (August 6). http://www.doi.org/10.1029/2009GL039191.

Fishing trawlers berthed at Brixham Harbor in Devon, England. The cod fishery in Great Britain has collapsed as this cold water species migrates northward. (Barry Batchelor/PA Images/Getty Images)

UNITED KINGDOM

Waters Warm, Fish Move

Cod, a cold-water fish similar in some ways to salmon, has been declining and migrating northward as habitats warm. In mid-July 2000, the World Wildlife Fund (WWF) placed North Sea cod, a staple of British fish and chips (contributing to a $60 million annual fishery), on its endangered-species list. Cod stocks in the North Sea have been in nearly continuous decline since the early 1970s; they fell below safe biological limits after 1984 (Smith 2001, 7). By 2017, cod in English waters were mainly a memory.

Populations of North Sea cod declined 90 percent in 30 years from 1970 to 2000, according to the World Wildlife Fund (Brown 2000, 3). Cod have been overfished, and their population is falling because they do not breed well in warmer water. The North Sea in the year 2000 was as much as 3°C warmer than it had been in 1970 (Brown 2000, 3). During the 1930s, British fishermen harvested roughly 300,000 tons of cod annually; by 1999, the catch was down to 80,000 tons ("Fished" 2000, 18). Spawning cod in the North Sea fell from about 277,000 English tons during the early 1970s to 54,700 English tons in 2001. Global warming and predation (overfishing) by humans and seals are major factors in the decline of the cod fishery. Another problem advanced by fishermen is dredging and laying of pipes (Smith 2001, 7).

Less Plankton, Fewer Fish

Research by Gregory Beaugrand and colleagues supports the idea that cod are declining in the North Sea not only because of overfishing, but also because larval cod feed on plankton, which also have declined because of rising temperatures and ocean acidity caused by rising levels of carbon dioxide. They have written that "variability in temperature affects larval cod survival; [we conclude] that rising temperature since the mid-1980s has modified the plankton ecosystem in a way that reduces the survival of young cod" (Beaugrand et al. 2003, 661). Such observations have been supported by North Sea fishermen, who have been dragging rolls of silk behind their boats for 70 years to monitor the density of plankton populations. Given the plankton's decline, even a complete ban on cod fishing is unlikely to restore the fishery. Cod have declined not only in population, but also in size. The peak of plankton abundance also now occurs later in the year, after cod larvae experience their greatest need for them. This mismatch means that fewer larval cod develop into adults.

Cod is only one of several fish species that are moving north as waters warm. William W. L. Cheung and colleagues wrote in *Nature* that they had developed an index, the mean temperature of the catch (MTC), calculated "from the average inferred temperature preference of exploited species weighted by their annual catch," which "increased at a rate of 0.19°C per decade between 1970 and 2006, and non-tropical MTC increased at a rate of 0.23°C per decade" (Cheung et al. 2013, 365–368). The index, which covers 52 large marine ecosystems, including many of the Earth's coastal and shelf areas, "shows that ocean warming has already affected global fisheries in the past four decades, highlighting the immediate need to develop adaptation plans to minimize the effect of such warming on the economy and food security of coastal communities, particularly in tropical regions" (Cheung et al. 2013, 365).

"As the world warms, the only way for wildlife species to live in the temperature they prefer is to move their ranges slowly poleward," according to a study published in the *Journal of the Marine Biological Association* (Connor 2002, 12). Fish cannot regulate their body temperature in the water they are in, and therefore, they continue to swim toward water that is a more comfortable temperature. As water temperatures change, the distribution of fish will change as well, reflecting increasing temperatures.

Warming Affects Sea Creatures

Roughly 300 surviving North Atlantic right whales may be threatened by a decline in their main food source, plankton, as a result of shifts in ocean circulation. At the same time, the common dolphin, *Delphinus delphis*, a warm-water species, has been increasing its range, while the cold-water range of the white-beaked dolphin, *Lagenorhynchus albirostris*, is shrinking. Predators are following their prey as species of fish change their latitude or depth in response to a warming climate. Exotic southern fish species, such as red mullet, anchovy, sardine, and poor cod, are now being found in the North Sea. Fish species are unable to regulate their body temperature, and their distribution and abundance are temperature dependent ("Climate Change Dislocates" 2006).

As the North Sea has grown warmer, the Pacific oyster, *Crassostrea gigas*, which was brought to Europe for commercial reasons, has been able to breed in the wild and is now displacing native oysters in the Wadden Sea. Previously, these oysters were unable to survive outside artificial pens ("Climate Change Dislocates" 2006).

Ecological "Meltdown" in the North Sea

Global warming is contributing to changes that some scientists describe as an "ecological meltdown," with devastating implications for fisheries and wildlife. The meltdown begins at the base of the food chain, as increasing sea temperatures reduce plankton populations. The devastation of the plankton then ripples up the food chain as fish stocks and seabird populations decline as well.

Scientists at the Sir Alister Hardy Foundation for Ocean Science in Plymouth, England, which has been monitoring plankton growth in the North Sea for more than 70 years, have said that unprecedented warming of the North Sea has driven plankton hundreds of miles to the north. They have been replaced by smaller, warm-water species that are less nutritious (Sadler and Lean 2003, 12).

Overfishing of cod and other species has played a role, but fish stocks have not recovered after cuts in fishing quotas. The number of salmon returning to British waters is now half of what it was 20 years previously. A decline of plankton stocks is a major factor in this decline.

"A regime shift has taken place, and the whole ecology of the North Sea has changed quite dramatically," said Dr. Chris Reid, the foundation's director.

> We are seeing a collapse in the system as we knew it. Catches of salmon and cod are already down, and we are getting smaller fish. We are seeing visual evidence of climate change on a large-scale ecosystem. We are likely to see even greater warming, with temperatures becoming more like those off the Atlantic coast of Spain or further south, bringing a complete change of ecology. (Sadler and Lean 2003, 12)

According to a report by Richard Sadler and Geoffrey Lean in *The Independent*, research by the British Royal Society for the Protection of Birds (RSPB) has determined that seabird colonies off the Yorkshire coast and the Shetlands during 2003 "suffered their worst breeding season since records began, with many simply abandoning nesting sites" (Sadler and Lean 2003, 12). The seabirds are starving because sand eels are declining. The sand eels feed on plankton, which have declined as water temperatures have risen. This survey concentrated on kittiwakes, one breed of seabirds, but other species that feed on the eels, including puffins and razorbills, also have been seriously affected. Dr. Euan Dunn of the RSPB commented: "We know that sand eel populations fluctuate, and you do get bad years.

But there is a suggestion that we are getting a series of bad years, and that suggests something more sinister is happening" (Sadler and Lean 2003, 12).

Sand eels also comprise a third to half of the North Sea catch, by weight. They have heretofore been caught in huge quantities by Danish factory ships, which turn them into food pellets for pigs and fish. During the summer of 2003, the Danish fleet caught only 300,000 English tons of its 950,000-ton quota, a record low (Sadler and Lean 2003, 12). The situation is "unprecedented in terms of its scale and the number of species it's affecting," said ornithologist Eric Meek of RSPB (Kaiser 2004, 1090).

Further Reading

Beaugrand, Gregory, Keith M. Brander, J. Alistair Lindley, Sami Souissi, and Philip C. Reid. 2003. "Plankton Effect on Cod Recruitment in the North Sea." *Nature* 426 (December 11): 661–664.

Brown, Paul. 2000. "Overfishing and Global Warming Land Cod on Endangered List." *The Guardian* (London), July 20: 3.

Cheung, William W. L., Reg Watson, and Daniel Pauly. 2013. "Signature of Ocean Warming in Global Fisheries Catch." *Nature* 497 (May 16): 365–368.

"Climate Change Dislocates Migratory Animals, Birds." 2006. Environment News Service, November 17. Accessed November 30, 2018. http://www.ens-newswire.com/ens/nov2006/2006-11-17-01.html.

Connor, Steve. 2002. "Strangers in the Seas; Exotic Marine Species Are Turning Up Unexpectedly in the Cold Waters of the North Atlantic." *The Independent*, August 5: 12–13.

"Fished to the Point of Ruin, North Sea Cod Stocks So Low as to Spell Disaster." 2000. *The Herald* (Glasgow, Scotland), November 7: 18.

Healy, Patrick. 2002. "Warming Waters: Lobstermen on Cape Cod Blame Light Hauls on Higher Ocean Temperatures." *Boston Globe*, August 30: B-1.

Kaiser, Jocelyn. 2004. "Reproductive Failure Threatens Bird Colonies on North Sea Coast." *Science* 305 (August 20): 1090.

Sadler, Richard, and Geoffrey Lean. 2003. "North Sea Faces Collapse of Its Ecosystem." *The Independent*, October 19: 12.

Smith, Craig S. 2001. "One Hundred and Fifty Nations Start Groundwork for Global Warming Policies." *New York Times*, January 18: 7.

UNITED STATES

Coastlines Sink as the Ocean Rises

United States coastal land is sinking from southern Maine to Florida, as well as along the Gulf of Mexico; this, coupled with actual sea-level rise, is making apparent sea-level rise much larger than what is being caused by melting ice and thermal expansion of warming water alone. The rate of subsidence, as Justin Gillis wrote in the *New York Times*,

> is fastest in the Chesapeake Bay region, [where] whole island communities that contained hundreds of residents in the 19th century have already disappeared. Holland

Island, where the population peaked at nearly 400 people around 1910, had stores, a school, a baseball team, and scores of homes. But as the water rose and the island eroded, the community had to be abandoned. (Gillis 2014)

Eventually, wrote Gillis, "Just a single, sturdy Victorian house, built in 1888, stood on a remaining spit of land, seeming at high tide to rise from the waters of the bay itself" (Gillis 2014).

Some soils compress and sink over time. The New Jersey coast is sinking in this manner. In addition, removal of subsurface water for human use can cause land to sink. At times, this subsidence expresses itself suddenly and violently when a surface collapses in sinkholes that may swallow cars and even houses. With a modest actual sea-level rise of eight inches by 2050, the apparent rise will be 14 inches at the Battery (on the southern tip of Manhattan Island) and 15 inches on the New Jersey shore, a team from Rutgers University projected in 2013. By 2100, their analysis indicates, there will be an average rise in the global ocean of 28 inches, with 36 inches at the Battery and 39 inches on the New Jersey coastal plain (Miller et al. 2013). In 2009, the New York City Panel on Climate Change issued a prophetic report. "In the coming decades, our coastal city will most likely face more rapidly rising sea levels and warmer temperatures, as well as potentially more droughts and floods, which will all have impacts on New York City's critical infrastructure," said William Solecki, a geographer at Hunter College and a member of the panel (Atlas 2012).

James Hansen, former director of NASA's Goddard Institute for Space Studies, described the problem:

> The East Coast of the United States, including many major cities, is particularly vulnerable, and most of Florida would be under water with a 25-meter sea-level rise. Most of Bangladesh and large areas in China and India also would be under water. . . . The population displaced by a 25-meter sea-level rise, for the population distribution in 2000, would be about 40 million people on the East Coast of the United States and 6 million on the West Coast. More than 200 million people in China occupy the area that would be under water with a sea-level rise of 25 meters. In India it would be about 150 million and in Bangladesh more than 100 million. (Hansen 2006, 22–23)

Effects of sea-level rise would be felt most acutely in times of weather emergencies, according to Hansen: "The effects of a rising sea level would not occur gradually, but rather they would be felt mainly at the time of storms. Thus, for practical purposes, sea-level rise being spread over one or two centuries would be difficult to deal with. It would imply the likelihood of a need to continually rebuild above a transient coastline" (Hansen 2006, 23).

Florida's Sea-Swept Future

Most of Florida's real-estate assets (and tax base) are on or near its coasts, with waterfront condos, businesses, and other homes perilously vulnerable to sea-level rise. Florida is by far the riskiest state in the United States for prospective flooding in a warming world, with almost half of the country's at-risk population. Three-quarters of Florida's 18 million people live in coastal counties, which generate 80 percent of the state's economy. A five-foot sea-level rise would flood 1 million homes in the state. In the metropolitan Miami area, the shoreline will move inland 500 to 2,000 feet for every foot of sea-level rise (Parker 2015, 114–116).

In Miami, Florida, "sunny-day flooding" has become common with lunar high tides. "During high tide one recent afternoon," reported Coral Davenport of the *New York Times*,

> Eliseo Toussaint looked out the window of his Alton Road laundromat and watched bottle-green salt water seep from the gutters, fill the street, and block the entrance to his front door. "This never used to happen," Mr. Toussaint said. "I've owned this place eight years, and now it's all the time." Down the block at an electronics store, it is even worse. Jankel Aleman, a salesman, keeps plastic bags and rubber bands handy to wrap around his feet when he trudges from his car to the store through ever-rising waters. (Davenport 2014)

Even with densely populated land within a few feet of sea level, construction of seawalls is nearly useless because the area is underlain by porous limestone, "allowing the rising seas to soak into the city's foundation, bubble up through pipes and drains, encroach on fresh water supplies, and saturate infrastructure. County governments estimate that the damages could rise to billions or even trillions of dollars" (Davenport 2014). Miami ranks first worldwide among cities by number of residents (4.8 million in 2014) who will risk coastal flooding by 2070, according to a report issued in 2012 by the Organisation for Economic Co-operation and Development.

Donald Trump's Mar-a-Lago Is Going Under

One prominent example of sinking land and rising sea levels involves President Donald J. Trump's Mar-a-Lago estate, which sits roughly three to six feet above the high-tide line on a barrier island at Palm Beach, Florida. It's not simply rising seas that are going to put Trump's prized estate under water. Like most of Florida, Mar-a-Lago sits on land that is slowly sinking, accelerating relative sea-level rise.

Thermal inertia—the amount of sea-level rise already "in the pipeline" due to fossil fuels already burned—guarantees a dunking for Mar-a-Lago. The amount of time all of this will require is subject to some debate. The best estimate may be 50 to 100 years—sooner, perhaps, if the area is struck by one or more major hurricanes during that period. Mar-a-Lago was built about 90 years ago. In 90 more years, it will be below sea level.

Trump tweeted in 2012 that global warming is a hoax concocted by the Chinese to make U.S. industry less competitive. In early December 2016, he said several times that nobody really knows whether climate change is real. He also chose Oklahoma attorney general Scott Pruitt, a staunch denier of climate change, to run the U.S. Environmental Protection Agency, where ignorance of climatic reality is now a litmus test.

Even Trump denied that climate change is a problem, officials in Palm Beach had overhauled 12 pumping stations to vastly increase the amount of seawater that they can pump into the Intracoastal Waterway with a 20-year, $120 million plan that eventually will be able to suck up almost 1 million gallons per minute. "I just deal with the reality that sea levels are rising," said Palm Beach town manager Thomas Bradford. "I don't want to rile people up about it" (Johansen 2017). While the president ignores climate change, brackish water is already bubbling out of the ground near the Trump estate during "king tides," when the sun and moon align.

According to Palm Beach County's online climate-change mapping tool, "The back quarter or so of Mar-a-Lago's verdant, palm-tree-lined grounds would flood if sea levels rise two to three feet" (Johansen 2017). The town also has required higher seawalls around homes built on the adjacent seacoast. Palm Beach County's own internal documents call for a two-foot sea-level rise within the next 40 years.

Suzanne Goldenberg wrote in the *Guardian* that

> the water is already creeping up bridges and advancing on access roads, lawns, and beaches because of sea-level rise. . . . In 30 years, the grounds of Mar-a-Lago could be under at least a foot of water for 210 days a year because of tidal flooding along the Intracoastal Waterway, with the water rising past some of the cottages and bungalows, [an] analysis by Coastal Risk Consulting found. . . . Parts of the estate are already at high risk of flooding under heavy rains and storms, the analysis found. By 2045, the storm surge from even a category-two storm would bring waters crashing over the main swimming pool and up to the main building. (Goldenberg 2017)

Mar-a-Lago is only one example of how rising oceans will affect much of the U.S. Atlantic and Gulf of Mexico coastlines. Within 50 years, much of Florida's coastline will be in peril, along with many large cities, among them New Orleans, Miami, Charleston, Washington, D.C., New York City, and Boston. The West Coast of the United States is less threatened because many of its cities are at least partially elevated. Nor is the West Coast subject to subsiding land that threatens much of the Atlantic and Gulf coasts.

Further Reading

Atlas, James. 2012. "Is This the End?" *New York Times*, November 24. Accessed October 9, 2018. http://www.nytimes.com/2012/11/25/opinion/sunday/is-this-the-end.html.

Davenport, Coral. 2014. "Miami Finds Itself Ankle-Deep in Climate Change Debate," *New York Times*, May 8. Accessed October 9, 2018. http://www.nytimes.com/2014/05/08/us/florida-finds-itself-in-the-eye-of-the-storm-on-climate-change.html.

Gillis, Justin. 2014. "The Flood Next Time." *New York Times*, January 13. Accessed October 9, 2018. http://www.nytimes.com/2014/01/14/science/earth/grappling-with-sea-level-rise-sooner-not-later.html.

Goldenberg, Suzanne. 2017. "Water World: Rising Tides Close in on Trump, the Climate Change Denier." *The Guardian* (U.K.), March 1. Accessed October 9, 2018. https://www.theguardian.com/us-news/2016/jul/06/donald-trump-climate-change-florida-resort.

Hansen, James E. 2006. "Declaration of James E. Hansen." *Green Mountain Chrysler-Plymouth-Dodge-Jeep, et al., Plaintiffs v. Thomas W. Torti, Secretary of the Vermont Agency of Natural Resources, et al., Defendants.* Case Nos. 2:05-CV-302 and 2:05-CV-304, Consolidated. United States District Court for the District of Vermont. August 14.

Johansen, Bruce E. 2017. "Trump's Mar-a-Lago Is Due for a Dunking." *News-Times* (Danbury, CT), June 14. Accessed October 9, 2018. http://www.newstimes.com/opinion /article/Bruce-E-Johansen-Trump-s-Mar-a-Lago-is-due-11217139.php.

Miller, Kenneth G., Robert E. Kopp, Benjamin P. Horton, James V. Browning, and Andrew C. Kemp. 2013. "A Geological Perspective on Sea-Level Rise and its Impacts along the U.S. Mid-Atlantic Coast." *Earth's Future* 1, no. 1 (December 5). Accessed October 9, 2018. http://onlinelibrary.wiley.com/doi/10.1002/2013EF000135/abstract.

Owen, David. 2017. "Mar-a-Lago's Destiny as Trump's Presidential Retreat." *New Yorker,* January 20. Accessed October 9, 2018. http://www.newyorker.com/culture/culture-desk /mar-a-lago-trump-presidential-retreat.

Parker, Laura. 2015. "Treading Water." *National Geographic*, February: 106–128.

Smith, Michael, and Jonathan Levin. 2016. "Trump Rejects Climate Change, but Mar-a-Lago Could Be Lost to the Sea." Bloomberg News, December 16. Accessed October 9, 2018. https://www.bloomberg.com/news/articles/2016-12-16/trump-rejects-climate -change-but-mar-a-lago-could-be-lost-to-the-sea.

Chapter 10: Water Issues

OVERVIEW

We sometimes forget that the substance that is most basic to human life is water in liquid form. Our bodies, like the Earth's surface, are two-thirds water. We can live without oil or products manufactured from it. We can exist for several days without food, but two days without water (especially in desiccating heat) can be fatal.

Cleanliness of water is vital as well. In this section, we will visit Chinese villages whose people have risen up in protest over foul water. Few of us miss water until it is unavailable. Likewise, few of us can live with water in damaging, flooding superabundance.

Changes in the hydrological cycle wrought by rising temperatures are having major effects on weather patterns over land masses worldwide. While a warming atmosphere generally increases the amount of moisture available in the atmosphere, this is not always the case. Enduring droughts also have intensified. These have scorched an entire continent (Australia) and have damaged areas that historically have been wet, such as the Amazon Valley of South America. Always a desert continent, Australia has become even drier and hotter during recent years. The people there have come to realize that climate change is the culprit, but evidence had to convince them.

Climate change is a global issue, ipso facto; all national manifestations cross borders. The effect of atmospheric circulation changes does not stop at any line on a map. Consider, for example, Hadley Cells, atmospheric circulation patterns that strongly influence where air ascends (low pressure) and where it descends (high pressure), with major influences on what parts of the planet are routinely wet and dry.

The Indian Ocean monsoon affects the daily lives of hundreds of millions of people on the Indian subcontinent. In the past, a switch in upper-level winds, causing abnormally heavy rain or drought, has played a role in floods, droughts, and famines that have killed hundreds of thousands of people. With half of India's 1.3 billion people working in agriculture, a drought during monsoon season can have a major effect on millions of peoples' lives. The monsoon rains of the Indian subcontinent and eastern Africa also have become more erratic, playing a role in floods some years and droughts in others. Human practices, including deforestation, also play a role in spreading droughts that turn lands once subject to occasional drought into deserts.

All of these patterns are likely to intensify as temperatures continue to rise in coming years. As with many processes related to climate change, various feedbacks

A dust storm swallows a farm on the U.S. Great Plains during the 1930s Dust Bowl, an environmental disaster that combined overworked farmland soils and severe drought. (NOAA)

come into play that reinforce each other. A rain forest that had been a major "sink" (absorber) of carbon dioxide can become a net source of the gas once it has been subject to drought for an extended period. Areas subject to human development, including deforestation (as in parts of the Amazon Valley) compound this problem. Trees are roughly 50 percent carbon. Alive, they remove carbon dioxide from the atmosphere and replace it with oxygen. Dying and dead trees (and those being burned as fuel) become sources, not absorbers of carbon dioxide and other greenhouse gases.

By 2007, about one-fifth of the human-caused greenhouse gases being released into the atmosphere came from deforestation, during which carbon stored in trees entered the air. Indonesia has been clearing more forests than any other country. In some areas, such as the province of Riau (on the island of Sumatra), more than half the forests have been felled in a decade, many for palm oil plantations.

The Mato Grosso region of the Amazon, formerly rain forest, is becoming a dryland savanna. The destruction of the forest reinforces itself because rain clouds form more easily above moist forests. Deforestation also degrades soil quality, because most of the Amazon's nutrients come from decaying vegetation. Removing the forests removes the nutrients. The deforestation of the Mato Grosso ("Great Forest" in Portuguese) is being aggravated by logging; about 17 percent of this region's forest already has been cleared

Africa has been plagued with expanding deserts, as the Sahara pushes the populations of Morocco, Tunisia, and Algeria northward toward the Mediterranean. At the southern edge of the Sahara, in countries from Senegal and Mauritania in the west to Sudan, Ethiopia, and Somalia in the east, demands of growing human populations and livestock are converting land into desert. Iran also is engaged in a battle with deserts. Mohammad Jarian, who directs Iran's Anti-Desertification Organization, reported in 2002 that sandstorms had buried 124 villages in the southeastern province of Sistan-Baluchistan. Here, as elsewhere, people are losing their livelihoods on lands that have become too hot, too dry—and, by turns, sometimes also too wet—in a world of diverging extremes that are becoming worse.

In this chapter, we'll examine Australia's alternating battles with drought and floods, extreme droughts in Brazil, the Navajo Nation, and other areas of the United States, rising toxic waters in Italy, and the search for clean water in China, India, and Indonesia.

Further Reading

Economy, Elizabeth C. 2013. "China's Water Pollution Crisis." *The Diplomat,* January 22. Accessed October 10, 2018. http://thediplomat.com/2013/01/forget-air-pollution-chi nas-has-a-water-problem/.

Johansen, Bruce E. 2018. "Global Warming Savages India's Poor." *Nebraska Report*, July–August: 8.

AUSTRALIA

Drought and Atmospheric Circulation

Australia's southern two-thirds is a desert continent, so people who live there are accustomed to periodic drought. Few of them were prepared, however, for the epic droughts of the 21st century. By 2010, Australia's extended drought was being called "The Big Dry."

Some of Australia's enduring drought stems from changes in atmospheric circulation compelled by worldwide climate change. Even though warmer air generally holds more moisture, not everyone will see more precipitation in a globally warmed world. Many deserts already are expanding in a worldwide pattern influenced by atmospheric circulation patterns that meteorologists call "Hadley Cells." Near the equator, warm, moist air rises, cools, and unleashes downpours. In the upper troposphere, the air spreads north and southward toward both poles, descending at about 30° north and south latitude and creating deserts. For reasons that are not yet fully understood, as temperatures rise, the Hadley Cells reach further north and south of the equator. Most deserts around the world range between 20° and 40° north and south latitude, precisely where Australia lies.

There is little doubt that Australia has been heating up. Global warming was on stage in 2014, during the Australian Open tennis tournament, when temperatures as high as 111°F (44° C) in Melbourne forced officials to invoke an "extreme heat policy" that shut down play on outside courts. The year 2013 turned out to be

Australia's hottest year on record, as the calendar year started and ended with record heat across much of the continent. Temperatures reached 120.7°F in Moomba, Queensland, and 118° at several other locations. The heat wave was accompanied by intense drought and frequent wildfires that included Australia's agricultural heartland in the Murray-Darling river valley. Fires blazed on all sides of Sydney, Australia's largest city. The heat has worsened as time has elapsed. By the summer of 2016–2017, record highs were notched at 127°F at some locations in the outback.

"A four-month heat wave during the Australian summer culminated in January in bushfires that tore through the eastern and southeastern coasts of the country, where most Australians live," according to a *New York Times* report. The report continued, "Those record-setting temperatures were followed by torrential rains and flooding in the more densely populated states of New South Wales and Queensland that left at least six people dead and caused roughly $2.43 billion in damage along the eastern seaboard" (Siegel 2013). The country's natural cycles of drought and deluge, already extreme compared to much of the world, have become worse as temperatures rise, according to a federal government report titled *The Angry Summer*. Thousands of cattle and sheep died in the fires that whipped through parts of Australia's agricultural heartland, which had expanded and prospered during more tranquil times with irrigation.

"Drought-Proof" Reservoirs No Longer Suffice

During the 1950s and 1960s, Australia built multiyear reservoirs that were supposed to protect its urban areas as well as farming regions against recurring droughts. These gave the country the highest storage capacity per capita in the world. Together with hundreds of miles of irrigation conduits, Australia was said, at the time, to be "drought-proof." However, the post-2000 droughts exceeded the capacity of this system. Melbourne's water storage was 28 percent of capacity by mid-2007; Sydney's was 37 percent, and Perth's, 15 percent. Inflows behind Sydney's dams in 1991–2006 were 71 percent less than their averages from 1948 to 1990 (Pincock 2007, 336).

Perth's first desalination plant was completed in 2006 with a wind farm meant to provide the 24 megawatts required to operate it. Perth soon was drawing 17 percent of its water from that plant. Sydney and Melbourne added desalination plants soon thereafter. Some industries, such as BHP Billiton's copper and uranium mines in the South, also have plans to build their own desalination facilities. The plants require a great deal of energy to operate and produce a salty mush that will render whatever land or water is used for disposal useless for other purposes. Brisbane's government is considering recycling sewage after its main water supply runs dry (Nowak 2007, 11).

Drought Alternates with Deluges

During the summers of 2002 and 2003, wildfires pushed by raging hot, dry winds from Australia's interior seared parts of Canberra, charring hundreds of homes, killing four people, and forcing thousands to flee the area. "I have seen a lot of bushfire scenes in Australia . . . but this is by far the worst," Australia's prime minister, John Howard, said ("Australia Assesses" 2003, 6-A). Flames spread through undergrowth and exploded as they hit oil-filled eucalyptus trees. The 2002–2003 drought knocked 1 percent off Australia's gross domestic product and cost $6.8 billion in exports. It reduced the size of Australia's cattle herd by 5 percent and its sheep flock by 10 percent (Macken 2004, 61). Recovery was impeded during 2004 by continuing drought. Such fires have continued for many years, up to and including 2017.

Even as drought aggravated wildfires in Australia, some parts of the country, especially in the subtropical north, were dealing with deluges. Speaking in Sydney, Tomihiro Taniguchi, vice chairman of the Intergovernmental Panel on Climate Change (IPCC), said that recent Australian flooding and similar weather conditions around the world were further evidence that the impact of global warming was beginning to be felt. While he admitted it was difficult to directly link deluges in New South Wales to global warming, he added: "We can say with high confidence that the likelihood of flooding will increase in the future" (Maynard 2001, 17). The higher temperatures create greater seawater evaporation, which in turn produces more precipitation in coastal areas.

With the advent of a strong La Niña cycle in the Pacific during 2010, Australia's wheat belt's three-year drought broke dramatically. Record-setting precipitation drenched sections of Victoria and New South Wales in August, followed by the wettest September on record in Australia as a whole. By the end of December 2010, 200,000 people had been stranded by floods in parts of northeastern Australia's province of Queensland, in an area the size of France and Germany combined. In Rockhampton, near the coast, the Fitzroy River reached a level not seen since 1918 (Goodman 2010). Rain was falling at four times the average, producing the worst floods in nearly a century. In mid-January 2011, floodwaters were surging toward Brisbane, Australia's third-largest city, with 2 million people. The floodwaters killed 10 people, including five children in Toowoomba and other sections of the Lockyer Valley on January 10, ripping buildings from the ground and sweeping away many cars. On February 2, 2011, flooding was topped off by Cyclone Yasi, the strongest tropical storm to hit Australia in a century, a Category 5 with wind gusts up to 200 miles per hour.

After the floods submerged large parts of Brisbane, the rebuilding effort was compared to that which would follow a war. More than 30,000 homes and businesses were inundated in Australia's northeast at the peak of the flood, as the area remained under water for several days while the Brisbane and Bremer rivers crested. "We now face a reconstruction task of postwar proportions," the state premier, Anna Bligh, said, choking back tears (Belford and Foley 2011). The floods also nearly paralyzed Queensland's suppliers of coking coal, which is used in steel production.

A Chronic Drought?

Drought may become chronic in Australia if modeling done by the Commonwealth Scientific and Industrial Research Organisation (CSIRO) proves accurate.

According to Kevin Hennessey of the agency's Atmospheric Research Climate Impact Group, "There is a consistency between our modeling and the reality of Australia's weather. Our modeling suggests Australia will become warmer and drier in the future as a result of global warming. By 2030, most of Australia will be between 0.5° and 2° warmer and potentially 10 percent drier" (Macken 2004, 61).

Australia's decade-long drought, provoked by climate change and natural variability, also has been aggravated by deforestation, according to Clive McAlpine of the University of Queensland in Brisbane and colleagues, who used a climate model that simulated Australian climatic conditions from the 1950s to 2003, compared to ground conditions before European immigration began in 1788. The drought is most intense in southeast regions of the continent, where less than 10 percent original vegetative cover remains, increasing the number of days with temperatures over 35°C by 300 percent (six to 18 a year) and the number of dry days by a similar ratio, five to 15 ("Land Clearances" 2009).

Six years of drought aggravated by warming temperatures from 2002 to 2008 reduced Australia's rice crop 98 percent, playing a major role in rapidly rising worldwide prices, including a doubling in three months during the first half of 2008. The price rise contributed to food riots as far away as Haiti, where rice shortages "spurred panicked hoarding in Hong Kong and the Philippines and set off violent protests in countries including Cameroon, Egypt, Ethiopia, Haiti, Indonesia, Italy, Ivory Coast, Mauritania, the Philippines, Thailand, Uzbekistan and Yemen" (Bradsher 2008).

Further Reading

"Australia Assesses Fire Damage in Capital." 2003. Associated Press in *Omaha World-Herald*, January 20: A-4.

Belford, Aubrey, and Meraiah Foley. 2011. "Australian Floods Rage through Brisbane." *New York Times*, January 13. Accessed October 10, 2018. http://www.nytimes .com/2011/01/14/world/asia/14australia.html.

Bradsher, Keith. 2008. "A Drought in Australia, a Global Shortage of Rice." *New York Times*, April 17. Accessed October 10, 2018. http://www.nytimes.com/2008/04/17/business /worldbusiness/17warm.html.

Goodman, J. David. 2010. "Australia Floods Show No Signs of Retreating." *New York Times*, December 31. Accessed October 10, 2018. http://www.nytimes.com/2011/01/01/world /asia/01australia.html.

"Land Clearances Turned Up the Heat on Australian Climate." 2009. *New Scientist*, May 16. Accessed September 18, 2018. http://www.newscientist.com/article/mg2022 7084.700-land-clearances-turned-up-the-heat-on-australian-climate.html.

Macken, Julie. 2004. "The Double-Whammy Drought." *Australian Financial Review*, May 4: 61.

Maynard, Roger. 2001. "Climate Change Bringing More Floods to Australia." *Straits Times* (Singapore), March 14: 17.

Nowak, Rachel. 2007. "Australia: The Continent That Ran Dry." *New Scientist*, June 16: 8–11.

Pincock, Steve. 2007. "Climate Politics: Showdown in a Sunburnt Country." *Nature* 450 (November 15): 336–338.

Siegel, Matt. 2013. "Report Blames Climate Change for Extremes in Australia." *New York Times*, March 4. Accessed September 18, 2018. http://www.nytimes.com/2013/03/05/world/asia/australian-government-blames-climate-change-for-angry-summer.html.

BRAZIL

Drought in the Amazon

Changes in the hydrological cycle wrought by rising temperatures are having major effects on weather patterns over land masses worldwide. While a warming atmosphere generally increases the amount of moisture available in the atmosphere, this is not always the case. Devastating droughts also have intensified. These have scorched an entire continent (Australia) and damaged areas that historically have been wet, such as the Amazon Valley of South America.

The swing from deluge to drought for large parts of the Amazon is important over and above the devastation of its own flora and fauna, discussed in this work in chapter 2. It affects worldwide climate patterns, most notably El Niño and La Niña. Prevailing winds over South America reverse as the Amazon dries, from east-west to west-east, promoting floods on South America's West Coast and drought east of the Andes, the reverse of the usual pattern. The pattern tends to reinforce itself as the drought intensifies, promoting changes in global air circulation that also intensify drought in such places as Australia and southern Africa. Thus, a water shortage in São Paulo, Brazil, is related to low reservoirs in Cape Town and Perth.

A drying Amazon forest may "accelerate climate change through carbon losses and changed surface energy balances" (Phillips et al. 2009, 1344). Oliver Phillips and his colleagues used records from several long-term monitoring plots across Amazonia to evaluate the response of the forest to the intense 2005 drought, which they take as a possible analog of future events. The forest in these areas lost biomass, "reversing a large long-term carbon sink, with the greatest impacts observed where the dry season was unusually intense. . . . Amazon forests therefore appear vulnerable to increasing moisture stress, with the potential for large carbon losses to exert feedback on climate change" (Phillips et al. 2009, 1344).

Severe Drought Plagues the Amazon Valley

During 2005, a severe drought spread through the Amazon Valley at the same time that new evidence was being assembled indicating that damage from logging had been more than previously reported. "We think this [additional logging] adds 25 percent more carbon dioxide to the atmosphere" from the Amazon than previously estimated, said Michael Keller, an ecologist with the U.S. Department of Agriculture's Forest Service and coauthor of an Amazon logging inventory published in *Science* (Naik and Samor 2005, A-12; Asner et al. 2005, 480–481).

This study differed from others that measured only the clear-cutting of large forest areas. The study by Gregory Asner and colleagues included these measures of deforestation and added trees cut selectively, while much of the surrounding forest was left standing in five Brazilian states (Mato Grosso, Para, Rondonia, Roraima, and Acre) that account for more than 90 percent of deforestation in the Brazilian Amazon (Asner et al. 2005, 480). In addition, the Amazon Valley's worst drought in about 40 years was causing several tributaries to evaporate, probably contributing even more carbon dioxide via wildfires.

A report of the United Nations Intergovernmental Panel on Climate Change, issued during April 2007, projected continued drying in the Amazon Valley. "By mid-century, increases in temperature and associated decreases in soil water are projected to lead to gradual replacement of tropical forest by savanna in eastern Amazonia," it anticipated, warning that "crop productivity is projected to decrease for even small local temperature increases" in tropical areas, "which would increase risk of hunger" (Rohter 2007).

By 2008, the drought in some areas was the worst since record keeping had begun a century previously. This drought may be only an early indication of a new weather regime in the Amazon Valley, which holds nearly a quarter of the world's fresh liquid water. The Amazon Valley could be caught in a double vise as the world warms, as rising Atlantic Ocean temperatures combine with El Niño events to provoke more frequent droughts. El Niño events tend to reverse the air circulation over the Amazon from east-west to west-east, setting up drying, down-slope winds.

In the Amazon basin, lush forests have become recognized by the government as assets for their value as sources of carbon sequestration as well as harvestable products. Police, called *Forca Verde* ("Green Police"), now pursue men who burn logs into charcoal in ragged kilns, a "gritty, hellish business," as well as deforestation traffickers who sell high-end hardwoods for $1,000 or more per tree (Schapiro 2014, 77, 81). Jennifer Balch wrote in *Nature* that "aircraft have captured the 'breath' of the Amazon forest—carbon emissions over the Amazon basin. The findings raise concerns about the effects of future drought and call for a reassessment of how fire is used in the region" (Balch 2014).

The Kamayurá Lose Their Fish

Members of the Kamayurá tribe, who live in the Amazon's Xingu National Park, say that hotter, drier weather has imperiled the fish that are at the basis of their diet and has played a role in deforestation and wildfires that have threatened their homeland for the first time in memory. "Us old monkeys can take the hunger, but the little ones suffer—they're always asking for fish," said Kotok, the tribe's chief (Rosenthal 2009). Fishermen haul in nearly empty nets while people swim in water that was once rife with piranhas. Some Kamayurá children have been substituting ants for fish that they eat on flatbread made from cassava flour.

The Kamayurás' homeland in the formerly isolated national park has become a forested island in a sea of farms and ranches, while deforestation decreases the amount of rainfall, making rainy seasons less dependable. Elisabeth Rosenthal reported in the *New York Times*:

> Last year, for the first time, the beach on the lake that abuts the village was not covered by water in the rainy season, rendering useless the tribe's method of catching turtles by putting food in holes that would fill up, luring the animals. The tribe's agriculture has suffered, too. For centuries, the Kamayurá planted their summer crops when a certain star appeared on the horizon. "When it appeared, everyone celebrated because it was the sign to start planting cassava since the rain and wind would come," Chief Kotok recalled. But starting seven or eight seasons ago, the star's appearance was no longer followed by rain, an ominous divergence, forcing the tribe to adjust its schedule. It has been an ever-shifting game of trial and error since. Last year, families had to plant their cassava four times—it died in September, October, and November because there was not enough moisture in the ground. It was not until December that the planting took. The corn also failed, said Mapulu, the chief's sister. "It sprouted and withered away," she said. (Rosenthal 2009)

The forest, once too wet to burn, has become dangerously flammable. During 2007, fire burned several thousand acres in the park. "'The whole Xingu was burning—it stung our lungs and our eyes,' Chief Kotok said. 'We had nowhere to escape. We suffered along with the animals'" (Rosenthal 2009).

Further Reading

"Amazon Fire Destroying Rare Forest Home of Uncontacted Tribe." 2015. Survival International Press Release by e-mail, December 1.

Asner, Gregory P., David E. Knapp, Eben N. Broadbent, Paulo J. C. Oliveira, Michael Keller, and Jose N. Silva. 2005. "Selective Logging in the Brazilian Amazon." *Science* 310, no. 5747 (October 21): 480–482.

Balch, Jennifer. 2014. "Atmospheric Science: Drought and Fire Change Sink to Source." *Nature* 506 (February 6): 41–42.

Brahic, Catherine. 2009. "Parts of Amazon Close to Tipping Point." *New Scientist*, March 5. Accessed October 10, 2018. http://www.newscientist.com/article/dn16708-parts-of-amazon-close-to-tipping-point.html.

Joyce, Christopher. 2011. "'Alarming' Amazon Droughts May Have Global Fallout." National Public Radio in NASA Earth Observatory, February 7. Accessed October 10, 2018. http://www.npr.org/2011/02/07/133462608/alarming-amazon-droughts-may-have-global-fallout.

Naik, Gautam, and Geraldo Samor. 2005. "Drought Spotlights Extent of Damage in Amazon Basin." *Wall Street Journal*, October 21: A-12.

Phillips, Oliver L., Aragao, Luiz Eduardo O. C., Lewis, Simon L., Fisher, Joshua B., Lloyd, Jon, Lopez-Gonzalez, Gabriela, …Armando Torres-Lezama. 2009. "Drought Sensitivity of the Amazon Rainforest." *Science* 323, no. 5919: 1344–1347. https://doi.org/10.1126/science.1164033.

Rohter, Larry. 2007. "Brazil, Alarmed, Reconsiders Policy on Climate Change." *New York Times*, July 31. Accessed October 10, 2018. http://www.nytimes.com/2007/07/31/world/americas/31amazon.html.

Romero, Simon. 2012. "Swallowing Rain Forest, Cities Surge in Amazon." *New York Times,* November 24. Accessed October 10, 2018. http://www.nytimes.com/2012/11/25 /world/americas/swallowing-rain-forest-brazilian-cities-surge-in-amazon.html.

Rosenthal, Elisabeth. 2009. "An Amazon Culture Withers as Food Dries Up." *New York Times*, July 25. Accessed October 10, 2018. http://www.nytimes.com/2009/07/25 /science/earth/25tribe.html.

Schapiro, Mark. 2014. *Carbon Shock: A Tale of Risk and Calculus on the Front Lines of the Disrupted Global Economy; How Carbon Is Changing the Cost of Everything*. White River Junction, VT: Chelsea Green Publishing.

CHINA

Killer Water

Water pollution from industrial waste is still a major problem, as is soil contamination. Many Chinese rivers are so seriously polluted that they contain no fish, and by one estimate 78 percent of their water was unfit for human consumption. Chemical runoff from agricultural fertilizer and factories in eastern China also contributes to off-shore dead zones, waters starved of oxygen in which nearly all life has died (Hays 2014). China's three great rivers—the Yangtze, Pearl, and Yellow—are so filthy that it is dangerous to swim or eat fish caught in them (Hays 2014).

Chinese president Xi Jinping said that "the standard that Internet users apply for lake water quality is whether the mayor dares to jump in and swim." Xi was

---

Ocean Acidity and Dead Zones

In addition to the increasing use of nitrogen-based fertilizers and increasing sewage runoff, accelerating emissions of carbon dioxide into the atmosphere (and its absorption by the oceans) are contributing to the rapid expansion of "dead zones"—desolate regions where pollution-fed algae deprive other marine life of oxygen. The number of such areas worldwide had increased to about 200 by 2008, a 34 percent rise in two years. Given carbon dioxide's long life in the atmosphere, this trend could persist for centuries or longer, even after greenhouse-gas emissions end. A 100,000-year computer simulation indicates that severe ocean oxygen depletion could last for thousands of years. Dead zones that cover about 2 percent of ocean surface today could expand to 20 percent by that time (Shaffer et al. 2009, 105). In the simulation, business-as-usual carbon emissions increase to present-day rates until 2100, as atmospheric levels increase 400 percent and climate warms by about 5°C.

Further Reading

Shaffer, Gary, Steffen Malskær Olsen, and Jens Olaf Pepke Pedersen. 2009. "Long-Term Ocean Oxygen Depletion in Response to Carbon Dioxide Emissions from Fossil Fuels." *Nature Geoscience* 2 (January): 105–109.

referring to news that residents in Zhejiang Province had challenged local government officials to swim in polluted waterways (Schmitz 2016). Beijing has been funding sustainable irrigation practices and requiring mandatory water recycling systems in new factories. China's prime minister, Li Keqiang, said, "Poverty and backwardness in the midst of clear waters and verdant mountains is no good, nor is it to have prosperity and wealth while the environment deteriorates" (Schmitz 2016).

Water Pollution Ubiquitous in Eastern China

Water pollution is ubiquitous in much of China's thickly populated eastern and southern coastlines. Liang Bo works at the Shenzhen Mangrove Wetlands and Conservation Foundation, north of Hong Kong, and monitors local mangrove swamps, where, according to a report in the *New York Times*, garbage washes up on the rocks from murky, gray water in a bay that is part of the Pearl River delta.

> "It's worse at low tide," Liang said. "You really see how filthy it has become." Because of all the landfill and new development, she said, water no longer flows in and out of the bay as it once did. So garbage gets trapped, stagnation gets worse, and fish are killed off. That scenario is repeated throughout the delta, where small rivers and tributaries have been filled in and paved over to make way for new highways, office parks, and housing developments, adding to the strains on an already inadequate system of drains and sewers. "The sea, the wetlands and mangroves used to be part of people's lives here," Liang pointed out. "But most of the people who live here now weren't around when the mangroves were still here. They see this park, which makes us more vulnerable to rising seas and typhoons, as they do all the tall buildings and highways. They equate it with progress." (Kimmelman and Haner 2017)

According to another report, "up to 40 percent of China's rivers were seriously polluted" and "20 percent were so polluted their water quality was rated too toxic even to come into contact with" (Economy 2013). An estimated 10,000 petrochemical plants have been sited along the Yangtze River and 4,000 along the Yellow—and these are not even among China's seven most polluted major waterways. The Chinese government has reported that 60,000 premature deaths in China result from water pollution (Economy 2013). A Chinese newspaper, the *Southern Weekly*, said that a married couple, both known as water experts in Beijing, said that they had avoided drinking tap water for 20 years, having observed water quality deteriorate significantly at an accelerating rate.

Half of Chinese Lack Safe Drinking Water

China was ranked 116th of 132 countries on water quantity for human consumption, including industrial, agricultural, and household uses, according to Yale University's 2012 Environmental Performance Index.

Jonathan Kaiman wrote in *The Guardian,* "The head of China's ministry of water resources said in 2012 that up to 40 percent of the country's rivers are 'seriously polluted' . . . China's lakes are often affected by pollution-induced algae blooms, causing the surface of the water to turn a bright iridescent green. Yet even greater threats may lurk underground. A recent government study found that groundwater in 90 percent of China's cities is contaminated, most of it severely." (Hays 2014)

According to an account by teacher Jeffrey Hays:

About one third of the industrial wastewater and more than 90 percent of household sewage in China is released into rivers and lakes without being treated. Nearly 80 percent of China's cities (278 of them) have no sewage treatment facilities and few have plans to build any. Underground water supplies in 90 percent of Chinese cities are contaminated. Water shortages and water pollution in China are such a problem that the World Bank warns of "catastrophic consequences for future generations." . . . Nearly two thirds of China's rural population—more than 500 million people—use water contaminated by human and industrial waste. (Hays 2014)

Hays continued:

Drinking water in China contains dangerous levels of arsenic, fluorine, and sulfates; an estimated 980 million of China's 1.3 billion people drink water every day that is partly polluted. More than 600 million Chinese drink water contaminated with human or animal wastes and 20 million people drink well water contaminated with high levels of radiation. A large amount of arsenic-tainted water has been discovered. China's high rates of liver, stomach, and esophageal cancer have been linked to water pollution. (Hays 2014)

After several large protests, the central government pledged to spend more than 70 billion yuan (US$11.25 billion) "to implement a clean water action plan, strengthen the protection of drinking water sources, and prevent water pollution in key river basins" ("Nine Biggest" 2014).

Hays wrote:

One-sixth of China's population is threatened by seriously polluted water. One study found that eight of 10 Chinese coastal cities discharge excessive amounts of sewage and pollutants into the sea, often near coastal resorts and sea farming areas. Water pollution is especially bad along the coastal manufacturing belt. Despite the closure of thousands of paper mills, breweries, chemical factories, and other potential sources of contamination, the water quality along a third of the waterway falls far below even the modest standards that the government requires. Most of China's rural areas have no system in place to treat wastewater. (Hays 2014)

Reports have surfaced, some of them on clandestine social media, that "many rivers are filled with garbage, heavy metals, and factory chemicals. There have been devastating fish kills caused by the release of chemicals into the Haozhongou River in Anhui Province and Min Jiang River in Sichuan Province" (Hays 2014).

Some villages have reported clusters of cancers induced by water pollution. In one of many examples, Jennifer Duggan reported in Great Britain's *Guardian* that "in September 2013, Xinhua reported on a village in Henan where the groundwater has been badly polluted. The news agency said that locals claimed the deaths of 48 villagers from cancer are linked to the pollution" (Duggan 2013). As many as 100 such villages were located by 2012 along the Huai River and its tributaries in Henan Province alone.

"Cancer Villages"

There are said to be around 100 cancer villages, especially along the Shaying River. Death rates on the Huai River are 30 percent higher than the national average. In 1995, the government declared that water from a Huai tributary was undrinkable and the water supply for 1 million people was cut off. The military had to truck in water for a month until 1,111 paper mills and 413 other industrial plants on the river were shut down.

Rob Schmitz reported from one such "cancer village" in 2016:

> Three years ago, I sat beside the bed of Zhang Runxiang as the 42-year-old lay dying of uterine cancer. She and dozens of others in Liuchong Village, a tiny hamlet in central China's Hubei Province, have petitioned China's government and complained on state-run television that they've been poisoned by Dasheng chemical, a company that manufactures phosphate fertilizer on a hill above the village. Two days after I visited Zhang, she died. (Schmitz 2016)

The fatal pollutant in this case was phosphogypsum, a by-product of fertilizer manufacture, which is radioactive (containing arsenic, uranium, and radium) and therefore carcinogenic. A 200-foot high waste pile of phosphogypsum looms over the village, as radioactive dust fills the air on windy days.

The Huai river in Anhui Province, west of Shanghai, has become so polluted that all fish have died. Some places have water that is too toxic to touch. Furthermore, "The Qingshui River, a tributary of the Huai whose names means 'clear water,' has turned black with trails of yellow foam from pollution from small mines that have opened up to meet the demand for magnesium, molybdenum, and vanadium used in the booming steel industry" (Hays 2014).

Between 2010 and 2014, China experienced a wave of toxic chemical accidents that leached into water supplies. Incidents continued after that, but at a slower rate. For example, on April 11, 2014, according to the Chinese newspaper *Global Times*, "Excessive levels of benzene, a carcinogenic chemical, almost 20 times the national limit in tap water affected more than 2.4 million people in downtown Lanzhou, [in] Northwest China's Gansu Province" ("Nine Biggest" 2014). More than 34 tons of the chemical was absorbed into the ground and into an underground water duct.

On March 9, 2012, Peter Smith wrote in the *Times* (London):

> Beyond the brick cottages of Tongxin runs Lou Xia Bang, once the soul of the farming village and a river where, until the digital revolution, children swam and mothers washed rice. Today it flows black: a chemical mess heavy with the stench of China's

high-tech industry—the hidden companion of the world's most famous electronics brands [including Apple Computer] and a reason the world gets its gadgets on the cheap. (Hays 2014)

Dead Pigs Afloat

Reports of dead pigs floating in some rivers and others running red as blood surfaced frequently, often with lurid photographs. The "blood-red" river was Maozhou in Guangdong Province (near Hong Kong), where the Fuan textile factory had spilled dyes from the manufacture of T-shirts and other clothing. Rapid growth of industry in China, along with lax environmental enforcement of overused farm chemicals, contaminated waterways and the countryside with lead, cadmium, and pesticides, among other toxic chemicals. Local news media contain photographs that demonstrate how severe China's water pollution has become. Half of China's rivers and lakes are severely polluted by industrial effluent and chemical fertilizers (Sim 2017).

In Shanghai, during March and April of 2013, according to the *Global Times*, almost 16,000

> dead pigs [were] found in the Huangpu River, which supplies over a fifth of Shanghai's drinking water. . . . More than 10,000 carcasses have been found in the Huangpu River in Shanghai in March 2013, and another 4,600 in Jiaxing in Zhejiang Province, 100 kilometers southwest. Despite authorities' claims that the river water was not contaminated and Shanghai's tap water was safe, laboratory tests found porcine circovirus in one water sample. The virus can spread among pigs, though not to human beings. ("Panic over Dead Pigs" 2013)

Authorities were unable to find any reason for the pigs' appearance, which provoked many jokes on the Internet because "there is an old saying in Chinese culture that the appearance of a fat pig at the front door augurs abundance and good fortune. The sight of a bloated one floating dead down the nearest river portends something else entirely," reported the *South China Morning Post* (Garland 2013).

Authorities did assure residents that none of the river-borne dead pigs had shown up in local markets and that "Yu Kangzhen, China's chief veterinarian with the Chinese Ministry of Agriculture (MOA) . . . said [that] water in the Huangpu River, where a great number of dead pigs were found, meets national standards for drinking water" ("Nine Biggest" 2014). Tags affixed to their ears later were traced to farmers in nearby Jiaxing, Zhejiang Province. More dead pigs surfaced in the same river a month later.

Satiric social media posts indicated that many residents did not trust official assurances that the dead pigs posed no threat to their health: "Public reaction toward the scandal has turned from panic and anger to satire during the first week. While studious Shanghai residents carried out their own water-quality tests and posted results online to show tap water was dirty and unsafe, many others joked about the scandal" ("Nine Biggest" 2014). Beginning about 2008, China began to spend enormous amounts of money to clean up some of its rivers, only to see pollution return:

> China wanted to reduce water pollution discharges by 10 percent between 2008 and 2010. More than $8 billion was spent on cleaning up the Huai River basin in Henan and Anhui Province in the 1980s and 1990s. Great progress was made. In the mid-1990s, the cleanup was heralded as a great success and much of the work stopped. By the mid-2000s, the river was polluted again, in many cases worse than it was before. (Hays 2014)

Further Reading

Duggan, Jennifer. 2013. "China Hit by Another Airpocalypse as Air Pollution Cancer Link Confirmed." *The Guardian,* October 24. Accessed November 30, 2018. https://www.theguardian.com/environment/chinas-choice/2013/oct/24/china-airpocalypse-harbin-air-pollution-cancer.

Economy, Elizabeth C. 2013. "China's Water Pollution Crisis." *The Diplomat,* January 22. Accessed October 11, 2018. http://thediplomat.com/2013/01/forget-air-pollution-chinas-has-a-water-problem/.

Garland, Matthew. 2013. "China's Deadly Water Problem." *South China Morning Post,* March 26. Accessed October 11, 2018. http://www.scmp.com/comment/insight-opinion/article/1199574/chinas-deadly-water-problem.

Hays, Jeffery. 2014. "Water Pollution in China." Facts and Details.com. Accessed October 10, 2018. http://factsanddetails.com/china/cat10/sub66/item391.html.

Kimmelman, Michael, and Josh Haner. 2017. "Rising Waters Threaten China's Rising Cities." *New York Times,* April 7. Accessed October 11, 2018. https://www.nytimes.com/interactive/2017/04/07/world/asia/climate-change-china.html.

"Nine Biggest Water Pollution Disasters in China (Since 2010)." 2014. *Global Times,* April 15. Accessed October 11, 2018. http://www.globaltimes.cn/content/854711.shtml.

"Panic over Dead Pigs in Shanghai River Prompts Satire." 2014. *Global Times.* March 14, Accessed November 30, 2018. http://www.globaltimes.cn/content/768137.shtml.

Schmitz, Rob. 2016. "Life and Death Inside a Chinese 'Cancer Village.'" *The Marketplace,* May 19. October 11, 2018. http://www.marketplace.org/2016/05/03/world/life-and-death-inside-chinese-cancer-village.

Sim, David. 2017. "Dead Pigs and Rivers of Blood: Shocking Photos of Water Pollution in China." *IBTimes* (United Kingdom), February 2. Accessed October 11, 2018. http://www.ibtimes.co.uk/dead-pigs-rivers-blood-shocking-photos-water-pollution-china-1459222.

INDIA

Too Many People, Too Little Water

India, with one of the hottest climates on Earth, has 1.3 billion people packed into an area roughly one-third the size of the United States. With increasing population, rising levels of affluence, and a short monsoon that supplies much of its moisture made more erratic by global warming, periodic shortages of water have become chronic. India has only 4 percent of the world's fresh water but 16 percent of its people.

More than 600 million Indian people depend on agriculture for their livelihood; about 60 percent of cultivated land is not irrigated and so depends on an adequate, timely monsoon between June and September, when about 80 percent of India's rain occurs.

Underground Water Pumped at Unsustainable Rates

One report anticipates that India will need to double its capacity to generate water by the year 2030 to meet the needs of a growing population (Harris 2013). Many major cities, including Delhi, Mumbai, Chennai, and Hyderabad, are drawing down groundwater at unsustainable rates that will be exhausted well before 2030. Judges have been ordering halts in new construction until builders can supply evidence that they will be using recycled water.

The National Aeronautics and Space Administration has conducted satellite surveys of India that show a sharp decline in groundwater levels in irrigated fields of rice, wheat, and barley under farm fields in the densely populated Ganges Valley. Wells have been dug deeper, making the problem worse. At the same time, temperatures soared to the highest levels in recorded history, intensifying evaporation.

Rapid depletion of groundwater is accelerating as farmers using subsidized electricity pump up more groundwater than China and the United States combined. In 50 years, roughly 1965 to 2015, the number of boreholes increased from fewer than 100,000 to 20 million ("*The Economist* Explains" 2016). With 85 percent of India's drinking water coming from aquifers, their decline is a source of deep anxiety. "Water-starved regions often cultivate water-hungry crops like paddy [rice], cotton, and sugarcane. Punjab in the north and Tamil Nadu and Karnataka in the south continue to squabble over the ownership of rivers. The problem is not lack of adequate water, but its reckless overuse. China, with a larger population, uses 28% less fresh water than India," according to an analysis in *The Economist* (2016).

Water Scarcity and Overpopulation

An analysis by the Water Project asserted that "water scarcity in India is expected to worsen as the overall population is expected to increase to 1.6 billion by year 2050. To that end, global water scarcity is expected to become a leading cause of national political conflict in the future, and the prognosis for India is no different" (Snyder 2017).

Even the village of Cherrapuni, India, in the Himalayas, which receives nearly 40 feet of rain a year (one of the wettest places on the planet), suffers periodic drought during the dry season from November through May. Some villagers must walk several miles to fill jugs with water for drinking, cooking, and bathing. Some of them make the trip four or five times a day. Water taps in Shillong, the capital of Meghalaya State (one of the few places in the area with piped-in water), supply it for only a few hours each day. What comes out of the taps is usually not drinkable, a widespread problem throughout India. About 600,000 Indian children in an average year die from diarrhea or pneumonia caused or aggravated by tainted water and poor sanitation, according to the United Nations Children's Fund (UNICEF) (Harris 2013).

Two-thirds of India's people live in rural areas, where most water is contaminated. According to the World Bank, about one-fifth of communicable diseases in India in an average year are caused at least in part by unsafe water and the lack of sanitation. Almost half of India's population, 569 million people, lack access to toilets and practice open defecation ("India's Water Crisis" 2017). Water in India is so frequently contaminated that anyone who drinks it (or even eats raw food washed in it) is taking a health risk for "Delhi belly" or worse. Like the ubiquitous mosquitoes that spread malaria and dengue fever, the tap water is best avoided, even for brushing teeth.

Urban water supplies often are plagued by corruption and theft. "Water plants in New Delhi, for instance, generate far more water per customer than many cities in Europe, but taps in the city operate on average just three hours a day because 30 percent to 70 percent of the water is lost to leaky pipes and theft," reported Gardiner Harris in the *New York Times* (Harris 2013). Sewage treatment is rudimentary in much of India, adding to problems with water contamination.

> Even as towns and cities increase water supplies, most fail to build the far more expensive infrastructure to treat sewage. So as families connect their homes to new water lines and build toilets, many flush the resulting untreated sewage into the nearest creek, making many of the less sophisticated water systems that much more dangerous. "As drinking water reaches more households, all the resulting sewage has become a huge problem," said Tatiana Gallego-Lizon, a principal urban development specialist at the Asian Development Bank. (Harris 2013)

Drought, Deluges, and Water Scarcity

When the monsoon rains fail to arrive on time or precipitation is erratic, India's water scarcity becomes more intense. Climate change has been playing a role in delaying the seasonal rains—and, in some cases, producing deluges alternating with droughts. Chennai, for example, had record rains and floods late in 2015, in the midst of drought that afflicted much of the country. Mumbai experienced 37 inches of rain in one day.

In 2016, roughly 330 million people—more than one quarter of India's population—suffered drought conditions. The drought

> turned vast areas of the subcontinent into a dust bowl, withering crops and forcing farmers from their lands. Coal-fired power plants—the major source of India's electricity—have had to suspend output because there is not enough water in nearby rivers to generate steam. Armed guards are being posted at dams to prevent desperate farmers from stealing water. ("India's Water" 2016)

Several farmers, faced with ruin, committed suicide.

Three Feet of Rain in One Day

India, with its annual dry season that usually alternates with heavy monsoon rains, has adapted to a drought-deluge cycle. About 80 percent of India's precipitation falls between June and September during an average year, so heavy rain in Mumbai (Bombay) during late July is hardly unusual. On July 26 and 27, 2005, however, 37.1 inches of rain fell in Mumbai during 24 hours, the heaviest on record for an Indian city in one diurnal cycle. The deluge contributed to more than 1,000 deaths in and near Mumbai and surrounding Maharashtra State ("Record Rainfall" 2005, A-12). Metropolitan Mumbai, a city of 17 million, was shut down by the rain, and several people drowned in their cars. Mass transit and telephone services stopped. Other people were electrocuted by wires falling onto flooded streets. Tens of thousands of animals also died. Two years later, some of the worst monsoon rains in memory killed at least 2,800 people in India, Bangladesh, Nepal, and Pakistan during 2007. Several million people lost their homes.

Further Reading

"Record Rainfall Floods India." 2005. *New York Times*, July 28: A-12.

Power Cut for Lack of Water

On March 11, 2016, at one of India's largest coal-fired power stations, on the banks of the Ganges, the water level in an adjoining canal sank so low that operations had to be suspended. (Water is used to run turbines and for cooling at the plant.) India's power grid was snarled. A day later, housing for plant workers ran out of water. The plant was closed for 10 days, the first lengthy outage in its history. "'Never before have we shut down the plant because of a shortage of water,' said Milan Kumar, a senior plant official. 'We are being told by the authorities that water levels in the river have receded and that they can do very little'" (Biswas 2016).

At the power plant, Kumar said he was "afraid that this can happen again. We are being told that water levels in the Ganges have declined by a fourth. Being located on the banks of one of the world's largest rivers, we never thought we would face a scarcity of water. The unthinkable is happening" (Biswas 2016). Downriver, sand bars emerged from the Ganges and ferry service was halted. The 1,533-mile-long Ganges supplies water for one-quarter of India's people, about 300 million. Its water is supplied mainly by melting snow and ice from the Himalayas, which has been declining due to rising temperatures, and sometimes by the increasingly erratic nature of the annual monsoon rains.

Emmanuel Theophilus and his son, Theo, kayaked on the Ganges during an 87-day trip along several of India's rivers in 2015. They told BBC World News that they asked fishermen and people living on the river what had changed most about it. "All of them said there had been a reduction in water levels over the years. Also, when we were sailing on the Ganges, we did not find a single turtle. The river was so dirty that it stank. There were effluents, sewage, and dead bodies floating," said Theophilus (Biswas 2016). Before the 2016 drought ended with monsoon rains, India's reservoirs contained only 29 percent of capacity, according to the Central Water Commission.

The BBC World News reported that a fisherman on the Ganges, Balai Haldar, scanned his dwindling catch of prawns. "The river has very little water these days. It is also running out of fish. Tube wells in our village have run out of water," he says. "There's too much of uncertainty. People in our villages have moved to the cities to look for work" (Biswas 2016).

Further Reading

Biswas, Soutik. 2016. "Is India Facing Its Worst-ever Water Crisis?" British Broadcasting Corp. (BBC) World News, March 27. Accessed October 11, 2018. http://www.bbc .com/news/world-asia-india-35888535.

"*The Economist* Explains: Why India Has a Water Crisis." *The Economist* (U.K.), May 25, 2016. Accessed October 11, 2018. http://www.economist.com/blogs/economist-ex plains/2016/05/economist-explains-11.

Harris, Gardiner. 2013. "Rains or Not, India Is Falling Short on Drinkable Water." *New York Times*, March 12. Accessed October 11, 2018. http://www.nytimes.com/2013/03/13 /world/asia/rains-or-not-india-is-falling-short-on-drinkable-water.html.

"India's Water Crisis." 2016. Editorial in the *New York Times*, May 3. Accessed October 11, 2018. https://www.nytimes.com/2016/05/04/opinion/indias-water-crisis.html.

"India's Water Crisis." 2017. Water.org. Accessed October 11, 2018. http://water.org /country/india/.

Snyder, Shannyn. 2017. "Water in Crisis—India." The Water Project. Accessed November 30, 2018. https://thewaterproject.org/water-crisis/water-in-crisis-india.

INDONESIA

Gold Mining and Water Supplies in East Kalimantan

Since 1992, Rio Tinto's Kelian Gold Mine in the East Kalimantan (Borneo) has produced more than 400,000 ounces of gold per year using the cyanide heap-leaching process, producing cyanide-laced tailings that are ruining the lives of local people, whose water is laced with several poisons. The tailings are held in a dam and treated in a polishing pond near the Kelian River. Water from the polishing pond pours into the river through an outlet. The company claims that the water is clean, while the community says that people cannot drink or bathe in the water because it causes skin lesions and stomach ailments, at the very least ("Rio Tinto Kelian" 2000).

Rio Tinto and PT KEM signed a contract with the Indonesian government in 1985 and began exploration for gold in 1987. Since that time, the community has alleged numerous human-rights violations, destruction of property, and pollution of their lands and rivers. The community began the blockade in frustration over Rio Tinto's continued efforts to stall negotiations, as well as its violations of conditions of the negotiations.

Workers scan a small part of Rio Tinto's former Kelian gold mine in Borneo, which closed in 2005 and left contaminated water over a broad area. (Fairfax Media via Getty Images)

The Water Is Too Dangerous to Drink

This mine pollutes a tributary of the Mahakam River on which thousands of indigenous people rely for water. "The Kelian mine has consistently manipulated environmental reports and has wiped out without recognition the local community's traditional mining rights," asserted Mohammed Ramli, speaking for the Indonesian Mining Advocacy Network (JATAM), a group that represents affected communities, according to an analysis by the Mineral Policy Institute ("Rio Tinto's Shame" 2000).

The East Kalimantan Kelian gold mine, which is 80 percent owned by international mining giant Rio Tinto, has been causing serious pollution problems that endanger the local community's health, according to accounts from the scene. "Locals suffer from skin rashes when they bathe in the river. They can no longer catch the fish they rely upon as a protein source, and the water is so contaminated with insufficiently treated mine wastes that it's too dangerous to drink," said Ramli. In addition, the first five kilometers of the river near the mine has been artificially diverted without taking into consideration biological effects, leaving the previous watercourse devoid of life ("Rio Tinto's Shame" 2000). Community protests of the mine's pollution have been met with fire from police.

Drinking Poison

Rio Tinto's 1997 Health, Safety, and Environment Report described a massive acid-drainage problem at the Kelian mine. Levels of manganese in water discharged from the mine during 1997 averaged 800 micrograms per liter. This level would not be legal in European or North American drinking water; it also exceeds the World Health Organization's recommended limits of 100 to 500 micrograms per liter. On nine occasions in 1997 (and 105 occasions in 1996), manganese levels were more than 200 times the amount permitted in drinking water under European Union Directives (50 micrograms per liter). Rio Tinto's research unit pledged to investigate methods of reducing manganese pollution but seemed to be in no hurry. The company said that a new method currently under trial would not be implemented at Kelian until 1999 at the earliest ("Rio Tinto's Environmental" 1999).

Wastewater from the same mine also contained more than 500 kilograms of cyanide in 1997. While this level was about half the cyanide emissions for the previous year, the Kelian mine still was responsible for the worst cyanide pollution levels of any Rio Tinto gold or copper mine. (Cyanide compounds are used to extract gold from ore.) Rio Tinto implies that high cyanide levels are not a problem because "any residual-free cyanide breaks down rapidly in the presence of sunlight and does not persist in the environment" ("Rio Tinto's Environmental" 1999).

Gold or Water?

The Kelian mine also releases large amounts of suspended solids into the river. These are fine particles of soil and rock produced during the processing of ore. At 1,600 tons, the amount of suspended solids in the water discharged by PT KEM is the second highest of all Rio Tinto's operations worldwide. In 1996, levels of suspended solids were even higher—at 4,700 tons—as PT KEM diverted part of the River Kelian. Nevertheless, Rio Tinto puts much of the blame for the high turbidity of the river water on the operations of small-scale community miners ("Rio Tinto's Environmental" 1999).

PT KEM takes more than 6 million cubic meters of fresh water per year from the Kelian River for its mining operations. Only about 4 million cubic meters is recycled within the mine. Wastewater containing high levels of manganese, cyanide compounds, and mud has been discharged into the river.

Because it has polluted local drinking-water supplies, PT KEM has been required to provide drinking water for the local population since the start of its operations in the early 1990s. However, not all of the indigenous people or more recent settlers have access to piped drinking water. Water from the Kelian River is used for all other household and agricultural needs, including bathing, laundry, and preparing food, regardless of its pollution ("Rio Tinto's Environmental" 1999).

Local indigenous peoples also have questioned Rio Tinto's assertions that it has rehabilitated more than 500 hectares of forest that had been cleared in connection with mine operations. Instead, say local peoples, "rehabilitation" has consisted largely of "planting non-native tree species several hundred kilometers away in the Bukit Suharto Park—much of which went up in flames in . . . forest fires [during 1999]" ("Rio Tinto's Environmental" 1999).

Mine Operations Blockaded

Several thousand Dayak people blockaded operations at the Kelian mine in East Kalimantan, Indonesia, after a breakdown in negotiations between them and Rio Tinto. The blockade began on April 29, 2000, and lasted 10 days, forcing the company to maintain only minimal operations as its stockholders met in London. *Drillbits and Tailings* reported that local police arrested a dozen community representatives and held them overnight in an attempt to force them to lift the blockade. Rio Tinto, which owns 90 percent of its Indonesian subsidiary PT KEM, and the mine have stalled negotiations with the community meant to address concerns over compensation for land despoiled by mining ("Rio Tinto Kelian" 2000).

According to a statement from the community read at Rio Tinto's Annual General Meeting in Britain:

> In the name of the Kelian community of West Kutai district, East Kalimantan, Indonesia, we state that PT Kelian Equatorial Mining (PT KEM) has not been genuinely committed to settling the issues and demands raised by the people. The company has only paid lip service to various activities—community development projects, recruitment of local workers, environmental management, and mine closure plans—as a form of propaganda. ("Rio Tinto Kelian" 2000)

Toxic Tailings Flood Farm Fields

During 1998, heavy monsoon rains caused toxic tailings to overflow in Rapak Lama Village, Marangkayu subdistrict. The flood mixed toxic tailings and chemicals from Unocal's processing plant. Another similar flood occurred on February 11, 2000. The floods caused the company's tailing pipe to overflow, washing its toxic contents into local rice fields, which contaminated more than 400 hectares of indigenous peoples' rice fields, wiping out many villages' rice crops. According to an on-the-scene report, local people "noticed milky-white water with foams and brownish blobs on the surface around the plantation, in its irrigation gutters, and the mouth of the pipe close to Unocal factory fence" ("Unocal Tailing" 2000). Company officials took water samples but did little else to stem the flow of putrid water that was destroying the fields.

"In the evening," according to a report by local people in *Minergy News*, "Unocal workers removed the brownish blobs on the surface of the water, yet the water remained milky white with foams" ("Unocal Tailing" 2000). Later, the head of Marangkayu District Police (Kapolsek Marangkayu) and one of Unocal's security personnel visited the mouth of the pipe where the tailing-carrying water emerged. They were heard to have said, "If only the security had covered the . . . pipe-mouth with sand earlier in the morning, none of the locals would have been able to discover them" ("Unocal Tailing" 2000). Samples were again taken, but local people were not informed of the results.

A report in *Minergy News* described the damage to the local environment:

> [Many] hectares of rice fields, fish/shrimp embankments, and plantations were severely contaminated by the waste of the oil-processing plants. Tailings were

dumped in a 100-meter distance from the beach, causing the local fishing people to lose their livelihood since their daily catch smell of oil and are unfit for human consumption. Their previously white-sand beach is now brown, and the air is filled with the suffocating stench of crude oil. ("Unocal Tailing" 2000)

The Indonesian government's National Human Rights Commission investigated allegations of human-rights abuses by Rio Tinto, publishing a report during 2000 that documented "egregious violations" ("Rio Tinto Kelian" 2000). According to *Drillbits and Tailings*,

> The report revealed that the Indonesian military and company security forcibly evicted traditional gold miners, burned down their villages, and arrested and detained protestors. PT KEM employees have also been named in a number of incidents of sexual harassment, rape, and violence against women between 1987 and 1997. Local people have lost homes, lands, gardens, fruit trees, forest resources, family graves, and the right to mine for gold in the river. ("Rio Tinto Kelian" 2000)

In a statement addressed to Rio Tinto shareholders at their 2000 annual meeting, the Kelian community demanded fair compensation for land, crops, and property; restoration of small-scale mining rights; remediation of contamination of the rivers by cyanide, metals; and dust; legal action against PT KEM staff and local authorities for human rights violations; and genuine community development.

Further Reading

Abrash, Abigail. 2001. "The Amungme, Kamoro, and Freeport: How Indigenous Papuans Have Resisted the World's Largest Gold and Copper Mine." *Cultural Survival Quarterly* 25, no. 1 (Spring): 38–43.

"Freeport-McMoRan Divests of Papua Operation." 2018. Radio New Zealand, July 13. Accessed October 11, 2018. https://www.radionz.co.nz/international/pacific-news/361775/freeport-mcmoran-divests-of-papua-operation.

"Rio Tinto Kelian Mine Shut Down by Community Blockade." 2000. *Drillbits & Tailings* 5 (May 16): 8.

"Rio Tinto's Environmental Record in East Kalimantan." 1999. Down to Earth: International Campaign for Ecological Justice in Indonesia, September. Accessed November 30, 2018. http://www.downtoearth-indonesia.org/old-site/Cklpl.htm.

"Rio Tinto's Shame File: Indonesian Landowners' Discontent Represented at Rio Tinto AGM." 2000. Mineral Policy Institute, May 22.

"Unocal Tailing Pipe Flooded; Rice Fields in Marangkayu Contaminated." 2000. *Minergy News*, Indonesia.

ITALY

Venice's Acqua Alta

Floods have plagued Venice for most of its history, but subsidence and slowly rising seas due to global warming have worsened flooding during the late 20th and early 21st centuries. Venice, which sits atop several million wooden pillars pounded into marshy ground, has sunk by about 7.5 centimeters per century for the past 1,000 years. That rate is now accelerating.

Flooding Becomes Routine

Increased floods have provoked plans for movable barriers across the entrance to Venice's lagoon. In Venice, the water level rose to 125 centimeters above sea level in June 2002, a record for the month. At the beginning of the 20th century, St. Mark's Square, the center of the city, was flooded an average of nine times a year. During 2001, it flooded almost 100 times. Venice flooded 111 times during 2003, more than any other year in its lengthy history, to that time. In another century, it will flood on a permanent basis ("Heavy Rains" 2002).

Venice has lost two-thirds of its population since 1950; 60,000 people remain in the city, which hosts 12 million tourists a year, who make their way over planks into buildings with foundations rotted by perennial flooding. At the Danieli, one of Venice's most luxurious hotels, tourists often arrive on wooden planks raised two feet above the marble floors amid a suffocating stench from the high water (Poggioli 2002).

Venice residents and visitors have become accustomed to high-water drills for "acqua alta," high water. A system of sirens much like the ones that convey tornado warnings in the U.S. Midwest sounds when the water surges. Restaurants have stocked Wellington boots and moved their dining rooms upstairs. Venetian gondoliers ask their passengers to shift fore and aft—and watch their heads—as they pass under bridges during episodes of high water (Rubin 2003). Some of the gondoliers have hacked off their boats' distinctive tail fins to clear the bridges brought closer by rising waters.

Dikes and Controversy

Faced with rising waters, Venice has proposed construction of massive retractable dikes in an attempt to hold the water at bay, amid considerable controversy. After 17 years of heated debate, the Venice's MOSE (*Modulo Sperimentale Elettromeccanico*) project will cost about US$1 billion.

Some environmentalists assert that the barriers will destroy the tidal movement required to keep local lagoon waters free of pollution and thereby damage marine life. Water quality near Venice is already precarious because pollution has leached into the lagoon from industry, homes, and motor traffic. The Italian Green Party favors shaping the lagoon's entrances to reduce the effects of tides, along with raising pavements as much as a meter inside Venice.

As proposed, the barriers will be constructed at the three entrances to Venice's natural lagoon from the Adriatic Sea. Each barrier is planned to house 79 "flippers" that can be adjusted like the flaps of an aircraft. Installed below the waterline, they will be raised when the sea level rises by more than one meter, which at the turn of the millennium was occurring about a dozen times a year (Watson 2001, 18).

During normal tides, according to an account in *Scotland on Sunday*, "The hollow barriers will sit within especially constructed trenches in the bed of the channels connecting the lagoon to the open sea. When a dangerously high tide is forecast, compressed air will be forced into the flippers, which will have the effect of squeezing seawater out. As they rise, more water will trickle out to be replaced by air" (Watson 2001, 18). The project is expected to provide as many as 10,000 jobs during 10 years of construction.

By mid-century, the flood-control system may be running almost all the time, severing the city from the ocean, transforming its neighboring lagoon, according to one observer, "into a stagnant pond with devastating effects on marine life and health" (Poggioli 2002). Many in Venice say that Project MOSE will not help much because it will operate only when water rises at least 43 inches.

A Toxic Bathtub?

Environmentalists have argued that the MOSE flood-control system is a construction boondoggle that will turn Venice into a toxic bathtub in which the city's canals will be laced with sludge from surrounding heavy industry, as well as the urban area's human waste. Environmentalists have focused attention on bacteria from animal and human waste in the waters surrounding the city (Petrillo 2003). "Venice has no sewage system; they just dump the stuff right out into the canals. It's not pretty," said Rick Gersberg, a microbiologist. "Normally, the tides come in and flush everything out. But when you cut off the tide, it just sits there" (Petrillo 2003). Venice's deputy mayor, Gianfranco Bettin, has called MOSE "expensive, hazardous, and probably useless" (Nosengo 2003, 608).

James Atlas (2012) commented in the *New York Times*:

> Is the *Modulo Sperimentale Elettromeccanico*—the project's official name—some engineer's fantasy? It was scheduled for completion this year [2012], but that has been put off until 2014. Even if, by some miracle, the gates materialize, they will be only a stay against the inevitable. Look at the unfortunate Easter Islanders, who left behind as evidence of their existence a mountainside of huge blank-faced busts, or the Polynesians of Pitcairn Island, who didn't leave behind much more than a few burial sites and a bunch of stone tools. Every civilization must go.

In other words, sinking land and a rising sea cannot be cured with some stilts and a water-wall. Venice is becoming an historical artifact as we watch.

Further Reading

Atlas, James. 2012. "Is This the End?" *New York Times*, November 24. Accessed October 11, 2018. http://www.nytimes.com/2012/11/25/opinion/sunday/is-this-the-end.html.

"Heavy Rains Threaten Flood-Prone Venus." 2002. *Straits Times* (Singapore), June 8.

Nosengo, Nicola. 2003. "Venice Floods: Save Our City!" *Nature* 424 (August 7): 608–609.

Petrillo, Lisa. 2003. "Turning the Tide in Venice." Copley News Service, April 28. (LEXIS).

Poggioli, Sylvia. 2002. "Venice Struggling with Increased Flooding." National Public Radio *Morning Edition*, November 29. (LEXIS).

Rubin, Daniel. 2003. "Venice Sinks as Adriatic Rises." Knight-Ridder News Service, July 1. (LEXIS).

Watson, Jeremy. 2001. "Plan to Hold Back Tides of Venice Runs into Flood of Opposition from Greens." *Scotland on Sunday*, December 30: 18.

UNITED STATES

Drought and (Occasional) Deluges

Since the 1990s, an epic drought has seized the western United States, from the Navajo reservation (which has been inundated by sand dunes) to California, where the continent's most productive agricultural valleys were parched for five consecutive years. With a few exceptions, the dominant jet stream pattern, locked in place by advancing Hadley Cells in the upper troposphere, moved rain-bearing storms to the north, even during the usual winter rainy season.

As many as 80 million piñon trees (the state tree of New Mexico) died in that state and Arizona between 2001 and 2005 because of intense drought. That represents about 90 percent of the area's piñon trees (Carlton 2006, A-1). Four million

Hector Ramirez works a field of tomato plants outside of Huron, California, where drought has ravaged towns in the western San Joaquin Valley and many farm workers have moved away. (Michael Robinson Chavez/Los Angeles Times via Getty Images)

have died in Santa Fe alone. The trees also have fallen victim to a rice-grain-sized bark beetle that feasts on dying diseased trees.

One very notable exception was 2017, when record rainfall swamped the entire U.S. West Coast. During January of that year, the drought unexpectedly and suddenly abated, at least for a time, with flooding rains throughout California and record-setting snows in the Sierra Nevada. Seattle had its wettest winter on record, after several years of drought throughout the Pacific Northwest. The hydrological cycle had swung wildly from withering drought to damaging floods. No one was sure what to expect next, as extremes of both drought and deluge had become the new "normal."

Drought and Wildfires

"Widespread annual droughts, once a rare calamity, have become more frequent and are set to become the 'new normal,'" wrote Christopher Schwalm, Christopher A. Williams, and Kevin Schaefer in the *New York Times* (2012). Schwalm is a research assistant professor of earth sciences at Northern Arizona University, Williams is an assistant professor of geography at Clark University, and Schaefer is a research scientist at the National Snow and Ice Data Center.

With a scarcity of snow in western North America, as well as abnormally warm temperatures, intense early wildfires broke out in May 2015 as far north as the Canadian Northwest Territories. The fires were so numerous that the NASA Earth Observatory noted that "heavy smoke from the growing blazes merged with lingering smoke from previous days, making air quality a major concern" ("Intense Fires" 2015). "Fires this early are very unusual," said Mike Flannigan of the University of Alberta. He noted that the Northwest Territories and northern Alberta usually see the most active part of the fire season in July, and British Columbia's fire season typically runs from the end of July through August ("Intense Fires" 2015). Some of the fires were so large and intense that pyrocumulonimbus clouds (thunderstorms generated by the heat from fire rather than by the sun-warmed ground) from them generated thunderstorms. NASA said that in Alberta, more than 60 fires displaced thousands of residents and led some energy companies to suspend oil operations and evacuate staff, reducing tar-sands production ("Intense Fires" 2015; "Alberta Wildfires" 2015). By 2017, intense fires were ravaging Kansas, Oklahoma, and Texas in mid-March, at the end of calendar winter, until flooding rains arrived during April and May.

Wildfires across Canada during the summer of 2015 exported smoke to urban areas. Across the western half of North America, from Alaska to California, record heat and drought provoked a record fire season. During the first week of July, smoke that had risen into the stratosphere from fires in Alberta and Saskatchewan obscured the sun across the United States' upper Midwest, including Omaha, Des Moines, and other cities. Smoke was so thick some days in Denver that flights were delayed at its international airport. Colorado State health officials warned people with asthma to stay indoors on the days when smoke was thickest.

Hundreds of Wildfires at Once

By the end of July 2015, as many as 300 wildfires, most started by lightning, were burning at one time across Alaska as record temperatures prompted early snowmelt and drought turned forests to tinder. Nearly 5 million acres already had burned, making the season among Alaska's worst, with several weeks to go until the first snows of early fall would snuff residual cinders. During the record year, 2004, 6.59 million acres burned (Mooney 2015).

During 2015, more land burned in Alaska than in the rest of the United States combined, producing smoke so dense on some days that many outdoor events were canceled in Fairbanks. The intense burning was adding carbon dioxide to the atmosphere in two ways: the torched trees themselves, as well as the smoldering of what was once frozen ground (permafrost), much of which is laced with carbon-rich organic peat. "The more severe the fire, the deeper that it burns through the organic layer, the higher the chance it will go through this complete conversion," said Ted Schuur, a Northern Arizona University ecologist and permafrost specialist. "What happens in the summer of 2015 has the potential to change the whole trajectory of [the burned area] for the next 100 years or more" (Mooney 2015). Worldwide, across the Arctic, permafrost in North America, Northern Europe, and Siberia contains twice as much carbon as Earth's atmosphere. "The permafrost that we degrade now in these forest fires might never return in our lifetimes," Schuur said (Mooney 2015).

"Orange Is the New Blue"

Smoke from hundreds of fires burning across British Columbia swept into Vancouver, British Columbia, and nearby areas. NASA detected a thick pall of smoke July 5 and 6 that "settled over Vancouver and adjacent areas of British Columbia, leading some residents to wear face masks and health officials to warn residents and World Cup tourists against outdoor activities. . . . Tan and gray smoke almost completely obscures the Strait of Georgia and southern Vancouver Island. Winds shifted abruptly between July 5 and 6, driving the smoke plume toward the east, dispersing it in some places while fouling the air in areas to the east, such as the Fraser Valley" ("Smoke Blankets" 2015). For a few days, officials in Vancouver compared its air quality to that of Beijing ("Metro" 2016).

"When it comes to the skies overhead, it appears that orange is the new blue," reported Yvonne Zacharias in the *Vancouver Sun*.

> Get ready for more of the apocalyptic haze as scientists warn that climate change is blazing a whole new trail for the Earth. [John Innes, dean of forestry at the University of British Columbia] predicted global warming will not only increase the length and intensity of the forest-fire season, but will also affect sockeye salmon runs, ski resorts like Whistler, which is changing to an all-season resort, and the inaccessibility of remote areas in the north because of an early breakup that makes transportation routes impassable. And as for forestry, "we can no longer rely on the trees we are

planting today to be there in 80 or 100 years' time," he said. "I hope this is a wake-up call," said UBC forestry professor Lori Daniels. In other words, by the time today's youth grow into adulthood, "this is what is going to become the new norm." (Zacharias 2015)

As smoke from at least 110 fires in Saskatchewan poured southward into large areas of the United States, about 13,000 indigenous people of the Lac La Ronge and other bands were being forced to evacuate their homes during the first week of July 2015. About 100 miles to the south of this evacuation, the Montreal Lake Cree Nation declared a state of emergency as fire surrounded its community in a wall so fierce that several hundred firefighters were forced to withdraw. Residents were forced out of their homes as several structures burned to the ground.

"There was no stopping it—four water bombers, two helicopters bucketing, crews on the ground doing whatever they could, but we just couldn't do anything when the wind picked up," said Chief Edward Henderson, describing the wall of flames in La Ronge. "I've never seen anything like this," Lac La Ronge Indian Band Chief Tammy Cook-Searson told CBC News. "We've had to evacuate all six of our communities" ("Saskatchewan" 2015). At their height, about 2,500 firefighters and Canadian armed forces troops from Ontario, New Brunswick, Newfoundland, Alberta, and Saskatchewan battled the flames.

A New Climate Reality?

"California is facing a new climate reality," wrote Noah S. Diffenbaugh and Christopher B. Field in the *New York Times*. "Extreme drought is more likely. The state's water rights, infrastructure, and management were designed for an old climate, one that no longer exists." Their research, they said,

> has shown that global warming has doubled the odds of the warm, dry conditions that are intensifying and prolonging this drought, which now holds records not only for lowest precipitation and highest temperature, but also for lowest spring snowpack in the Sierra Nevada in at least 500 years. These changing odds make it much more likely that similar conditions will occur again, exacerbating other stresses on agriculture, ecosystems, and people.
>
> At the same time, the intensity of extreme wet periods will increase even during droughts because a warming atmosphere can carry a larger load of water vapor. El Niño conditions during 2015 and 2016 forced Californians to face both flooding and drought simultaneously. The more rainfall there is, the more water will be lost as runoff or river flow, increasing the risk of flooding and landslides. Add in the fact that the drought and wildfires have hardened the ground, and a paradox arises wherein the closer El Niño comes to delivering enough precipitation to break the drought this year, the greater the potential for those hazards. (Diffenbaugh and Field 2015)

An analysis of blue oak tree rings in California's Central Valley released August 14, 2015 indicated that snowfall in nearby mountains was lower by 2015 than it had been in 500 years. By that time, the Sierra Nevada was entirely bereft of snow for the first time in 75 years of recorded history. "The results were astonishing," said Valerie Trouet, an associate professor at the University of Arizona, an author of the analysis in *Nature Climate Change* (Belmecheri et al. 2015). "We knew it was an all-time low over a historical period, but to see this as a low for the last 500 years, we didn't expect that. There's very little doubt about it," she said ("Fears" 2015).

Agricultural Losses to Drought

California's drought caused $2.2 billion in agricultural losses as fields were left fallow in 2014 alone. More than 12 million trees had died in the five-year drought by 2015. This may be a foretaste, as "future droughts will be compounded by more intense heat waves and more wildfires. Soaring temperatures will increase demand for energy just when water for power generation and cooling is in short supply. Such changes will increase the tension between human priorities and nature's share" (AghaKouchak et al. 2015).

During the winter of 2014–2015, rain teased California for a few December days but evaporated in January with record heat and low rainfall. By March, temperatures in Southern California had risen to nearly 100°F in some places as mountain snowpacks eroded to the lowest on record. On March 3, statewide snowpack was only one-fifth of usual. Rising temperatures were shortening the period in and altitude at which snow could accumulate. What little precipitation did fall came more often as rain, which melted existing snow and then soaked into the ground before reaching reservoirs. Winter 2013–2014 was California's warmest on record, at 45.6°F, until the winter of 2014–2015 averaged 47.4°. Higher temperatures increase evaporation and human demand for water.

"The normal cyclical conditions in California are different now from what they used to be, and that's not because the long-term annual precipitation changed," said Noah Diffenbaugh, a senior fellow at the Stanford Woods Institute for the Environment. "What is really different is there has been a long-term warming in California," he said. "And we know from looking at the historical record that low precipitation years are much more likely to result in drought conditions if they occur with high temperatures" (Nagourney 2015).

Further Reading

AghaKouchak, Amir, David Feldman, Martin Hoerling, Travis Huxman, and Jay Lund. 2015. "Water and Climate: Recognizing Anthropogenic Drought." *Nature* 524 (August 27): 409–411. Accessed October 11, 2018. http://www.nature.com/news/water-and -climate-recognize-anthropogenic-drought-1.18220.

"Alberta Wildfires Prompt Oil Firms to Suspend Production and Evacuate Staff." 2015. Reuters in *The Guardian,* May 26. Accessed October 11, 2018. http://www.theguardian .com/world/2015/may/26/alberta-wildfires-oil-production-suspended-evacuations.

Belmecheri, Soumaya, Flurin Babst, Eugene R. Wahl, David W. Stahle, and Valerie Trouet. 2015. "Multi-Century Evaluation of Sierra Nevada Snowpack; California Snowpack Lowest in Past 500 Years." *Nature Climate Change,* September. Accessed October 11, 2018. http://www.nature.com/nclimate/journal/vaop/ncurrent/full/nclimate2809.html.

Carlton, Jim. 2006. "Some in Santa Fe Pine for Lost Symbol, But Others Move On." *Wall Street Journal*, July 31: A-1, A-8.

Diffenbaugh, Noah S., and Christopher B. Field. 2015. "A Wet Winter Won't Save California." *New York Times*, September 18. Accessed October 11, 2018. http://www.nytimes.com/2015/09/19/opinion/a-wet-winter-wont-save-california.html.

Fears, Darryl. 2015. "Scientists Say California Hasn't Been This Dry in 500 Years." *Washington Post,* September 14. Accessed November 30, 2018. https://www.washingtonpost.com/news/energy-environment/wp/2015/09/14/scientists-say-its-been-500-years-since-california-was-this-dry/.

"Intense Fires in Northern Canada." 2015. NASA Earth Observatory, June 3. Accessed October 11, 2018. http://earthobservatory.nasa.gov/IOTD/view.php?id=85972&src=eoa-iotd.

"Metro Vancouver Air Quality Comparable to Beijing." 2016. *Huffington Post.* July 6. Accessed November 30, 2018. https://www.huffingtonpost.ca/2015/07/07/metro-vancouver-air-quality-comparable-to-beijing_n_7740900.html.

Mooney, Chris. 2015. "Alaska's Terrifying Wildfire Season and What It Says About Climate Change." *Washington Post,* July 26. Accessed October 11, 2018. http://www.washingtonpost.com/news/energy-environment/wp/2015/07/26/alaskas-terrifying-wildfire-season-and-what-it-says-about-climate-change/.

Nagourney, Adam. 2015. "As California Drought Enters 4th Year, Conservation Efforts and Worries Increase." *New York Times*, March 17. Accessed October 11, 2018. http://www.nytimes.com/2015/03/18/us/as-california-drought-enters-4th-year-conservation-efforts-and-worries-increase.html.

"Saskatchewan First Nations Evacuate 13,000, Declare Wildfire State of Emergency." 2015. Indian Country Today Media Network, July 6.

Schwalm, Christopher, Christopher A. Williams, and Kevin Schaefer. 2012. "Hundred-Year Forecast: Drought." *New York Times*, August 12. Accessed October 11, 2018. http://www.nytimes.com/2012/08/12/opinion/sunday/extreme-weather-and-drought-are-here-to-stay.html.

"Smoke Blankets British Columbia." 2015. NASA Earth Observatory. Accessed November 30, 2018. https://earthobservatory.nasa.gov/images/86190/smoke-blankets-british-columbia.

Zacharias, Yvonne. 2015. "Global Warming Exacerbates B.C. Wildfire Severity, Scientist Says." *Vancouver Sun*, July 6. Accessed October 11, 2018. http://www.vancouversun.com/technology/global+warming+exacerbates+wildfire+severity+scientist+says/11192869/story.html.

UNITED STATES: NAVAJO NATION

A Sea of Sand Dunes

Persistent drought in the U.S. Southwest is forcing Navajos who have no indoor plumbing to travel several miles for water as their wells run dry, while also forcing early sale of livestock as formerly scanty pastures turn to naked dirt. "Perhaps among the worst of those impacts," wrote Terri Hansen in the Indian Country Today Media Network,

> are the runaway sand dunes it has unleashed, which extend over one-third of the 27,000-square-mile reservation. During the 1996–2009 drought period, the extent

of dune fields increased by some 70 percent. These dunes are moving at rates of approximately 35 meters per year, covering houses, burying cars, and snarling traffic, degrading grazing and agricultural lands, contributing to the loss of rare and endangered native plants, and when they occur, contributing to poor air quality, a serious health concern for many of the reservation's 173,667 residents. (Hansen 2014)

The Difference Between Little Rain and None

The 25 to 40 percent of Navajos who haul their own water pay 20 times per volume than non-Navajos who have piped in supplies on per capita income that is less than half of the U.S. average—before adding the expense of round-trips that average 28 miles (Cozzetto et al. 2013, 569). During droughts, which are becoming more frequent, both the cost of water and the distances required to acquire it increase.

The drought also has become pervasive in other parts of the U.S. Southwest. "Our 30,000-acre reservation is pretty dry because of drought," said Lawrence Snow, land resources manager for Utah's Shivwits Band of Paiutes. "Wildfires in the last decade have burned half our acreage and changed the landscape. We've got less trees, and bark beetles are trying to kill off the ones we do have. Once the fires happened and took out the ground cover, major storms brought big flooding" (Allen 2012). When Native areas in the U.S. Southwest do receive rain and snowfall, it increasingly comes in flooding deluges. Arizona's Havasupai Tribe between 2008 and 2010 endured several damaging floods.

Navajo and Hopi lands in Arizona have always been relatively dry, but climate change in recent years often has made matters worse. The Navajo and other Southwestern Native peoples have made a fine art of surviving on little water for centuries. The ways they farm and the animals they herd are used to it. There is a difference, however, between little rain and nearly none, and that's what they've been dealing with for 20 years. Cindy Dixon's sheep, for example, used to forage scrub on the desert near Farmington, New Mexico. By 2014, however, even the scrub had died, as Dixon turned to expensive bales of hay. "The landscape around her Navajo Reservation homestead," wrote Bobby Magill in Climate Central, "was as brown and bleak as the open-pit coal mine a few miles to the west and well within earshot" (Magill 2014).

Dixon lives without electricity or running water, but her sheep cannot eat sand. "Since it's all dry and bare and deserted—no vegetation—I have to constantly buy hay and grain to keep the sheep fed," Dixon said, looking at the land around her trailer. "This is a bad, bad area for livestock" (Magill 2014). The lack of forage is compounded by coal dust blowing in from the mine on stiff winds that are now pushing sand dunes across the brown, desiccated land. Sometimes, Dixon cuts her own grocery spending so that she can buy hay for the sheep.

Some Sand Dunes Reach 30 to 40 Feet

By 2014, sands dunes were "covering housing, causing transportation problems, and contributing to loss of endangered native plants and grazing land" (Cozzetto et al. 2013, 569). Rainfall in some parts of the Navajo Nation fell to three inches a year during the latest drought, wrote Margaret Redsteer (Redsteer et al. 2011a, 2011b). Because of the enduring drought, "more than one-third of Native lands on the Colorado Plateau (Navajo Nation and Hopi tribal lands) are covered with sand dunes and sand sheets," according to the U.S. Geological Survey (USGS) (Redsteer et al. 2011a, 2011b). Lands that once were marginally productive for grazing of sheep and dry-land agriculture (a long-time practice among the Navajo and Hopi) are becoming true water-starved deserts. Or, as the USGS phrases it: "Reactivation of inactive dunes could have serious consequences on human and animal populations, agriculture, grazing, and infrastructure on the Navajo Nation and similar areas in the Southwest" (Redsteer et al. 2011a, 2011b). Wind and drought have been worst in the spring.

Dunes are migrating faster across the landscape at speeds heretofore unknown in Navajo Country. During 2009, the USGS measured dune migration as fast as 112 to 157 feet per year. Some dunes moved more than 3.3 feet in a single windstorm. The Grand Falls dune field has grown in areal extent by 70 percent (laterally and downwind) in 15 years (1992 to 2007). The drought has continued since then, punctuated by a very occasional deluge that quickly runs off the cracked, parched, and increasingly sandy earth.

Streams that once were sources of water have dried up, feeding the wind with plumes of gritty, irritating sand. The USGS found that "the formation and movement of active dunes on the downwind side of streambed sand sources is presently endangering housing and transportation, potentially jeopardizing native plants and grazing lands, increasing health hazards to humans and animals, and affecting regional air quality" (Redsteer et al. 2011a, 2011b).

The Navajo Nation has experienced several decades of rising temperatures, declining snowfall, and decreased (or, in some cases, nonexistent) stream flow, resulting in water scarcity that has "magnified the impacts of drought that began in 1996 and continues today" (Redsteer et al. 2013, 390).

Like heavy snowdrifts, dunes move across roads and block them with increasing regularity. "As we're packing up and preparing to leave Navajo elder Chee Willie's house, Tohannie tells me that yesterday's windstorm shifted a nearby dune, causing it to cover part of a road used by the handful of families in the area, including Tohannie's parents," Kathy Ritchie wrote (2014). "'I got stuck with my son,' he says. 'We had to shovel our way out'" (Ritchie 2014). These are not small drifts, and they can't be moved, like snow, with a plow. And, of course, they never melt. Ritchie described one dune as a "magnificent sculpture shaped from mostly eroded Navajo and Entrada sandstone... Many of the dunes in the area are unexpectedly tall, some measuring anywhere from 30 feet to 40 feet high. Begay tells me that near Preston Mountain, some 45 minutes north of where I'm standing, the dune field is even higher, possibly 60 feet to 80 feet in places" (Ritchie 2014). The dominant plant species in some areas, where any survive, has become tumbleweed, which has evolved to move with the wind, anchoring nothing.

Wise Navajo Drivers Carry Shovels

Across the reservation, persistent winds have been driving sand into homes and across roads. Most of the roads are not paved, and their surfaces can become parts of the moving dunes. Even major paved roads have been blocked. On April 16, 2013, driven by winds gusting to 60 miles an hour, drifts of sand closed parts of Interstate 40 and traffic backed up 12 miles. A NASA satellite photographed the dust plume from space.

Wise Navajo drivers carry shovels: "Kee Tohannie, Begay's grandfather and Huskie Tohannie's father, always carries a shovel, chains, and sometimes a hatchet in case he or one of his neighbors is marooned in the sand. 'I've been stuck in the sand many times—it's a lot of digging,' he says. You just have to know how to drive in sand. Like you learn to drive in snow" (Ritchie 2014).

Redsteer, who is of Crow descent, was raised in their homeland near the Montana-Wyoming border, but during the 1970s, having married a Navajo, she moved to his homeland and mothered three children. Many people told her how much plant life in the area had changed over the years. Only later did she begin to associate these changes with climate change. In the meantime, Redsteer and her family had moved to Flagstaff, Arizona, when she was 29 years of age in 1986; there, she studied for a PhD. Shortly after 2000, Redsteer, employed by the U.S. Geological Survey, switched her focus of study from volcanic deposits near Yellowstone National Park to the effects of climate change on the Navajo Nation, as intensifying drought was provoking growth and migration of sand dunes there. She wrote several academic papers and had a key role in the National Climate Assessment, released by President Barack Obama in 2013. She has become known for linking elders' recollections with weather records to trace the evolution of climate change.

Snow Used to Be Like Money in the Bank

Bobby Magill wrote in Climate Central: "Navajo elders remember wetter times, when winter snows were knee-deep, water always ran in springs and arroyos, and the rangeland among the canyons, mesas, and volcanic hills could support large herds of livestock, a mainstay of the Navajo economy" (Magill 2014). Some elders recalled a time when they were children, with moist ground until the Fourth of July. Climate data on the Navajo Nation indicates a marked drying trend since the middle 1940s and a warming of 4°F in many areas since the 1960s. The decrease in snowfall has long-term implications. "Snow is like water in the bank," Redsteer said. "It takes a long time to melt. It soaks into the ground slowly" (Redsteer et al. 2011b). Over the past 60 years, the Southwest has experienced swings between very wet and very dry, but the current drought has dominated the past 20, with brief wet periods in 2004, 2005, and 2010 doing little to alleviate that long-term trend.

A technical report for the National Climate Assessment, issued in 2013, said that the Four Corners area probably will continue to endure warmer weather on average during the coming decades as soil continues to dry, with droughts becoming more intense and frequent (Redsteer et al. 2013). The drought—and spread of sand dunes—is worst in the southwestern quarter of the Navajo reservation, where many families may be forced to move away, according to Redsteer (Magill 2014), who has used the recall of elders as well as weather data to trace climatic changes. Average snowfall across the Navajo Nation has dropped from about 31 inches in 1930 to about 11 inches by 2010, according to a United Nations case study "Every tribal elder mentioned the lack of snowfall," Redsteer said. "They describe winters where the snow was 'chest high on horses.' The snowfall snows a significant decline over the 20th century and is still declining in recent years" (Magill 2014). Elders' memories are especially important in recent years as heat and drought have accelerated, because many U.S. government weather stations were shut down during the early 1980s to save money.

The Navajo drought continued to intensify into the summer of 2018, as much of the U.S. Southwest was "heading into summer in the throes of a persistent and ever-more severe drought," with very little precipitation, low snowpack, and temperatures much above average ("Intensifying Drought" 2018). According to the U.S. Drought Monitor, all areas bordering the Navajo Nation were in severe drought, including 97 percent of Arizona and 88 percent of New Mexico ("Intensifying Drought" 2018). Satellite measurements from NASA indicated that the Navajos were in the center of the worst conditions. The severity of the drought was indicated by the discovery on Navajo lands of 200 wild horses dead of heat and thirst during May of 2018.

Further Reading

Allen, Lee. 2012. "Southwest Tribes Struggle with Climate Change Fallout." Indian Country Today Media Network, June 14.

Cozzetto, K., K. Chief, K. Dittmer, M. Brubaker, R. Gough, K. Souza, . . . P. Chavan. 2013. "Climate Change Impacts on the Water Resources of American Indians and Alaska Natives in the U.S." *Climatic Change* 120 (September): 569–584.

Hansen, Terri. 2014. "Climate Disruptions Hitting More and More Tribal Nations." Indian Country Today Media Network, May 7. Accessed November 30, 2018. http://www.tulalipnews.com/wp/2014/05/12/climate-disruptions-hitting-more-and-more-tribal-nations/.

"Intensifying Drought in the American Southwest." 2018. NASA Earth Observatory, June 12. Accessed October 11, 2018. https://earthobservatory.nasa.gov/IOTD/view.php?id=92274&src=eoa-iotd.

Magill, Bobby. 2014. "The Navajo Nation's Shifting Sands of Climate Change." Climate Central, May 28. Accessed October 11, 2018. http://www.climatecentral.org/news/navajo-nation-climate-change-17326.

Redsteer, M. H., K. Bemis, K. Chief, M. Gautam, B. R. Middleton, and R. Tsosie. 2013. "Unique Challenges Facing Southwestern Tribes." In *Assessment of Climate Change in the Southwest United States: A Report Prepared for the National Climate Assessment*, by G. Garfin, A. Jardine, R. Merideth, M. Black, and S. LeRoy, 385–404. Washington, DC: Island Press. Accessed October 11, 2018. http://www.swcarr.arizona.edu/sites/default/files/ACCSWUS_Ch17.pdf.

Redsteer, M. H., R. C. Bogle, and J. M. Vogel. 2011a. "Monitoring and Analysis of Sand Dune Movement and Growth on the Navajo Nation, Southwestern United States." Fact Sheet Number 3085, U.S. Geological Survey, Reston, VA.

Redsteer, M. H., K. B. Kelley, and H. Francis. 2011b. "Increasing Vulnerability to Drought and Climate Change on the Navajo Nation." Paper GC43B-0928, delivered at American Geophysical Union Annual Meeting, San Francisco, December 5–9.

Ritchie, Kathy. 2014. "Dune and Gloom." *Arizona Highways*.

Selected Bibliography

Abboud, Leila. 2006. "Sun Reigns on Spain's Plains: Madrid Leads a Global Push to Capitalize on New Solar-Power Technologies." *Wall Street Journal*, December 5: A-4.

Abrash, Abigail. 2001. "The Amungme, Kamoro, and Freeport: How Indigenous Papuans Have Resisted the World's Largest Gold and Copper Mine." *Cultural Survival Quarterly* 25, no. 1 (Spring): 38–43.

Anand, Geeta. 2017. "India's Air Pollution Rivals China's as World's Deadliest." *New York Times*, February 14. Accessed October 12, 2018. https://www.nytimes.com/2017/02/14/world/asia/indias-air-pollution-rivals-china-as-worlds-deadliest.html.

Archer, David. 2009. *The Long Thaw: How Humans Are Changing the Next 100,000 Years of Earth's Climate*. Princeton, NJ:Princeton University Press.

Benton, Michael J. 2003. *When Life Nearly Died: The Greatest Mass Extinction of All Time*. London: Thames and Hudson.

Braasch, Gary. 2007. *Earth under Fire: How Global Warming Is Changing the World*. Berkeley: University of California Press.

Bradbury, Roger. 2012. "A World without Coral Reefs." *New York Times*, July 13. Accessed October 12, 2018. http://www.nytimes.com/2012/07/14/opinion/a-world-without-coral-reefs.html.

Brienen, R. J. W., O. L. Phillips, T. R. Feldpausch, E. Gloor, T. R. Baker, J. Lloyd, . . . R. J. Zagt. 2015. "Long-Term Decline of the Amazon Carbon Sink." *Nature* 519 (March 19): 344–348.

Brown, Lester R. 2003. *Plan B: Rescuing a Planet under Stress and a Civilization in Trouble*. New York: Earth Policy Institute/W. W. Norton.

Caldeira, Ken, and Philip B. Duffy. 2000. "The Role of the Southern Ocean in the Uptake and Storage of Anthropogenic Carbon Dioxide." *Science* 287 (January 28): 620–622.

Cardwell, Diane. 2014. "Copenhagen Lighting the Way to Greener, More Efficient Cities." *New York Times*, December 8. Accessed October 12, 2018. http://www.nytimes.com/2014/12/09/business/energy-environment/copenhagen-lighting-the-way-to-greener-more-effecient-cities.html.

Ceballos, Gerardo, Paul R. Ehrlich, and Rodolfo Dirzo. 2017. "Biological Annihilation via the Ongoing Sixth Mass Extinction Signaled by Vertebrate Population Losses and Declines." *Proceedings of the National Academy of Sciences* 114, no. 30 (July 10). Accessed October 1, 2018. http://www.doi.org/10.1073/pnas.1704949114.

Clarkson, M. O., S. A. Kasemann, R. A. Wood, T. M. Lenton, S. J. Daines, S. Richoz, . . . E. T. Tipper. 2015. "Ocean Acidification and the Permo-Triassic Mass Extinction." *Science* 348 (April 10): 229–232.

Cone, Marla. 2005. *Silent Snow: The Slow Poisoning of the Arctic*. New York: Grove Press.

Eichstaedt, Peter. 1994. *If You Poison Us: Uranium and American Indians*. Santa Fe, NM: Red Crane Books.

Estrada, Alejandro, Paul A. Garber, Anthony B. Rylands, Christian Roos, Eduardo Fernandez-Duque, Anthony Di Fiore, . . . Baoguo Li. 2017. "Impending Extinction Crisis of the World's Primates: Why Primates Matter." *Science Advances* 3, no. 1 (January 18). Accessed October 12, 2018. http://www.doi.org/10.1126/sciadv.1600946.

Ewen, Alexander. 1994. *Voices of Indigenous Peoples: Native People Address the United Nations.* Santa Fe, NM: Clear Light.

Fialka, John. 2016. "Why China Is Dominating the Solar Industry." *Scientific American,* December 19. Accessed October 8, 2018. https://www.scientificamerican.com/article/why-china-is-dominating-the-solar-industry/.

Flannery, Tim. 2005. *The Weather Makers: How Man Is Changing the Climate and What It Means for Life on Earth.* New York: Atlantic Monthly Press.

Freytas-Tamura, Kimiko de. 2017. "A Push for Diesel Leaves London Gasping Amid Record Pollution." *New York Times,* February 17. Accessed October 7, 2018. https://www.nytimes.com/2017/02/17/world/europe/london-smog-air-pollution.html.

Gedicks, Al. 2001. *Resource Rebels: Native Challenges to Mining and Oil Corporations.* Boston: South End Press.

Gillis, Justin. 2011. "The Threats to a Crucial Canopy." *New York Times,* October 1. Accessed October 12, 2018. http://www.nytimes.com/2011/10/01/science/earth/01forest.html.

Gillis, Justin. 2014. "Sun and Wind Alter Global Landscape, Leaving Utilities Behind." *New York Times,* September 13. Accessed October 8, 2018. http://www.nytimes.com/2014/09/14/science/earth/sun-and-wind-alter-german-landscape-leaving-utilities-behind.html.

Gomez, C. D. Harvell, P. F. Sale, A. J. Edwards, K. Caldeira, N. Knowlton, C. M. Eakin, . . . M. E. Hatziolos. 2007. "Coral Reefs Under Rapid Climate Change and Ocean Acidification." *Science* 318 (December 14): 1737–1742.

Gordon, Anita, and David Suzuki. 1991. *It's a Matter of Survival.* Cambridge: Harvard University Press.

Guillette, Elizabeth A., Maria Mercedes Meza, Maria Guadalupe Aquilar, Alma Delia Soto, and Idalia Enedina Garcia. 1998. "An Anthropological Approach to the Evaluation of Preschool Children Exposed to Pesticides in Mexico." *Environmental Health Perspectives* 106, no. 6 (June). Accessed December 3, 2018. https://www.ncbi.nlm.nih.gov/pmc/articles/PMC1533004/.

Hallam, Anthony, and Paul Wignall. 1997. *Mass Extinctions and Their Aftermath.* Oxford: Oxford University Press.

Hansen, J., D. Johnson, A. Lacis, S. Lebedeff, P. Lee, D. Rind, and G. Russell. 1981. "Climate Impact of Increasing Atmospheric Carbon Dioxide," *Science* 213 (1981): 957–966.

Hansen, James. 2012. "Game Over for the Climate." *New York Times,* May 9, 2012. Accessed September 30, 2018. http://www.nytimes.com/2012/05/10/opinion/game-over-for-the-climate.html.

Hansen, J., M. Sato, P. Hearty, R. Ruedy, M. Kelley, V. Masson-Delmotte, . . . K.-W. Lo. 2015. "Ice Melt, Sea Level Rise and Superstorms: Evidence from Paleoclimate Data, Climate Modeling, and Modern Observations That 2°C Global Warming Is Highly Dangerous." *Atmospheric Chemistry and Physics Discussions* 15 (July): 20059–20179. Accessed September 27, 2018. http://doi.org/10.5194/acpd-15-20059-2015.

Hansen, Terri. 2014. "Kill the Land, Kill the People: There Are 532 Superfund Sites in Indian Country!" Indian Country Today Media Network, June 17. Accessed December 3, 2018. https://newsmaven.io/indiancountrytoday/archive/kill-the-land

-kill-the-people-there-are-532-superfund-sites-in-indian-country-LpCDfEqzlkGEn
zyFxHYnJA/.

Harris, Gardiner. 2013. "Rains or Not, India Is Falling Short on Drinkable Water." *New York Times*, March 12. Accessed October 11, 2018. http://www.nytimes.com/2013/03/13 /world/asia/rains-or-not-india-is-falling-short-on-drinkable-water.html.

Hernández, Javier C. 2017. "'No Such Thing as Justice' in Fight over Chemical Pollution in China." *New York Times*, June 12. Accessed October 5, 2018. https://www.nytimes .com/2017/06/12/world/asia/china-environmental-pollution-chemicals-lead-poison ing.html.

Herrick, Thaddeus. 2002. "The New Texas Wind Rush: Oil Patch Turns to Turbines, as Ranchers Sell Wind Rights; a New Type of Prospector." *Wall Street Journal*, September 23: B-1, B-3.

Holland, Jennifer S. 2001. "The Acid Threat: As CO_2 Rises, Shelled Animals May Perish." *National Geographic*, November: 110–111.

Hughes, T. P., A. H. Baird, D. R. Bellwood, M. Card, S. R. Connolly, C. Folke, . . . J. Roughgarden. 2003. "Climate Change, Human Impacts, and the Resilience of Coral Reefs," *Science* 301 (August 15): 929–933.

Innis, Michelle. 2015. "Warming Oceans May Threaten Krill, a Cornerstone of the Antarctic Ecosystem." *New York Times*, October 19. Accessed October 12, 2018. http://www .nytimes.com/2015/10/20/science/australia-antarctica-krill-climate-change-ocean .html.

Jakob, Michael, and Jérôme Hilaire. 2015. "Climate Science: Unburnable Fossil-Fuel Reserves." *Nature* 517 (January 8): 150–152. Accessed September 27, 2018. http:// doi.org/10.1038/517150a.

Jamieson, Alan J., Tamas Malkocs, Stuart B. Piertney, Toyonobu Fujii, and Zulin Zhang. 2017. "Bioaccumulation of Persistent Organic Pollutants in the Deepest Ocean Fauna." *Nature Ecology & Evolution* 1 (February). Accessed October 3, 2018. http://www.doi .org/10.1038/s41559-016-0051.

Johansen, Bruce. 1972. "Ecomania at Home; Ecocide Abroad." University of Washington *Daily*, May 24: 4.

Johansen, Bruce E. 1993. *Life and Death in Mohawk Country*. Golden, CO: North American Press/Fulcrum.

Johansen, Bruce E. 2000. "Pristine No More: The Arctic, Where Mother's Milk Is Toxic." *The Progressive*, December: 27–29.

Johansen, Bruce E. 2001. "Arctic Heat Wave." *The Progressive*, October: 18–20.

Johansen, Bruce E. 2003. *The Dirty Dozen: Toxic Chemicals and the Earth's Future*. Santa Barbara, CA: Praeger.

Kelley, Colin, Shahrzad Mohtadi, Mark A. Cane, Richard Seager, and Yochanan Kushnir. 2015. "Climate Change in the Fertile Crescent and Implications of the Recent Syrian Drought." *Proceedings of the National Academy of Sciences* 112, no. 11 (March 2): 3241–3246. Accessed September 28, 2018. http://doi.org/10.1073/pnas.1421533112.

Kenneally, Christine. 2009. "The Inferno." *New Yorker*, October 26: 46–53.

Kimmelman, Michael. 2017. "Mexico City, Parched and Sinking, Faces a Water Crisis." *New York Times*, February 17. Accessed October 7, 2018. https://www.nytimes.com /interactive/2017/02/17/world/americas/mexico-city-sinking.html.

Kolbert, Elizabeth. 2007. "Unconventional Crude: Canada's Synthetic-Fuels Boom." *New Yorker*, November 12: 46–51.

Kolbert, Elizabeth. 2008. "The Island in the Wind: A Danish Community's Victory over Carbon Emissions." *New Yorker*, July 7. Accessed October 8, 2018. http://www.new yorker.com/reporting/2008/07/07/080707fa_fact_kolbert/.

Krauss, Clifford. 2008. "Move Over, Oil, There's Money in Texas Wind." *New York Times*, February 23. Accessed October 12, 2018. http://www.nytimes.com/2008/02/23/busi ness/23wind.html.

Lavers, Chris. 2000. *Why Elephants Have Big Ears*. New York: St. Martin's Press.

Lawton, R. O., U. S. Nair, R. A. Pielke Sr., and R. M. Welch. 2001. "Climatic Impact of Tropical Lowland Deforestation on Nearby Montane Cloud Forests." *Science* 294 (October 19): 584–587.

Liewer, Steve. 2016. "A Toxic Legacy." *Omaha World-Herald,* June 5: 1-A, 5-A.

Lynas, Mark. 2004. *High Tide: The Truth about Our Climate Crisis*. New York: Picador/ St. Martins.

Mauk, Ben. 2017. "States of Decay: A Journey through America's Nuclear Heartland." *Harper's*, October: 48–59.

McDowell, Bart. 1980. "The Aztecs." *National Geographic*, December: 704–752.

McGlade, Christophe, and Paul Ekins. 2015. "The Geographical Distribution of Fossil Fuels Unused When Limiting Global Warming to 2°C." *Nature* 517 (January 8): 187–190. Accessed September 27, 2018. http://doi.org/10.1038/nature14016.

McKibben, Bill. 2017. "Power Brokers: Africa's Solar Boom Is Changing Life beyond the Grid." *New Yorker*, June 26: 46–55.

Merchant, Brian. 2013. "The Nation's Top Climate Scientist Predicts an 'Ice-Free, Human-Free' Planet." Motherboard, April 17.

Molina Montes, Augusto F. 1980. "The Building of Tenochtitlan." *National Geographic*, December: 753–766.

Montaigne, Fen. 2010. *Fraser's Penguins: A Journey to the Future of Antarctica*. New York: Henry Holt.

Mooney, Chris. 2015. "Alaska's Terrifying Wildfire Season and What It Says about Climate Change." *Washington Post,* July 26. Accessed October 11, 2018. http://www.washing tonpost.com/news/energy-environment/wp/2015/07/26/alaskas-terrifying-wildfire -season-and-what-it-says-about-climate-change/.

Moyers, Bill. 2001. *Trade Secrets: A Moyers Report*. Program transcript. Public Broadcasting Service, March. Accessed October 4, 2018. http://www.pbs.org/tradesecrets/transcript .html.

Nicholls, Neville. 2004. "The Changing Nature of Australian Droughts." *Climatic Change* 63 (April): 323–336.

Nosengo, Nicola. 2003. "Venice Floods: Save Our City!" *Nature* 424 (August 7): 608–609.

Nowak, Rachel. 2007. "Australia: The Continent That Ran Dry." *New Scientist*, June 16: 8–11.

O'Reilly, Catherine M., Simone R. Alin, Pierre-Denis Plisnier, Andrew S. Cohen, and Brent A. McKee. 2003. "Climate Change Decreases Aquatic Ecosystem Productivity of Lake Tanganyika, Africa." *Nature* 424 (August 14): 766–768.

Pearce, Fred. 2016. "In Rural India, Solar-Powered Microgrids Show Mixed Success." *Yale Environment 360*, January 14. Accessed October 8, 2018. http://e360.yale.edu/features /in_rural_india_solar-powered_microgrids_show_mixed_success.

Pilkey, Orrin H., Linda Pilkey-Jarvis, and Keith C. Pilkey. 2016. *Retreat from a Rising Sea: Hard Choices in an Age of Climate Change*. New York: Columbia University Press.

Pincock, Stephen. 2007. "Climate Politics: Showdown in a Sunburnt Country." *Nature* 450 (November 15): 336–338.

Pollack, Henry. 2009. *A World without Ice*. London: Avery/Penguin.

Quammen, David. 1998. "Planet of Weeds: Tallying the Losses of Earth's Animals and Plants." *Harpers*, October: 57–69.

Redsteer, M. H., K. Bemis, K. Chief, M. Gautam, B. R. Middleton, and R. Tsosie. 2013. "Unique Challenges Facing Southwestern Tribes." In *Assessment of Climate Change in the Southwest United States: A Report Prepared for the National Climate Assessment*, by G. Garfin, A. Jardine, R. Merideth, M. Black, and S. LeRoy, 385–404. Washington, DC: Island Press. Accessed October 11, 2018. http://www.swcarr.arizona.edu/sites/default/files/ACCSWUS_Ch17.pdf.

Roberts, Leslie. 2017. "Nigeria's Invisible Crisis." *Science* 356 (April 7): 18–23.

Roberts, Paul. 2004. *The End of Oil: The Edge of a Perilous New World*. Boston: Houghton Mifflin.

Rohter, Larry. 2006. "With Big Boost from Sugar Cane, Brazil Is Satisfying Its Fuel Needs." *New York Times*, April 10. Accessed October 8, 2018. http://www.nytimes.com/2006/04/10/world/americas/10brazil.html.

Saro-Wiwa, Ken. 1992. *Genocide in Nigeria: The Ogoni Tragedy*. Port Harcourt, Nigeria: Saros International Publishers.

Schecter, A., O. Päpke, M. Ball, D. C. Hoang, C. D. Le, Q. M. Nguyen, . . . J. Spencer. 1992. "Dioxin and Dibenzofuran Levels in Blood and Adipose Tissue of Vietnamese from Various Locations in Vietnam in Proximity to Agent Orange Spraying." *Chemosphere* 25, no. 7–10 (October–November): 1123–1128.

Schiermeier, Quirin. 2016. "Solar on the Steppe: Ukraine Embraces Renewables Revolution." *Nature* 537, no. 7622 (September 29). Accessed October 12, 2018. http://www.doi.org/10.1038/537598a.

Service, Robert. 2014. "Perovskite Solar Cells Keep on Surging." *Science* 344 (May 2): 458.

Taub, Ben. 2017. "We Have No Choice." *New Yorker*, April 10: 36–49.

Tidwell, Mike. 2006. *The Ravaging Tide: Strange Weather, Future Katrinas, and the Coming Death of America's Coastal Cities*. New York: Free Press.

Toynbee, Arnold. 1973. "The Genesis of Pollution." *New York Times,* September 16. Reprinted from *Horizon*, n.p.

Walther, Gian-Reto. 2003. "Plants in a Warmer World." *Perspectives in Plant Ecology, Evolution, and Systematics* 6, no. 3: 169–185.

Weaver, Jace, ed. 1996. *Defending Mother Earth: Native American Perspectives on Environmental Justice*. New York: Maryknoll.

Weinberg, Bill. 1999. "Hurricane Mitch, Indigenous Peoples, and Mesoamerica's Climate Disaster." *Native Americas* 16, no. 3/4 (Fall/Winter): 50–59.

Wolfe, W. H., J. E. Michalek, and J. C. Miner. 1995. "Paternal Serum Dioxin and Reproductive Outcomes Among Veterans of Operation Ranch Hand." *Epidemiology* 6, no. 1 (January): 17–22.

Yablokov, Alexei, Sviatoslav Zabelin, Mikhail Lemeshev, Svetlana Revina, Galina Flerova, and Maria Cherkasova. 1991. "Russia: Gasping for Breath, Choking in Waste, Dying Young." *Washington Post*, August 18: C-3.

Yoon, Carol Kaesuk. 2001. "Something Missing in Fragile Cloud Forest: The Clouds." *New York Times*, November 20: F-5.

Zalasiewicz, Jan, and Mark Williams. 2016. *Ocean Worlds: The Story of Seas on Earth and Other Planets.* New York: Oxford University Press.

Zimmer, Carl. 2017. "Most Primate Species Threatened with Extinction, Scientists Find." *New York Times*, January 18. Accessed October 12, 2018. https://www.nytimes.com/2017/01/18/science/almost-two-thirds-of-primate-species-near-extinction-scientists-find.html.

Index

Page numbers in *italics* indicate photos.

About the Author

Bruce E. Johansen is Frederick W. Kayser Professor of Communication and Native American Studies at the University of Nebraska at Omaha, where he has been teaching and writing since 1982. He has authored 47 published books, most recently ABC-CLIO's *Climate Change: An Encyclopedia of Science, Society, and Solutions*. Johansen holds the University of Nebraska award for Outstanding Research and Creative Activity (ORCA). Johansen's writing has been published, debated, and reviewed in many academic venues. He also writes as a journalist in several national forums, including the *Washington Post* and the *Progressive*.